Transactions on Computer Systems and Networks

Series Editor

Amlan Chakrabarti, Director and Professor, A. K. Choudhury School of Information Technology, Kolkata, West Bengal, India

Transactions on Computer Systems and Networks is a unique series that aims to capture advances in evolution of computer hardware and software systems and progress in computer networks. Computing Systems in present world span from miniature IoT nodes and embedded computing systems to large-scale cloud infrastructures, which necessitates developing systems architecture, storage infrastructure and process management to work at various scales. Present day networking technologies provide pervasive global coverage on a scale and enable multitude of transformative technologies. The new landscape of computing comprises of self-aware autonomous systems, which are built upon a software-hardware collaborative framework. These systems are designed to execute critical and non-critical tasks involving a variety of processing resources like multi-core CPUs, reconfigurable hardware, GPUs and TPUs which are managed through virtualisation, real-time process management and fault-tolerance. While AI, Machine Learning and Deep Learning tasks are predominantly increasing in the application space the computing system research aim towards efficient means of data processing, memory management, real-time task scheduling, scalable, secured and energy aware computing. The paradigm of computer networks also extends it support to this evolving application scenario through various advanced protocols, architectures and services. This series aims to present leading works on advances in theory, design, behaviour and applications in computing systems and networks. The Series accepts research monographs, introductory and advanced textbooks, professional books, reference works, and select conference proceedings.

Srikrishnan Sundararajan

Multivariate Analysis and Machine Learning Techniques

Feature Analysis in Data Science Using Python

 Springer

Srikrishnan Sundararajan
Computer Science and Engineering
Adi Shankara Institute of Engineering
and Technology
Kaladi, India

ISSN 2730-7484   ISSN 2730-7492   (electronic)
Transactions on Computer Systems and Networks
ISBN 978-981-99-0352-8   ISBN 978-981-99-0353-5   (eBook)
https://doi.org/10.1007/978-981-99-0353-5

This Springer imprint is published by the registered company Springer Nature Singapore Pte Ltd.
The registered company address is: 152 Beach Road, #21-01/04 Gateway East, Singapore 189721,
Singapore

Preface

This book offers a comprehensive first-level introduction to data analytics. The book covers multivariate analysis, AI/ML, and other computational techniques for solving data science problems using Python. The topics covered include a working introduction to programming with Python for data analytics, probability and statistics, hypothesis testing, correlation and regression, factor analysis, classification (including logistic regression, linear discriminant analysis, decision tree, and support vector machines), cluster analysis, survival analysis, general computational techniques (market basket analysis, social network analysis, and recommendation systems), machine learning, and deep learning.

Many academic textbooks are available for teaching statistical applications using R, SAS, and SPSS. However, there is a need for textbooks that provide a comprehensive introduction to the emerging and powerful Python ecosystem, which is pervasive in data science and machine learning applications.

The book offers a judicious mix of theory and practice, reinforced by over 100 tutorials coded in the Python programming language. The tutorials and examples conceptualize real-world problems using data curated from public domain datasets. It is designed to benefit any data science aspirant with a primary (higher secondary school level) understanding of programming and statistics. This book may be a supplementary textbook for first-level courses in applied statistics, multivariate analysis, machine learning, deep learning, data mining, and business analytics. It can also be used as a reference book by data analytics professionals.

Chennai, India   Srikrishnan Sundararajan

Acknowledgements

The author extends profound gratitude to Mr. Sreejith Srikrishnan (Data Scientist at Tiger Analytics, Chennai (IIT Gandhinagar) for his invaluable contributions to the sections on 'social network analysis' and 'artificial intelligence and deep neural networks,' as well as for his meticulous review of the book. The author also wishes to express sincere appreciation to Mr. Sreenath Srikrishnan (Scientist at Cellarity, Boston | GaTech, IIT Madras) for providing insightful feedback on the chapters related to 'survival analysis,' 'Python for data analytics,' and 'artificial intelligence and deep neural networks'. The author would like to acknowledge the assistance of ChatGPT, an AI language model developed by OpenAI, which was instrumental in generating many questions based on the chapter content at the end of the chapters. ChatGPT's capabilities also played a valuable role in enhancing the content and engagement with the readers.

Furthermore, the author acknowledges with gratitude the efforts of Ms. T. Sobha, Associate Professor, CSE, and Ms. Remya Raveendran, Assistant Professor CSE-AI, Adi Shankara Institute of Engineering and Technology, Kaladi, India, for their diligent and thoughtful reviews of the chapters 'cluster analysis' and 'artificial intelligence and deep neural networks,' respectively.

The author expresses sincere gratitude to Ms. Kamiya Khatter, Associate Editor at Springer Nature India, for her pivotal role in making the book a reality and enhancing its overall presentation.

November 2023   Srikrishnan Sundararajan

Contents

About the Author

Srikrishnan Sundararajan is presently the Dean of Computer Science and Engineering at the Adi Shankara Institute of Engineering and Technology in Kaladi, Kerala, India. He has held tenured and visiting professor positions in the field of computer science and business analytics for over 12 years at various institutions, including the Kerala University of Digital Sciences, Innovation, and Technology in Thiruvananthapuram; Loyola Institute of Business Administration in Chennai; LM Thapar School of Management in Chandigarh; Agni College of Technology in Chennai; and SCMS-Cochin in Kerala.

Furthermore, he has over 25 years of experience as an IT consultant, leading multicultural teams engaged in information systems development across the USA, the UK, Japan, and India. His professional journey includes collaborations with prominent organizations like Tata Consultancy Services, Covansys Inc. USA, UST Global, HCL Technologies Ltd., and others.

His educational background comprises a Ph.D. degree from Cochin University of Science and Technology in Kerala, a Master of Technology degree from the Indian Institute of Technology in Kharagpur, and a Bachelor of Technology degree from the College of Engineering in Thiruvananthapuram.

List of Figures

List of Tutorials

Chapter 1
Introduction and Overview

Learning Objectives

- Understand the history and evolution of data analytics.
- Learn the basic concepts of data analytics.
- Outline data modeling for analytics.
- Acquire familiarity with job opportunities in analytics.
- Get familiar with the contents and the way to use this book.
- Get familiar with the features and access methods of the datasets used in the book.
- Understand the software installation process for Python, machine learning, and associated libraries.

Overview

The decade starting in 2011 is commonly called the age of the Fourth Industrial Revolution, or Industry 4.0. We are witnessing hitherto unknown disruptive innovation driven by the rapid growth of technology—impacting business and day-to-day social life—attributable to huge strides in interconnectivity and big data analytics.

In this chapter, we start with a formal and comprehensive introduction to the world of data analytics, through a discussion on the following topics—the history and evolution of data analytics, basic theoretical concepts, data modeling, and a glimpse of job opportunities. A quick tour of the book chapters follows this.

Supplementary Information The online version contains supplementary material available at https://doi.org/10.1007/978-981-99-0353-5_1.

Subsequently, you find a brief description of the datasets used in this book—the data features, sample data, and the code snippets to access the datasets. This book is intended to provide an applied orientation for using statistical, computational, and machine learning techniques through solving numerous problems in data analytics. Thus, the section on datasets used is expected to come in handy to all readers time and again. Finally, the guidelines for installing Python, TensorFlow, Keras, and other related software packages are described. The readers are requested to get familiar with current trends through self-learning. A set of exercises for this is provided. It is recommended that these exercises be undertaken as group projects.

Definitions

Big Data: Big Data is characterized by the amalgamation of factors such as the volume of data, velocity (e.g., streaming data), variety (diverse formats like audio, video, and text), and veracity (ambiguity, as observed in popular opinions emerging from social networks). These attributes have posed challenges to traditional data processing capabilities.

Business Analytics: Business analytics refers to tools and techniques (metrics and models) designed to empower managers with enhanced insights into their business operations, facilitating better fact-based decision-making. Coined in 2007, Business Analytics is a subset of data analytics.

Business Intelligence (BI): Business Intelligence (BI) gained prominence among business executives in the 1990s as a guiding principle for effective business decision-making. BI encompasses Online Analytical Processing (OLAP) and data mining.

Data Analytics: Data Analytics encompasses the entire spectrum of processes, including data gathering, transformation, storage, management, extraction, analysis, model building, and visualization. It is an interdisciplinary field that spans operations research, statistics, data science, and various business domains.

Data Mining: Data mining involves identifying patterns and trends in data that are not easily discernible using traditional techniques.

Data Models (as used in Analytics): Data Models can be classified into three categories based on objectives—descriptive, predictive, and prescriptive. Using statistical, computational, and machine learning techniques, these models represent the real world.

Data Visualization: In data visualization, users go through a three-step process: obtaining an overview of the data, identifying interesting patterns, and drilling down into the final details. Visualization technology for data exploration includes three components—the data to be visualized, visualization techniques, and interaction techniques.

Data Warehouse: A data warehouse is a historical repository of data collected from multiple heterogeneous sources, such as Operational/Transaction Processing Systems, ERP, CRM, Supply Chain, and external feeds like market prices. The data is organized in a Relational Database Management System (RDBMS) schema at a single site to facilitate management decision-making. Data warehouse processing involves Extract, Transform, Load (ETL) processes, Online Analytical Processing (OLAP), and data mining.

Data Stream Processing: Data stream processing involves using specialized algorithms for computations on large datasets arriving rapidly, requiring immediate processing without relying on data storage. Examples include satellite images and data from sensors.

Deep Learning: Deep Learning focuses on constructing large and complex neural networks.

Machine Learning: In machine learning, a subset of the data sample, known as the 'training set', is randomly chosen to train an algorithm. The remaining data is used to test the model. This test-train combination is iteratively repeated to develop an accurate model.

1.1 The World of Data Analytics

The terms data science and data analytics are used interchangeably, both in academics and industry. However, data science may be defined as a study of scientific principles to describe data and their interrelationship. A data engineer builds data models, and an analyst makes the best use of data from a business perspective to achieve business objectives. On a simpler note, data science deals mainly with theoretical concepts, whereas data analytics deals with practical applications. This book adopts an application-oriented approach, presenting a theoretical foundation for a topic followed by practical programming applications.

This section discusses the history and evolution of data analytics, basic concepts and terminology used in data analytics, data modeling for analytics, and job opportunities for analytics professionals.

1.1.1 History and Evolution

Let me start with an introduction to the cutting-edge AI technology of generative AI and one of its applications, 'ChatGPT'! I asked ChatGPT itself for an introduction. The answer is shown below.

'Generative AI, a remarkable field within artificial intelligence, is revolutionizing how we interact with and create content using machines. At its core, generative AI refers to the ability of computer systems to generate creative and contextually relevant content, such as text, images, or even music, without explicit human input. These systems can understand patterns, context, and style and produce content matching the desired criteria.

One of the most prominent examples of generative AI is ChatGPT, a creation of OpenAI. ChatGPT is a state-of-the-art language model built upon the GPT-3.5 architecture. It excels at understanding and generating human-like text, making it versatile for various applications, including natural language understanding, content creation, chatbots, and virtual assistants. By leveraging vast amounts of data and sophisticated algorithms, ChatGPT can engage in meaningful and contextually relevant conversations, assist with information retrieval, and generate coherent and contextually appropriate text, making it a powerful tool in the world of artificial intelligence and human–computer interaction' (OpenAI 2023).

As you know, today's world is witnessing rapid growth in data science and its flagship artificial intelligence (AI). However, it may be interesting to note that Academic programs in Data Analytics started only as recently as 2010 (Gorman and Klimberg 2014). 'By 2015, 2.35 million jobs were posted under the category of Data Science and Analytics in the United States alone' (Sundararajan 2021), and the demand is steadily increasing at 15% CAGR (Markow et al. 2017). Let us have a brief overview of the fascinating history of analytics.

With the introduction of microprocessor-based computing systems in the early 1980s and advancements in hardware, **Data Warehouse** (DWH) systems surfaced in the latter part of the decade. Structured historical data was gathered from diverse systems, transformed, and loaded into the DWH for Online Analytical Processing (OLAP) and management reporting (Han and Kamber 2014). OLAP operations, such as roll-up, drill-down, slicing, dicing, and pivoting, enable the examination of data from various perspectives. For instance, a report could showcase the quarter-to-quarter performance of a product in terms of sales, segmented by region, store, and quarterly sales for a retail company. Data warehouses offer functionalities for queries, periodic and ad-hoc reporting, and dashboards illustrating business health.

In the 1990s, **Business Intelligence** (BI) gained traction among executives as a guiding principle for effective business decision-making. BI encompasses OLAP and data mining. 'Data mining is the extraction of interesting (i.e., non-trivial, implicit, previously unknown, and potentially useful) patterns or knowledge from vast amounts of data. It is also referred to by various names such as Knowledge Discovery in Databases (KDD), knowledge extraction, data analysis, pattern analysis, data archaeology, and data dredging' (Sundararajan 2021; Han and Kamber 2014). Figure 1.1 illustrates a schematic diagram of data analytics.

The late 1990s witnessed significant disruptions in business models, shifting from brick and mortar to online and from local silos to global, fueled by the emergence of the World Wide Web and internet connectivity that bridged gaps across geographies and cultures. By the mid-2000s, digital mobility, social media, and cloud computing marked the onset of the big data era (Asamoah et al. 2017; Henry and Venkatraman 2015).

It's worth noting that big data lacks a precise definition; rather, it is characterized by a combination of factors such as data volume, velocity (e.g., streaming data), variety (diverse formats like audio, video, and text), and veracity (ambiguity, as seen in popular opinions from social networks) (Onofrei et al. 2004). These characteristics posed challenges to the capabilities of traditional data processing. The convergence of computing technologies—internet, web, social, mobile, cloud, and the Internet of Things (IoT)—along with existing information systems,

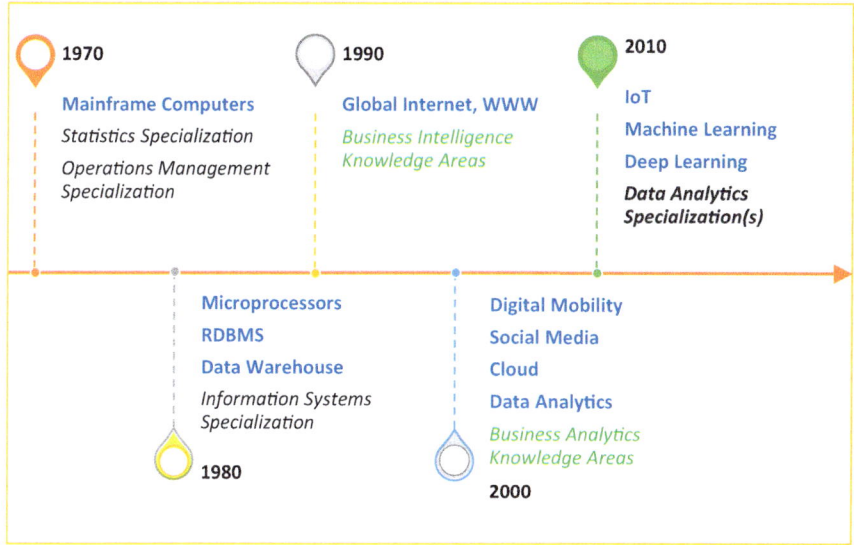

Fig. 1.1   The history and evolution of business analytics

open-source technology, and commodity hardware, led to an exponential data growth, doubling every 18 months. In 1986, the world had 0.4 zettabytes of optimally compressed information, growing to 1.9 ZB by 2007 (Sundararajan 2021). According to a conservative estimate by the International Data Corporation, the global data sphere reached 33 zettabytes by the end of 2018 and is projected to exceed 175 zettabytes by 2025. Figure 1.1 illustrates the history and evolution of information systems and the academic program 'business data analytics' (Sundararajan 2021).

Academic programs in statistics and operations management have been available since the 1970s. Academic programs in information systems emerged in the 1980s (Gorman and Klimberg 2014). Courses in data warehousing, data mining, and business intelligence were introduced in the 2000s (Schoenherr and Speier-Pero 2015), and business analytics courses became available in the 2010s. While academic programs in analytics have emerged, they are still evolving to meet industry demands, offering degrees such as Bachelor's (BA, BS), Master's (MS, MBA), and Ph.D.

1.1.2   Data Analytics—Basic Concepts and Terminology

Data Analytics includes all the processes from data gathering, transformation, storage, management, extraction, analysis, model building, and visualization (see Fig. 1.2). As discussed in the earlier section, data analytics is an interdisciplinary

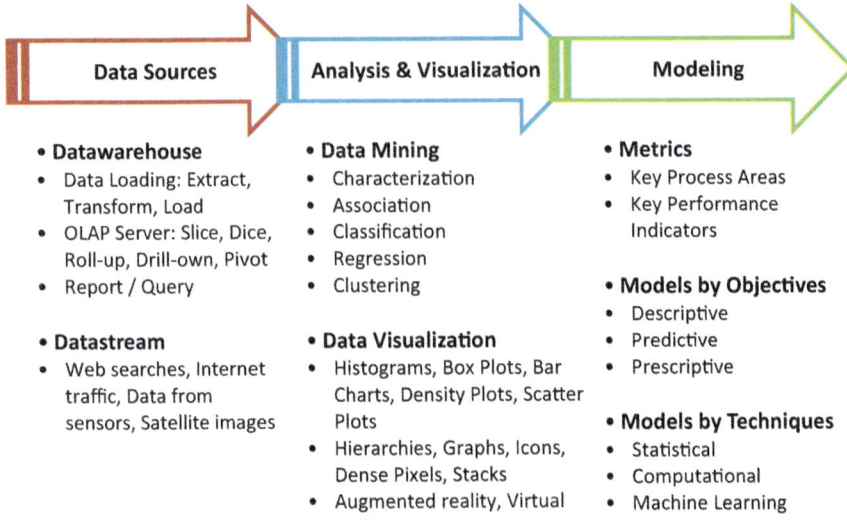

Fig. 1.2 Data analytics schema

field encompassing the broad academic areas of operations research, statistics, data science, and business domains.

The core of analytics is **Data Mining**, which offers tools and techniques for modeling and analyzing data. Data Models may be classified into three categories by way of objectives—descriptive, predictive, and prescriptive. These models depict the real world using statistical, computational, and machine-learning techniques (Sundararajan 2021; Leskovec et al. 2020).

Business Analytics is a subset of data analytics, a word coined in 2007. Business analytics can be defined as a set of tools and techniques (metrics and models) that help managers gain improved insight into their business operations and make better, fact-based decisions (Sorger 2013). The Business Analytics discipline focuses on business problem-solving ability, rather than the underlying statistical nuances, algorithms, or database technology. The domain in question varies; it can be any business, though the following domains offer relatively high job opportunities—Marketing (Liu and Levin 2018) Click or tap here to enter text., Financial Services, Healthcare (Kamble et al. 2019), Retail, Insurance, Supply Chain, Telecom, and HR—ranked based on the order of job postings in the job portals of the USA.

The data warehouse is a historical repository of data gathered from multiple heterogeneous data sources, such as Operational/Transaction Processing Systems, ERP, CRM, Supply Chain, and External feeds like market prices. Data is organized in a relational database management system (RDBMS) schema, at a single site, to facilitate management decision-making. Data warehouse processes include ETL processing, online analytical processing (OLAP), and data mining. The incoming data is subjected to the ETL process: extract, transform, and load. Data is cleaned by handling missing values, noise, outliers, and inconsistencies. Data in multiple formats and measurements are integrated into a coherent store. Data volume is reduced through data reduction techniques. Data is discretized, summarized, and standardized using data transformation techniques. The processed data is normalized and loaded into the data warehouse. OLAP supports multidimensional data analysis and decision-making, using tools such as roll-up, drill down, slice, dice, pivot, drill across, and drill through, providing an increasing granularity of information, as desired.

Data stream processing needs specialized algorithms for computations on large data that arrive fast so that it can be processed immediately without relying on data storage—for example, satellite images, sensor data, internet traffic, and web searches. Advanced algorithms that use high computing power, memory, data storage, and parallelization are necessary to achieve this.

Statisticians were the first to use the term '**data mining**'. Traditionally, data mining is defined as identifying patterns and trends in data that are not easily discernible. The statistical techniques include characterization, association, regression, classification, and clustering. The chapter 'Introduction to probability and Statistics' discusses data characterization using descriptive and inferential statistics. The topic 'association' gets specific treatment in the chapter 'computational techniques'. The topics of regression, classification, and clustering are discussed in separate chapters.

In **data visualization**, the user usually follows a three-step process: get an overview of the data, identify interesting patterns, and drill down final details. See Fig. 1.3. Visualization technology for data exploration involves three components—the data to be visualized, visualization techniques, and interaction techniques (Berthold et al. 2007). Visualizing data can be spatially structured in 1D/2D/Multi dimensions. Alternatively, the data may have the structure of a tree, graph, or text. Visualization techniques may include stacks, dense pixels, icons,

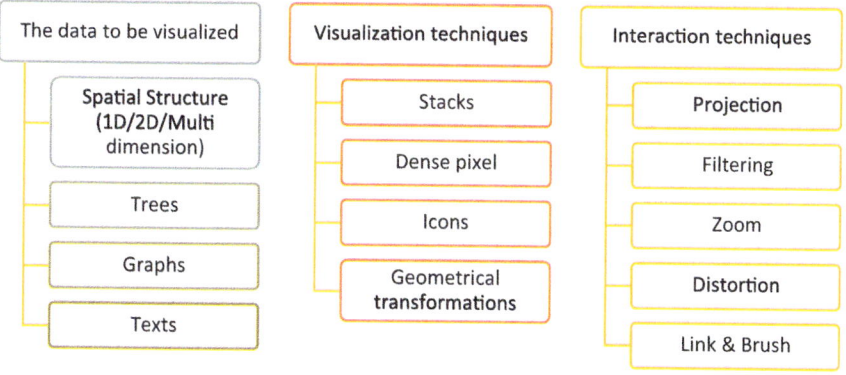

Fig. 1.3 Data visualization

and geometrical transformations. Users can interact with the display in multiple ways, such as projection, filtering, zoom, distortion, and link and brush.

R and Python provide excellent programming techniques through ever-growing and powerful software libraries available for free. At the same time, highly sought-after specialized tools (e.g., Tableau for data visualization) are available for a fee. Common statistical visualization techniques include histograms, box plots, bar charts, density plots, scatter plots, and multi-dimensional plots. Nevertheless, augmenting the perception of the tangible world can be achieved by applying audio-visual techniques or other sensory stimuli facilitated by technology. This method is commonly referred to as augmented reality. In contrast, virtual reality creates a computer-generated simulated environment, immersing users in a virtual space.

As we are aware, expressing the real world using a linguistic description, picture, or 3D model is customary. In data science, the term 'data model' typically refers to depicting real-world phenomena using mathematical models. These models may be classified into three categories by way of objectives—descriptive, predictive, and prescriptive. The descriptive model characterizes a phenomenon. Predictive models help to determine the likely outcomes based on given inputs. The prescriptive model helps to determine the best course of action under the given constraints to optimize an objective.

Data models may be classified based on the techniques employed—statistical, computational, or machine learning. Statisticians view a data model as the construction of a statistical model—the parameters of an underlying distribution to which the visible data belongs. We need computational methods to model data when the data does not conform to Gaussian or other well-defined statistical distributions. Examples include Google page rank, market basket analysis, product recommendations, and social network analysis. For example, the similarity of user preferences upon a few selected features is analyzed for product recommendation systems used by Amazon and Netflix.

Machine learning is used in recommendation systems, image processing, speech recognition, fingerprint identification, and medical diagnosis. Machine learning methods are broadly classified into supervised, unsupervised, and reinforcement learning. For example, let us consider supervised machine learning. This method randomly selects a subset of the data sample named 'training dataset' to train an algorithm such as Bayes nets, support vector machines, decision trees, and hidden Markov models. The rest of the data is used for testing the model. The test-train combination is repeated iteratively to develop an accurate model. The other machine learning methods also depend on the principle of learning underlying patterns from data and the method of train/test. However, the details differ—please see Chap. 11 for a better understanding.

Deep Learning (DL) techniques represent a recent advancement in machine learning within the past decade, revitalizing Artificial Intelligence. This resurgence is attributed to the abundant availability of high computing power, including GPUs, CPUs, memory, and parallel processing algorithms. DL focuses on constructing extensive and intricate neural networks, employing architectures like convolutional neural networks, recurrent neural networks, stacked autoencoders, deep Boltzmann machines, deep belief networks, and more. Despite the effectiveness of deep learning methods with substantial datasets, the resulting models can occasionally pose challenges in terms of interpretation.

1.1.3 Job Opportunities

The job profiles in the analytics area may be broadly categorized as data systems developers, data scientists, data analysts, functional analysts, and data-driven decision-makers, each requiring varied skill sets (Sundararajan 2021; Markow et al. 2017). Functional analysts and data-driven decision-makers require domain-specific knowledge and an understanding of analytics skills, which are important for organizations consuming and interpreting data.

The top disruptive skills in growing demand are big data, Hadoop, R, data visualization, and machine learning. The Spark framework is several times faster than the Hadoop framework in storing and manipulating big data. Built-in machine learning frameworks are associated with big data platforms—e.g., Spark ML for Hadoop and Spark. R boasts an excellent set of in-built statistical tools, while the world of Python is rapidly catching up. Python and associated frameworks such as Scikit Learn Tensorflow, Keras, and PyTorch provide an excellent environment for developing models for machine learning and artificial neural networks.

Most of the software stacks mentioned above belong to FOSS (free and open-source software) and are available for free. Several commercial systems (database and software tools) are available for a fee or as freemium. They are widely used in the industry, e.g., Tableau data visualization tool, Google Analytics for web data analytics, and SAS for statistical and data visualization, among others. Some of the traditional skill sets in demand are ETL (tools for data warehousing), data management, SQL, and business data analysis (financial analytics, marketing analytics, health care, telecom).

1.2 How to Use This Book

The book provides an applied orientation for using statistical and computational techniques to solve data analytics problems. This book will be helpful for students of data analytics, computer science, computer application, statistics, business analytics, biological sciences, and the healthcare domain. The hands-on approach envisages step-by-step learning with over 100 tutorials and exercises. In addition, each chapter provides a set of descriptive-type questions. Case studies are given at the end of some of the chapters.

The following is a note to those who need to become more familiar with Python programming. You may need to install an IDE, a toolset that helps you develop and test programs. You may use popular IDEs such as Spyder, PyCharm, Jupyter Notebook, etc. An IDE can be installed on your local system individually or under the package 'anaconda'. Google Colab provides internet-based Jupyter Notebook emulation. If you prefer to use Google Colab, you need not install any software; all you need is internet connectivity.

If you are new to programming, Spyder (local installation) or Google Colab (internet-based access) is recommended.

There are 12 chapters. If the reader has preliminary programming knowledge, Chap. 2, 'Python for Data Analytics' can be skipped. If the reader is familiar with statistics concepts, Chap. 3, 'Introduction to Probability and Statistics', and Chap. 4, 'Hypothesis Testing' can be skipped. Chapters 5–8 cover multivariate analysis techniques such as regression, classification, factor analysis, and cluster analysis. Chapter 9 discusses a topic of special interest in statistics—survival analysis. Chapter 10 discusses Computational Techniques (frequent itemset/market basket analysis, social network analysis, and recommendation systems). Chapter 11 introduces Machine Learning, and Chap. 12, Artificial Intelligence and Deep Neural Networks.

Python is a versatile programming language used for multiple purposes—common programming, solving statistical problems, and building machine learning/deep learning models. Chapter 2 provides a brief overview of programming in Python, which includes (a) basic programming—variables and operators, data structures, control flow, and functions, (b) libraries useful for data analytics—pandas, NumPy, matplotlib/Seaborn for data visualization, (c) statistical packages, and (d) scikit learn library for machine learning. If you work out the problems in this section, you will get a jump start on the Python programming techniques necessary to solve the problems in the rest of the book.

Chapter 3 provides a foundation in probability by introducing random numbers and probability distribution functions such as binomial, Poisson, normal (Gaussian), etc. An examination of descriptive statistics follows this. Skip this chapter if you have basic concepts of probability. Chapter 4 introduces hypothesis testing, tests for means, tests for variances, and tests for categories. Skip this chapter if you have basic concepts of statistics.

Chapter 5 covers the concepts of correlation, regression, validity and reliability, and polynomial regression. Chapter 6 introduces classification and proceeds to cover various classification techniques such as binary logistic regression, linear discriminant analysis, support vector machines, decision tree classification, and comparison of classifier performance. Chapter 7 describes techniques for dimension reduction, exploratory factor analysis, confirmatory factor analysis, validity, reliability of factor analysis, and use of factor scores for regression. Chapter 8 describes various cluster analysis techniques and explores hierarchical and k-means in-depth.

Chapter 9 covers survival analysis. Survival analysis is the study of the expected duration for an event to occur based on a study of censored data. The survival probability is the probability that an individual survives an event from the study's starting point. This is a specialized topic, particularly useful in the healthcare domain; therefore, the reader may skip this topic if he/she chooses to do so.

What about data mining algorithms that do not use statistical techniques, or where the use of statistical techniques is not the primary focus? Such techniques for data mining are introduced in Chap. 10 under the umbrella of computational techniques. Three of the commonly used computational techniques are covered here—(a) frequent item sets and market basket analysis, (b) social network analysis, and (c) recommendation systems.

Chapter 11 introduces machine learning. Several machine learning tutorials in regression and classification analysis are demonstrated. Chapter 12 introduces artificial intelligence and deep neural networks and deep learning. The chapter discusses feedforward networks, RNN, LSTM, and CNN and demonstrates them using the Keras framework.

Wish you an enjoyable read!

1.3 A Brief Description of the Datasets Used

The datasets used in this book are briefly described below for quick and easy reference. Please refer to Sundararajan (2023) for some of the datasets and their descriptions.

Tutorial 1.3.1 Diamonds

How to Access the Dataset?
Refer: R documentation

```
sb.get_dataset_names()
pdf  = sb.load_dataset('diamonds')
pdf.info()
```

Description

The diamonds dataset contains the prices and other attributes of almost 54,000 diamonds. The variables are as follows: -
price: price in US dollars   (326 - 18823)
carat: weight of the diamond (0.2 - 5.01)
cut: quality of the cut (Fair, Good, Very Good, Premium, Ideal)
color: diamond color, from D (best) to J (worst)
clarity: indicates how clear the diamond is (I1), SI2,..., IF)
x: length in mm   (0 -10.74)
y: width in mm   (0 - 58.9)
z: depth in mm   (0 -31.8)
depth: total depth % = Z / mean(X,Y)   (43 - 79)
table: width of to the p of the diamond relative to widest point (43-95)

Sample Data

```
       carat  cut          color clarity depth table price x     y     z
0      0.23   Ideal        E     SI2     61.5  55.0  326   3.95  3.98  2.43
1      0.21   Premium      E     SI1     59.8  61.0  326   3.89  3.84  2.31
2      0.23   Good         E     VS1     56.9  65.0  327   4.05  4.07  2.31
       ...    ...          ...   ...     ...   ...   ...   ...   ...   ...
53937  0.70   Very Good    D     SI1     62.8  60.0  2757  5.66  5.68  3.56
53938  0.86   Premium      H     SI2     61.0  58.0  2757  6.15  6.12  3.74
53939  0.75   Ideal        D     SI2     62.2  55.0  2757  5.83  5.87  3.64
[53940 rows x 10 columns]
```

Tutorial 1.3.2 Planets

How to Access the Dataset?
Refer: seaborn

```
import seaborn as sb
pdf = sb.load_dataset('planets')
pdf.info()
```

Description

```
RangeIndex: 1035 entries, 0 to 1034
Data columns (total 6 columns):
 #     Column         Non-NullCount      Dtype
---    ------         --------------     -----
 0     method         1035  non-null     object   Category with 10 labels
 1     number         1035  non-null     int64
 2     orbital_period 992   non-null     float64  * missing values
 3     mass           513   non-null     float64  * missing values
 4     distance       808   non-null     float64  * missing values
 5     year           1035  non-null     int64
```

Sample Data

```
             method    number  orbital_period  mass   distance  year
0      Radial Velocity      1    269.300000    7.10      77.40  2006
1      Radial Velocity      1    874.774000    2.21      56.95  2008
2      Radial Velocity      1    763.000000    2.60      19.84  2011
             ...      ...          ...       ...        ...   ...
1032          Transit      1      3.191524     NaN     174.00  2007
1033          Transit      1      4.125083     NaN     293.00  2008
1034          Transit      1      4.187757     NaN     260.00  2008
[1035 rows x 6 columns]
```

Tutorial 1.3.3 Penguins

How to Access the Dataset?
Refer: https://github.com/allisonhorst/penguins

```
import seaborn as sb
pdf = sb.load_dataset('penguins')
pdf.info()
```

Description

The dataset comprises information for 344 penguins, encompassing three distinct species. These penguins were gathered from three islands within the
Palmer Archipelago, Antarctica.

	Column	Non-Null Count	Dtype
0	species	344 non-null	['Adelie', 'Chinstrap', 'Gentoo']
1	island	344 non-null	['Biscoe','Dream','Torgersen']
2	bill_length_mm	342 non-null	float64 * 2 missing values
3	bill_depth_mm	342 non-null	float64 * 2 missing values
4	flipper_length_mm	342 non-null	float64 * 2 missing values
5	body_mass_g	342 non-null	float64 * 2 missing values
6	sex	333 non-null	['Female','Male'] * 9 miss.values

Sample Data

	species	island	bill_len…	...	flipper_len…	body_mass_g	sex
0	Adelie	Torgersen	39.1	...	181.0	3750.0	Male
1	Adelie	Torgersen	39.5	...	186.0	3800.0	Female
2	Adelie	Torgersen	40.3	...	195.0	3250.0	Female
..	...	...	...	...	...	...	...
341	Gentoo	Biscoe	50.4	...	222.0	5750.0	Male
342	Gentoo	Biscoe	45.2	...	212.0	5200.0	Female
343	Gentoo	Biscoe	49.9	...	213.0	5400.0	Male

[344 rows x 7 columns]

Tutorial 1.3.4 Chickweight

How to Access the Dataset?
Refer: R Documentation

```
import statsmodels.api as sm
pdf =sm.datasets.get_rdataset("ChickWeight").data
pdf.columns
pdf.info()
```

Description

Chick body weights were recorded at birth and subsequently every other day up to day 20, with an additional measurement on day 21. The chicks were grouped into four categories based on different protein diets. The dataset includes the following variables:

- Weight: A numeric vector representing the body weight of the chickens in grams.
- Time: A numeric vector indicating the number of days since birth when each measurement was taken.
- Chick: An ordered factor with levels of 18 … 48, serving as a unique identifier for each chick. The ordering of levels organizes chicks on the same diet and arranges them based on their final weight within the diet, from lightest to heaviest.
- Diet: A factor with levels of 1 … 4, indicating the experimental diet each chick received.

```
RangeIndex: 578 entries, 0 to 577
Data columns (total 4 columns):
 #    Column   Non-Null Count   Dtype
---   ------   --------------   -----
 0    weight   578 non-null     int64
 1    Time     578 non-null     int64   # 0 ... 21
 2    Chick    578 non-null     int64
 3    Diet     578 non-null     int64   # values 0,1,2,3
```

Sample Data

	weight	Time	Chick	Diet
0	42	0	1	1
1	51	2	1	1
2	59	4	1	1
..	...	...	...	...
575	234	18	50	4
576	264	20	50	4
577	264	21	50	4

[578 rows x 4 columns]

Tutorial 1.3.5 Iris

How to Access the Dataset?
Refer: R Documentation
(a) Common Method

```
import seaborn as sb
pdf = sb.load_dataset('iris')
pdf.columns
pdf.info()
```

(b) For Machine Learning

```
from sklearn import datasets
iris = datasets.load_iris()
X = iris.data
y = iris.target
```

Description

The dataset consists of measurements on 150 iris flowers belonging to three categories - setosa, versicolor, and virginica. Four features are measured. Length and width of sepals and petals.

	Column	Non-Null Count	Dtype
0	sepal_length	150 non-null	float64
1	sepal_width	150 non-null	float64
2	petal_length	150 non-null	float64
3	petal_width	150 non-null	float64
4	species	150 non-null	object - setosa, versicolor, virginica

Sample Data

	sepal_length	sepal_width	petal_length	petal_width	species
0	5.1	3.5	1.4	0.2	setosa
1	4.9	3.0	1.4	0.2	setosa
2	4.7	3.2	1.3	0.2	setosa
..	...	...	...	...	...
147	6.5	3.0	5.2	2.0	virginica
148	6.2	3.4	5.4	2.3	virginica
149	5.9	3.0	5.1	1.8	virginica

[150 rows x 5 columns]

Tutorial 1.3.6 Tips

How to Access the Dataset?
Refer: seaborn

```
import seaborn as sb
pdf=sb.load_dataset('tips')
pdf.info( )
```

Description

Many factors may influence food servers' tips in restaurants. The following
dataset shows customers served during two and a half months.
RangeIndex: 244 entries, 0 to 243

```
Data columns (total 7 columns):
 #    Column      Non-Null Count   Dtype
---   ------      --------------   -----
 0    total_bill  244 non-null     float64
 1    tip         244 non-null     float64
 2    sex         244 non-null     category ['Male', 'Female']
 3    smoker      244 non-null     category ['Yes', 'No']
 4    day         244 non-null     category ['Thur', 'Fri', 'Sat', 'Sun']
 5    time        244 non-null     category ['Lunch', 'Dinner']
 6    size        244 non-null     int64
```

Sample Data

	total_bill	tip	sex	smoker	day	time	size
0	16.99	1.01	Female	No	Sun	Dinner	2
1	10.34	1.66	Male	No	Sun	Dinner	3
2	21.01	3.50	Male	No	Sun	Dinner	3
..	...	...	...	...	...	...	...
241	22.67	2.00	Male	Yes	Sat	Dinner	2
242	17.82	1.75	Male	No	Sat	Dinner	2
243	18.78	3.00	Female	No	Thur	Dinner	2

[244 rows x 7 columns]

Tutorial 1.3.7 IT Projects

How to Access the Dataset?
Refer: {(Sundararajan, 2023)}
```
   import pandas as pd
   # Download the file from GitHub
   pdf=pd.read_csv('itprojects.csv')
   pdf.info()
```

Description

A study was conducted to explore the variables that influence software project performance, a float. Eleven variables were hypothesized to influence
project performance – they are measured on a scale of 0..6. 100 responses
were collected from IT consultants working on various projects. Project
Type is a category Variable with three labels - 0,1,2
RangeIndex: 100 entries, 0 to 99

Data	columns	(total 14 columns):		
#	Column	Non-Null Count	Dtype	
---	------	--------------	-----	
0	Case_No	100 non-null	int64	Serial Number
1	change_mgmt	100 non-null	int64	feature
2	project_plan	100 non-null	int64	feature
3	tech_mentoring	100 non-null	int64	feature
4	pm_tools	100 non-null	int64	feature
5	dev_process	100 non-null	int64	feature
6	system_arch	100 non-null	int64	feature
7	design_think	100 non-null	int64	feature
8	team_skills	100 non-null	int64	feature
9	core_team	100 non-null	int64	feature
10	prior_exp	100 non-null	int64	feature
11	rewards_recog	100 non-null	int64	feature
12	project_type	100 non-null	int64	[0, 1, 2]
13	project_perf	100 non-null	float64	target variable

dtypes: float64(1), int64(13)

Sample Data

	Case_No	change_mgmt	...	project_type	project_perf
0	1	5	...	1	0.374933
1	2	6	...	1	2.131779
2	3	5	...	1	1.679185
..	...	...	...	...	...
97	98	5	...	2	-1.192247
98	99	6	...	1	2.401460
99	100	5	...	1	0.383024

[100 rows x 14 columns]

Tutorial 1.3.8 Breast Cancer - Gene Prognostic Signature

How to Access the Dataset?

Refer: https://www.ncbi.nlm.nih.gov/geo/query/acc.cgi?acc=GSE7390

```
from sksurv.datasets import load_breast_cancer
X, Y = load_breast_cancer()
```

Description

A Survival analysis study was conducted on 198 breast cancer patients to predict distant metastases (dm). Distant metastasis (dm) refers to cancer that has spread from the original (primary) tumor to distant organs/lymph nodes. The study included 76-gene prognostic signatures in lymph node-negative (N-) breast cancer patients.

Features (X0 … X79): 80 features
- 76 gene characteristics
- age
- er (estrogen-receptor) categories {positive, negative}

Note:- About 80% of all breast cancers are er-positive
- grade (the abnormality of cancer cells) categories:- {intermediate, poorly differentiated, well differentiated}
- size

Survival Data (Y): structured array with 2 fields
- e.tdm: boolean - True indicates that metastasis has occurred; False indicates that the event time is right-censored
- t.tdm: time to distant metastasis in days

```
Features (X)
Data columns (total 80 columns):
 #     Column         Non-Null Count   Dtype
---    ------         --------------   -----
 0     X200726_at     198 non-null     float64
 1     X200965_s_at   198 non-null     float64
 2     X201068_s_at   198 non-null     float64
 3     X201091_s_at   198 non-null     float64
 4     X201288_at     198 non-null     float64


...    ...            ... ...          ...
 74    X221916_at     198 non-null     float64
 75    X221928_at     198 non-null     float64
 76    age            198 non-null     float64
 77    er             198 non-null     category     ['negative', 'positive']
 78    grade          198 non-null     category
['poorly differentiated', 'intermediate', 'well differentiated', 'unkown']
 79    size           198 non-null     float64
```

Sample Data

```
X
      X200726_at   X200965_s_at  ...                   grade   size
0      10.926361      8.962608   ...   poorly differentiated   3.0
1      12.242090      9.531718   ...   poorly differentiated   3.0
2      11.661716     10.238680   ...   poorly differentiated   2.5
3      12.174021      9.819279   ...   poorly differentiated   1.8
4      11.484011     11.489233   ...            intermediate   3.0
..        ...           ...      ...                     ...   ...
193    12.018292      8.323876   ...   poorly differentiated   2.2
194    11.711415     10.428482   ...   poorly differentiated   3.2
195    11.939616      9.615587   ...     well differentiated   2.5
196    11.848449     10.528911   ...            intermediate   1.2
197    11.425778      9.901486   ...   poorly differentiated   2.5

[198 rows x 80 columns]


Y
array([( True,   723.), (False, 6591.), ( True,   524.), (False, 6255.),
...            ...            ...                        ...
      (False, 2722.), (False, 1781.)]
```

Tutorial 1.3.9 German Breast Cancer Study Group 2

How to Access the Dataset?
Refer: https://ascopubs.org/doi/abs/10.1200/jco.1994.12.10.2086

```
from sksurv.datasets import load_gbsg2
x, y = load_gbsg2()
x.info()
```

Description

We are exploring a dataset with having 686 samples and 8 features. The end-point is recurrence-free survival, which occurred for 299 patients (43.6%). This dataset is the part of a study by the German Breast Cancer Study Group (GBSG) in 1984. This randomized clinical trial compared the effectiveness of different drug dosages on recurrence-free and overall survival.

X (features)

0	age	686 non-null	float64
1	estrec	686 non-null	float64
2	horTh	686 non-null	category -> ['no', 'yes']
3	menostat	686 non-null	category -> ['Pre', 'Post']
4	pnodes	686 non-null	float64
5	progrec	686 non-null	float64
6	tgrade	686 non-null	category -> ['I', 'II', 'III']
7	tsize	686 non-null	float64

Y (outcome)
  • 'cens', Boolean
  • 'time', float (survival in days)

Sample Data

X

	age	estrec	horTh	menostat	pnodes	progrec	tgrade	tsize
0	70.0	66.0	no	Post	3.0	48.0	II	21.0
1	56.0	77.0	yes	Post	7.0	61.0	II	12.0
2	58.0	271.0	yes	Post	9.0	52.0	II	35.0
..	...	...	...	...	...	...	...	...
683	51.0	0.0	no	Pre	5.0	43.0	III	25.0
684	52.0	34.0	no	Post	3.0	15.0	II	23.0
685	55.0	15.0	no	Post	9.0	116.0	II	23.0

[686 rows x 8 columns]

Y
array([(True, 1814.), (True, 2018.), (True, 712.), (True, 1807.),
       ...   .
       (True, 727.), (True, 1701.)]

Tutorial 1.3.10 West Virginia Student Fraternity

How to Access the Dataset?
Refer: {(Sundararajan, 2023)}

```
import pandas as pd
# Download the file from GitHub
d=pd.read_csv('WV-fraternity.csv')
d.shape          # (116, 58)
```

Description

This dataset consists of information about the interaction of students in a student fraternity in West Virginia - the number of times they interacted over a week. This dataset is used for social network analysis in Chapter 10.

Tutorial 1.3.11 The CIFAR-10 Dataset and CIFAR-100 Dataset

How to Access the Dataset?
Refer: https://www.cs.toronto.edu/~kriz/cifar.html

```
from tensorflow.keras import datasets
(X_train, y_train), (X_test, y_test) = datasets.cifar10.load_data()
```

Description

The CIFAR-10 dataset consists of 60,000 32x32 color images in 10 classes, with 6000 images per class. There are 50,000 training images and 10,000 test images. The image classes are: - airplane, automobile, bird, cat, deer, dog, frog, horse, ship, and truck.
The CIFAR-100 dataset has 100 classes containing 600 images each. The 100 classes are grouped into 20 superclasses.

Tutorial 1.3.12 MNIST Handwritten Character Recognition

How to Access the Dataset?
Refer: Refer:- http://yann.lecun.com/exdb/mnist/
This is described in Chapter 12, Artificial Neural Network

Description

The MNIST database, consisting of handwritten digits, includes 60,000 training and 10,000 test samples. These samples feature images of digits in black and white (0,1). The digits have been standardized and centered within a fixed-size image of 28x28 pixels. Construct a convolutional neural network to classify these images, and elucidate the network's structure and parameters.

Tutorial 1.3.13 PowerConsumption

How to Access the Dataset?
Refer - {(Sundararajan, 2023)}

```
import pandas as pd
# Download the file from GitHub
d=pd.read_csv('PowerConsumption_2013_19.csv')
d.shape          # (2161, 9)
```

Description

The dataset consists of data gathered from the electricity board for a state in India, reported from 2013 to 2019. The features include daily power demand, shortage, consumption, load, OD/UD, and date. These are typical time-series / sequences and can be used for RNN/ LSTM exercises.

Tutorial 1.3.14 Bank Marketing

How to Access the Dataset?

Refer - https://archive.ics.uci.edu/ml/datasets/bank+marketing

```
import pandas as pd
# Download the file from the url referred above
d = pd.read_csv ('BankMarketing.csv',delimiter=';')
d.shape #(45211, 17)
```

Description

The dataset consists of data gathered from a bank marketing survey. It has 21 variables and 45211 rows. Feature variables considered in the exercises are 'age', 'balance', 'day', 'duration', 'campaign', 'pdays', 'previous'. The target (independent variable) is 'Has the client in the training dataset, subscribed to a term deposit?' with binary answers ('yes', 'no').

X (features)

0	age	45211	non-null	int64
1	job	45211	non-null	object
2	marital	45211	non-null	object
3	education	45211	non-null	object
4	default	45211	non-null	object
5	balance	45211	non-null	int64
6	housing	45211	non-null	object
7	loan	45211	non-null	object
8	contact	45211	non-null	object
9	day	45211	non-null	int64
10	month	45211	non-null	object
11	duration	45211	non-null	int64
12	campaign	45211	non-null	int64
13	pdays	45211	non-null	int64
14	previous	45211	non-null	int64
15	poutcome	45211	non-null	object

Y (outcome)

16	y	45211	non-null	object

1.4 Python, Packages, TensorFlow, and Keras

Software libraries get revised frequently. Therefore, the installation instructions also change. Refer to the home page of the respective software projects for current usage. Alternatively, one may refer to installation instructions.

Conda is a package manager and environment management system that handles packages from multiple programming languages. Conda is typically associated with Anaconda, a Python and data science package distribution. We can use the conda command to create and manage Python environments and install packages. It is beneficial when managing dependencies for scientific computing and data science projects. It also helps in resolving and managing complex package dependencies.

Pip is the default package manager for Python and is explicitly used for Python packages. It is a package manager for installing, upgrading, and managing Python packages from the Python Package Index (PyPI). We can use the pip command to install Python packages on our system. In environments like Google Colab, Jupyter Notebooks, Spyder, and other IDEs, we can use the '!pip' command in Python program code cells to install Python packages. Note that the '!' character runs shell commands within a Python program in these IDEs. The '!pip' command(s) also helps in Checking and Resolving Path Dependencies.

Data Analytics in Action

A gazal evening with ChatGPT	
A gazal evening with ChatGPT (A narration of my interaction with ChatGPT (OpenAI 2023), a generative/conversational large language AI model from OpenAI) *[me]* Some for the Glories of This World; and some Sigh for the Prophet's Paradise to come; *[chat.openai]* This line is from the famous poem 'The Rubaiyat' by Omar Khayyam, a Persian poet, mathematician, and philosopher. The poem reflects on the themes of life, death, and the human condition ... Overall, the poem encourages readers to live in the present and enjoy the beauty of life, rather than getting caught up in worldly ambitions or religious dogma *[me...* *]* Ah, neither take the Cash or the Credit, Nor heed to the rumble of a distant Drum *[chat.openai]* Your response is a variation of the original stanza from Omar Khayyam's poem 'The Rubaiyat'. While the original stanza suggests taking the cash and enjoying the present moment, your variation suggests avoiding both cash and credit and not worrying about the distant future. This variation could be interpreted as a call to live in the moment and not be overly concerned with material possessions or future rewards. It suggests that true happiness can be found by living a simple, contented life and not being distracted by external pressures or desires...	*[me...* *In continuation... My own lines]* Fill your vital breath with the fragrance of April blossom, and enjoy the gentle melody of the river side breeze *[chat.openai]* Your statement is a poetic and evocative call to savor the beauty of nature and the present moment... The imagery in your statement is vivid and sensory, inviting the reader to imagine the sights, sounds, and smells of a tranquil spring day... It is a call to live in the present moment and to find joy in the small things that make life meaningful *[me]* *chat.openai, continue and write a few more stanzas* *[chat.openai]* Let your worries drift away on the river's gentle flow, And feel your troubles dissipate like the soft, drifting snow, As the world around you comes alive with vibrant glow Breathe in the fresh, crisp air and feel your spirit soar, As you revel in the beauty of nature's boundless store, And let your heart be filled with joy forevermore For life is but a fleeting moment in time, A precious gift that's yours and mine, To cherish and savor, like a sweet, fragrant wine So let us drink deeply from the cup of life, And revel in its beauty and all its strife, For in the end, it's the journey that counts, not the final strife

Summary

The history of data analytics reveals the evolution of hardware in the 1980s, the rise of data warehouse (DWH) systems, and business intelligence (BI) in the 1990s. The late 1990s witnessed a shift in business models from brick-and-mortar to online platforms, facilitated by the World Wide Web and internet connectivity, breaking geographical and cultural barriers. The twenty-first century sees disruptive innovation fueled by technology growth, impacting business and daily life, with big data analytics facing challenges in volume, velocity, variety, and veracity. Academic programs in data warehousing, data mining, and business intelligence emerged in the 2000s, with analytics programs still evolving.

Data analytics involves processes from gathering to visualization, incorporating operations research, statistics, data science, and business domains. The core is data mining, contributing to descriptive, predictive, and prescriptive models. Business analytics, coined in 2007, involves tools for better decision-making. The data warehouse stores historical data, utilizing ETL processing, OLAP, and data mining. Data stream processing requires specialized algorithms.

Data visualization follows a three-step process: overview, pattern identification, and detailed exploration. Statisticians construct statistical models, while machine learning trains algorithms using sample data. Deep Learning (DL) focuses on large neural networks. Job profiles include data systems developers, scientists, analysts, functional analysts, and decision-makers.

The book emphasizes applied learning with over 100 tutorials and exercises, addressing Python and frameworks like Scikit Learn, TensorFlow, Keras, and PyTorch. Most software used in data analytics is free and open source. Installation steps for the software stack are provided in the book.

Questions

Comprehension:

1. What is the distinction between data science and data analytics, and how do they differ in their focus and applications?
2. How has the field of data analytics evolved over the years, and what significant developments have contributed to its growth?
3. What are the primary categories of data models, and how do they differ regarding their objectives and techniques?
4. What are the essential skills and technologies in demand in data analytics, and how do they relate to job opportunities?
5. Can you explain the role of data warehousing and data mining in data analytics and how they contribute to decision-making?
6. How does data visualization enhance the understanding of data, and what are the essential components of data visualization technology?
7. What are the main challenges associated with big data, and what characteristics define it in terms of volume, velocity, variety, and veracity?
8. What are the different types of data analytics, and how do they contribute to solving practical problems in various domains?
9. Define business analytics.
10. Describe the basic principle of machine learning.

11. Provide three examples of computational techniques in data mining.
12. Compare and Contrast Deep learning with Machine learning.

Application:

13. Imagine you are working with a dataset of customer behavior for an e-com-
    merce website. How would you use descriptive analytics to gain insights into
    customer demographics and browsing patterns?
14. You are tasked with predicting customer churn for a subscription-based ser-
    vice. What type of analytics (descriptive, predictive, or prescriptive) would
    you use, and what data and techniques would be essential for this analysis?
15. A marketing team is looking to assess the effectiveness of an advertising cam-
    paign. What metrics and analytical methods would you use to measure the
    campaign's success, and how would you present the results to stakeholders?

Exercises

Note: The objective of the following exercises is to understand the current trends
in Data Analytics. This exercise may be given as a group assignment. The groups
shall share their learnings through scheduled dissemination.

Suggested topics:

1. AI in Business, AI and Society, Ethics & AI, The Future of AI (suggested
   source: articles from indexed and peer-reviewed academic journals)
2. Current Job Trends in Data Analytics/AI/ML (suggested source:—job portals,
   peer-reviewed articles from academic journals)
3. Technology Trends in AI
4. Disruptive Technologies—social media, mobility, data analytics, cloud, and
   IoT (suggested source: articles by IT analysts, Management Consultancies, as
   well as peer-reviewed academic journals)
5. Databases: RDBMS and Data warehousing, No-SQL Databases, Big Data
   Platform—Hadoop Ecosystem and Spark Ecosystem
6. Business Analytics—Financial, Marketing, HR, Supply Chain, Healthcare.

References

Asamoah DA, Sharda R, Hassan Zadeh A, Kalgotra P (2017) Preparing a data scientist: a ped-
    agogic experience in designing a big data analytics course. Decis Sci J Innov Educ 15(2).
    https://doi.org/10.1111/dsji.12125
Berthold N, Fricke H, Müller A (2007) Kleine Bundesländer—Achillesferse des Föderalismus?
    Zeitschrift Für Wirtschaftspolitik 56(2). https://doi.org/10.1515/zfwp-2007-0203
Gorman MF, Klimberg RK (2014) Benchmarking academic programs in business analytics.
    Interfaces 44(3). https://doi.org/10.1287/inte.2014.0739

Henry R, Venkatraman S (2015) Big data analytics is the next big learning opportunity. J Manag Inf Decis Sci 18(2)

Han J, Micheline Kamber, Pei J (2014). Data mining: concepts and techniques (The Morgan Kaufmann Series in Data Management Systems). In Proceedings—2013 international conference on machine intelligence research and advancement, ICMIRA 2013, 3rd ed

Kamble SS, Gunasekaran A, Goswami M, Manda J (2019) A systematic perspective on the applications of big data analytics in healthcare management. Int J Healthc Manag 12(3). https://doi.org/10.1080/20479700.2018.1531606

Leskovec J, Rajaraman A, Ullman JD (2020) Mining of massive datasets. In: Biometrics (Issue 4). Cambridge University Press, Cambridge. https://doi.org/10.1111/biom.12982

Liu X, Burns AC (2018) Designing a marketing analytics course for the digital age. Mark Educ Rev 28(1):28–40. https://doi.org/10.1080/10528008.2017.1421049

Liu Y, Levin MA (2018) A progressive approach to teaching analytics in the marketing curriculum. Mark Educ Rev 28(1). https://doi.org/10.1080/10528008.2017.1421048

Markow W, Braganza S, Taska B, Miller S, Hughes D (2017) The quant crunch: how the demand for data science skills is disrupting the job market. In: Burning Glass Technologies

Onofrei M, Hunt J, Siemienczuk J, Touchette DR, Middleton B (2004) A first step towards translating evidence into practice: heart failure in a community practice-based research network. Inform Prim Care 12(3). https://doi.org/10.14236/jhi.v12i3.119

OpenAI (Nov 2023) ChatGPT personal communication

Schoenherr T, Speier-Pero C (2015) Data science, predictive analytics, and big data in supply chain management: current state and future potential. J Bus Logist 36(1). https://doi.org/10.1111/jbl.12082

Sorger S (2013) Marketing analytics: strategic models and metrics

Sundararajan S (2021) Business analytics-overview, curriculum, opportunities and skills. Researchgate.Net

Sundararajan S (2023) MVA-ML. https://github.com/sun-sri/MVA-ML

Chapter 2
Python for Data Analytics

Learning Objectives

- Acquire Familiarity with the Python Programming Environment.
- Demonstrate the use of Python programming constructs—Variables, Operators, Data Structures, Control Flow, and Functions and Libraries.
- Examine Pandas software package functionality for data manipulation.
- Examine NumPy software package functionality for matrix algebra and numeric operations.
- Illustrate basic data processing operations.

Overview

In this chapter, we take a whirlwind tour of programming in Python, focusing on programming concepts necessary to complete the tutorials and exercises for applying statistical, computational, and machine-learning techniques.

Those already familiar with the technical nuances of the programming environment offered by Python can skip Sect. 2.1 and directly move to Sect. 2.2, where we start programming exercises. Section 2.1.1 introduces the Python software ecosystem—Python programming language and associated software packages. We need an environment to code and test Python programs—an Integrated Development Environment or IDE. Section 2.1.2 IDE discusses popular IDEs. Section 2.1.3 provides a formal view of the Python programming language.

Supplementary Information The online version contains supplementary material available at https://doi.org/10.1007/978-981-99-0353-5_2.

The programming constructs are discussed in Sects. 2.2–2.5, along with associated tutorials. The sections include Variables and Operators, Advanced Data Structures, Control Flow, and Functions and Libraries. This is followed by tutorials on the NumPy package, which is highly effective in numerical computations such as linear algebra, Fourier transforms, and matrix manipulation. Tutorials on the Pandas package for data analysis and manipulation follow this.

Understanding and preparing data for analysis is an essential step in data analytics. The last sub-section introduces the basic concepts related to data manipulation—data features, data types, and preprocessing techniques.

Definitions

Data discretization—A numeric variable can be grouped into categories. This method is called discretization.

Data preprocessing is an essential step before data mining and model building. Missing values, noise, outliers, inconsistencies, multiple data formats, and the like make data preprocessing challenging. Data preprocessing includes cleaning, integration, reduction (of features and data volume), transformation, discretization, standardization, and summarization.

Data standardization helps us compare variables measured on different scales, e.g., pounds and kilograms. There are many methods for data standardization—min–max, z-score, and decimal scaling. Data standardization helps in improving the performance of specific algorithms.

An integrated development environment (IDE) is a set of software tools that helps a programmer develop and test computer programs. Examples include Spyder IDE, Jupyter Notebook IDE, and Google Colab IDE.

Keras is a framework for deep learning.

Matplotlib and Seaborn are software packages for data visualization.

NetworkX is a software package for Graphs and Network Programming.

NumPy is a popular software package for matrix algebra and numerical operations.

Outliers are the data points that deviate considerably from the rest of the data in a dataset.

Pandas is a powerful, flexible, and easy-to-use data analysis and manipulation tool built on top of the Python programming language.

Python is an object-oriented programming language. It is also a structured language and a procedural language.

Scikit-learn is a software package for statistics and machine learning.

SciPy is a software package for statistics and optimization.

Statsmodels is a software package for statistics.

TensorFlow is a framework for deep learning.

2.1 The Python Environment

Python is one of the top two popular programming languages. It is a free and open-source project with numerous existing and emerging software libraries and is supported worldwide by researchers, students, programming enthusiasts, and numerous communities. Python was created in the early 1990s by Guido van Rossum at Centrum Wiskunde & Informatica, a mathematics and computer science research institute in Amsterdam. Python Software Foundation, a non-profit organization, owns Python-related intellectual property, with its principal office in Fredericksburg, Virginia.

This chapter is not a substitute for a regular programming course in Python. The chapter focuses on programming concepts necessary for doing the tutorials and exercises in this book oriented toward applying statistical, computational, and Python-based techniques for data analytics.

2.1.1 The World of Python

Python is a multipurpose programming language that can be used for developing business applications, websites, scientific applications, statistical models, and machine learning models. Python's standard libraries and ever-growing community contributions allow for endless possibilities of programming—https://docs.python.org/.

Technically, the capabilities of Python can be summarized as follows:

1. Pandas package for data access, analysis, and processing.
2. NumPy package (a Python library for multidimensional matrix-algebra operations and numerical computations).
3. NetworkX package for Graphs and Network Programming. For details, refer to https://networkx.org/.
4. Matplotlib graphics library for creating static, animated, and interactive visualizations in Python (the underlying modules of Matplotlib are written in C). For cheatsheets and handouts, please refer to https://github.com/matplotlib/cheatsheets.
5. Seaborn—a Python data visualization library based on matplotlib. It provides a high-level interface for creating rich statistical graphs quickly. For detailed information and sample codes, please refer to https://seaborn.pydata.org/.
6. SciPy algorithms for statistics, optimization, and other computational techniques written in C and wrapped in Python. Please refer to https://scipy.github.io/devdocs/ for details.
7. Statsmodels Python package. It complements SCiPy in statistical computations, including descriptive statistics, estimation, and hypothesis testing. For detailed information, you may refer to https://pypi.org/project/statsmodels/.

8. Scikit-learn Python library for statistics and machine learning, built on SciPy, NumPy, and matplotlib libraries.
9. TensorFlow framework for deep learning—a Python library (that invokes C++ modules to construct and execute dataflow graphs). For detailed information and code samples, please refer to https://www.tensorflow.org/.
10. Keras' deep learning API is written in Python and runs on the TensorFlow platform. Please refer to https://keras.io/guides/ for guidance.
11. Inbuilt libraries for Internet access.
12. Website development using Python-based frameworks such as Django, and Flask.
13. Desktop GUIs.
14. Packages for Business application Development.

2.1.2 IDE for Python Programming

An integrated development environment (IDE) is a set of software tools that helps a programmer to develop computer programs. The following three components form the basic building blocks of an IDE.

- Editor: A text editor for writing program code. It has added features such as automatic formatting, syntax check, visual prompts, auto code completion, and flagging out simple programming mistakes.
- Code Build: Commands for compiling program code into binary code, packaging it, and executing it.
- Debugger: A program that helps to trace the execution of program code and locate bugs if present. Graphical displays help a programmer trace the instructions and display the associated data (e.g., the value of variables).

Numerous IDEs exist for developing and testing Python programs—Jupyter Notebook, Google Colab, and Spyder. We can install Spyder IDE or Jupyter Notebook IDE in our local system and use them for programming without internet access. To work in Colab, internet access is necessary. The code snippets in this book are tested in Google Colab IDE and Spyder IDE.

The author recommends Google Colab IDE—https://Colab.research.google.com/. It is a product from Google Research. It comes with a set of utilities and interfaces that make programming easy. The data in Google Drive can be accessed, and the program code can be shared via email with others. It allows parallelism using GPUs. A programmer can open and use multiple Colab notebooks at the same time. Colab is a hosted Jupyter Notebook service that provides free access to computing resources, including GPUs.

The websites of these IDEs provide adequate help with their usage. If you are new to programming, start with the simple exercises provided by w3schools (https://www.w3schools.com/python/). You can try out code snippets in the lovely little pop-up window 'Try it Yourself'. Then, move on to the Google Colab introduction and exercises (https://Colab.research.google.com/).

2.1.3 Defining Python Programming Language

Primarily, Python is an object-oriented programming language. Python shares many other interesting features also. In Python, we are not allowed to use 'go to' statements that arbitrarily transfer the flow of control—therefore, Python is a structured programming language. Python is a procedural language since a Python program module consists of a set of instructions that are executed in sequence. We can define and use classes and objects in Python, supporting object-oriented programming. Python is a dynamically typed language—it is not necessary to declare the variable type before assigning a value to it. Python has high-level data types suitable for data manipulation and number crunching. The feature-rich Python offers numerous benefits over most other programming languages like Fortran, C, C++, Java, or R.

The Python program that we write is first compiled into byte code. The byte code is platform-independent. The interpreter converts bytecode to instructions specific to an operating system (Windows, Mac-OS, etc.). To the programmer, the above two steps are transparent. The programmer finds that his source code can be run on any platform. Therefore, it is believed that Python is an interpreted language.

2.2 Variables and Operators

In Python, variables are not explicitly declared. They are created from assignment statements. Integer (int), Decimal Numbers (float), Character String (str), True/False (bool), etc., are some of the basic data types available in Python, like any other programming language. List, Tuple, Set, and Dictionary are more advanced data types. The primary data types and operators are shown in Tables 2.1 and 2.2, respectively.

Let us work out some examples. Open a notebook in the IDE of your choice. Type the code below. Execute it with the 'run' command. The output displayed on the monitor is shown after '#>>'.

Table 2.1 Data types

Basic data types	Advanced data types
Integer (int)	List
Decimal numbers (float)	Range
Character string (str)	Tuple
Bool (True/False)	Set
Complex	Dictionary

Table 2.2 Operators

Operators	Symbols
Assignment operators	=, +=, −= etc.
Arithmetic operators	+, −, *, /, %, //, **
Comparison operators	==, ! =, >, <, <=, >=
Logical operators	and, or, not
Bitwise operators	&, |, ^, ~, <<, #>>
Membership operators	in, not in
Identity operators	is, is not

Tutorial 2.2.1 Variables and Operators

```
# Display "Hello" on the screen
```
In the Tutorial below, x is a variable. A string (of characters) is assigned to x. Therefore, x assumes the data type 'str'. '#' indicates that the rest of the line must be treated as a comment

```
x = "Hello"
print(x)
# Hello
#
# The print function
```
The 'print' is an inbuilt function defined for displaying the output on our display monitor.

```
x = "Hello"
y = 'Python!'
print(x,y)
# Hello Python!
```

```
# String concatenation
```
Here, the print function takes a set of parameters, x, and y, and concatenates them (appends y after x). Note that when we concatenate two strings, in the resultant string, there is no intervening space

```
x = "Hello"
print(x)
y = 'Python!'
print(x+y)
#HelloPython!
#
```

Tutorial 2.2.2 Variable Names

As we learned, in Python, variables of appropriate types are created when a value is assigned to them. A variable name must start with a letter or the underscore character. A variable name can have alpha-numeric characters and underscores only (A to Z, a to z, 0 to 9, _)

```
05September = 'Mother's Day'   #invalid variable name
September05 = 'Mother's Day'   #valid variable name
_05September = 'Mother's Day' #valid variable name
```

```
World Bicycle Day = '2021 June 3'   #invalid variable name
World-Bicycle-Day = '2021 June 3'   #invalid variable name
World_Bicycle_Day = '2021 June 3'    #valid variable name
```

Tutorial 2.2.3 Basic Data Types int, float, str, bool

Four variables with different data types are used below. The print function receives the four variables as input and displays them all (separated by a space)

```
i = 10              # int
f = 14.1            # float
b = True            # bool
s = 'pounds is equal to dollar' # str
print(i, s, f, b)
# 10 pounds is equal to dollar 14.1 True
```

Tutorial 2.2.4 Comments

We use '#' to indicate that the rest of the line is a comment text. We use ''' to enclose comments running into multiple lines (before and after the body of text).

```
# Bool takes one of the values - True or False.
# Note that the first letter 'T' of True and 'F' of False is in uppercase
'''

Basic Data Types are:    int, float, str, bool, complex,
Advanced Data Types are: list, tuple, range, set, dict
'''
```

Tutorial 2.2.5 Type Setting and Conversion

We noted earlier that we do not declare any variables in Python. However, Python allows us to set the data type if we want to do so

```
i = 1
# convert the integer 'i' to float and assign it to the variable f.
f = float(i)
print('i =', i, "  and i after conversion =",f)
# f = 1    and f after conversion = 2
#
print('data type of f = ',f, type(f))
# data type of f =  2.54 <class 'float'>
#
# convert the float 'f' to an integer and assign it to the variable i.
i = int(f)
```

```
print('f =', f, "  and f after conversion =",i)
# f = 1.0   and f after conversion = 1
# data type of i =  1 <class 'int'>
```

Tutorial 2.2.6 Assignment Operators

We saw the assignment operator '=' in the previous Tutorials. Assignments can be more elegantly done. For example, multiple variables can be assigned values, all in a single statement. Many more assignment operators, like +=, -=, *=, /=, etc., are not covered here.

```
#
i, f, b, s = 10, 14.1, True, 'pounds is equal to dollar'
print(i, s, f, b)
# 10 pounds is equal to a dollar 14.1 True
```

Tutorial 2.2.7 Arithmetic Operators

The arithmetic operations using Python are demonstrated below.

```
#
x = 10; y =3
x + y          # 13    : Addition
x - y          # 7     : Subtraction
x * y          # 30    : Multiplication
x / y          # 3.3333333333333335 : Division
x % y          # 1     : Modulus (Reminder)
x // y         # 3     : Floor division (Quotient)
x ** y         # 1000 : Exponentiation
```

Tutorial 2.2.8 Comparison Operators

The result of a comparison operation will be True or False

```
x =  2; y = 3
x == y # False : Equal?
x != y # True  : Not equal?
x >  y # False : Greater than?
x <  y # True  : Less than?
x >= y # False : Greater than or equal to?
x <= y # True  : Less than or equal to?
```

Tutorial 2.2.9 Logical Operators

Logical Operators are 'and', 'or', 'not'. They return True or False

```
x =  2; y = 3
x < 3 and y < 3
# False
```

```
# or:   Returns True if one of the statements is true
x < 3 or  x < 3
# True

# not: Reverses the result
not(x < 5 and y < 5)
# False
```

Tutorial 2.2.10 Strings

The Python compiler considers a string as an array of characters.

```
#
s = 'Hello, World!'
print(s)
# Hello, World!
#
print(len(s)) # returns the length of a string
# 13

print(s[0])   # returns the first character of the string
# H

print("lo" in s)  #returns true or false
# True

s[0:5]          # substring from position 0 to 2; 5th position not included
# Hello
```

Tutorial 2.2.11 Print Formatting - Insert Variables in Strings

This example shows one of the ways to structure the print output. The characters {} can be used as placeholders for variables while printing strings.

```
#
quantity    = 4
itemNo      = 1234     # Assume 1234 is the item number of Cup Cake
price       = 0.99
Order       = 'I want {} pieces of Cup Cakes Item# {} for $ {} each'
#                       ^                          ^       ^
print(Order.format(quantity,                  itemNo,  price))
# I want 4 pieces of Cup Cakes Item# 1234 for $ 0.99 each
```

2.3   Advanced Data Structures

This section will explore Lists, Tuple, Set, Range, and Dictionary. This book
extensively uses two of the above data types: list and range.

List

- List items that are ordered and changeable and allow duplicate values.
- Ordered and Indexed: The first item has an index [0], the second item has an index [1], etc.
- Changeable: We can update, add, and remove items in a list.
- Allow duplicate values (as they are Ordered and Indexed).

Tuple

- A tuple is a collection of items that are ordered and immutable (cannot be changed).
- Ordered and Indexed: The first item has an index [0], the second item has an index [1], etc.
- Allow duplicate values (as they are Ordered and Indexed).
- Tuples are immutable: we cannot change, add, or remove items.

Set

- A set is a collection that is unordered, immutable, and unindexed.
- We can remove and add new items using set class 'methods'.
- Does not allow duplicate values.

Range

- The range() construct returns a sequence of numbers.

Dictionary

- Dictionaries store data values in key: value pairs. Dictionaries can be nested.

Tutorial 2.3.1 Advanced Data Types

We will first get familiar with three similar data types - List, Tuple, and Set.
- LIST is ordered, changeable (allows duplicates)
- TUPLE is ordered, unchangeable (allows duplicates)
- SET is unordered. No duplicates allowed

```
#
list1 = ['tomato', 'beans', 'carrot', 'beans']
print(list1[0])
# 'tomato'
tuple1 = ('tomato', 'beans', 'tomato')
print(tuple1[1])
# 'beans'
set1   = {'tomato', 'beans', 'carrot'}
print(set1[0])
# TypeError: 'set' object is not subscriptable
```

Tutorial 2.3.2 List operations

The list is an important data structure.
Let us find out how to access a subset of items from a list. Note that we can use positive and negative indices for obtaining a subset.

```
# Counting the number of elements in a set
list1 = ['tomato', 'beans', 'carrot']
len(list1)
# 3

# list subsetting
list1 = ['tomato', 'beans', 'carrot']
# To print the last two items
print(list1[1:3])
# ['beans', 'carrot']
# Note that, list1[1:3] means print items 1..2. The 3rd item is not
printed

# list - 'append' function
# Adding an item to the end of a list
list1 = ['tomato', 'beans', 'carrot']
list1.append('pear')
print(list1)
# ['tomato', 'beans', 'carrot', 'pear']

# list - lookup for an item
# Lookup for an item. Returns True/False
'tomato' in list1
# True

# list - concatenate two lists
list1 = ['tomato', 'beans', 'carrot','milk']
list2 = ['bread', 'butter', 'milk']
listBig = list1 + list2
print(listBig)
# ['tomato', 'beans', 'carrot', 'milk', 'bread', 'butter', 'milk']

# list - insert: Insert item 'plums' at position 0
list1 = ['tomato', 'beans', 'carrot', 'milk', 'beans']
list1.insert(0,'plums')
print(list1)
# ['plums', 'tomato', 'beans', 'carrot', 'milk', 'beans']

# list - remove: Remove the first occurrence of the specified item
list1.remove('plums')
print(list1)
# ['tomato', 'beans', 'carrot', 'milk', 'beans']
```

```
# How many occurrences of 'beans' are there?
list1.count('beans')
# 2

# How many occurrences of 'cabbage' are there?
list1.count('cabbage')
# 0

# Find the position of the first occurrence of a particular item
list1.index('beans')
# 1
list1.index('carrot')
# 2
```

Tutorial 2.3.3 Tuple operations

Note: The triple data type is used only in very few places in this book.

```
# A tuple allows duplicates
tuple1 = ('tomato', 'beans', 'tomato')
print(tuple1)
# ('tomato', 'beans', 'tomato')

# the count of items in a tuple
len -> length -> number of items
len(tuple1)
# 3

# tuple items can be of different data types
tuple2 = (1, 5, 7, True, False, 4.75)
print(tuple2)
# (1, 5, 7, True, False, 4.75)

# one-item tuple
```
We must add a comma after the item to create a tuple with only one item.
Otherwise, Python will consider it to be a string.

```
tuple3 = ('tomato')    # this is not a tuple. It is a string
print(tuple3)
#tomato

tuple4 = ('tomato',)    #-> this is tuple; note the comma!
print(tuple4)
#('tomato',)

#a tuple is an ordered collection of items
tuple1 = ('tomato', 'beans', 'tomato')
print(tuple1[0])
# tomato
```

```
#a tuple is not mutable
tuple1[0] = 'cucumber'
# *** TypeError: 'tuple' object does not support item assignment

#tuple subsetting
```
Accessing a subset of items from a tuple is similar to that of a list.
We can use positive and negative indices to obtain a subset.

```
print(tuple1[0:2])
# ('tomato', 'beans')
# Note: tuple1[0:2] means items 0..1. The 2nd item onwards is discarded
```

Tutorial 2.3.4 Range construct

The range is a fundamental construct. Range object has just three components:
start, stop, and increment. The increment can be negative or positive. By
default, the increment is 1. Range helps to create a sequence of numbers in
ascending or descending order.

```
range1 = range (0,100,12)
type(range1)
# <class 'range'>

range1
# range(0, 100, 12)

listr = list(range (0,100,12)) # this constructs a list
print(listr)
# [0, 12, 24, 36, 48, 60, 72, 84, 96]

# Range (Start, Stop, Increment)
```
range(0,5,1) means to start with '0', increment by 1, and generate integers
sequentially up to 5 (but do not include 5)
```
list(range(0,5,1))
# [0, 1, 2, 3, 4]

# range(0,5,1) is same as range(0,5).By default, the increment is 1
list(range(0,5))
# [0, 1, 2, 3, 4]
```

Tutorial 2.3.5 Dictionary

Dictionaries store data values in key: value pairs. Dictionaries can be nested. This section is optional for doing the tutorials/exercises in this book.
```
#
#Create a dictionary with index 'brand', 'model', 'year'
dict1 = {'brand': 'hyundai', 'model': 'i10', 'year': 2011}
print(dict1)
# {'brand': 'hyundai', 'model': 'i10', 'year': 2011}
```

```
dict1['brand'] #here key is 'brand'; data value is 'hyundai'
# 'hyundai'

dict1['model']
# 'i10'

dict1['year'] = 2021    # dictionary is mutable
print(dict1)
# {'brand': 'hyundai', 'model': 'i10', 'year': 2021}


#Create a two-dimensional dictionary
vegPrices = {'itemNo':  [1, 3, 4, 8, 10],
'itemName': ['oranges','apples','bananas','peaches','grapes'],
             'unitPrice': [3.5, 2.5, 4, 4, 2.5],
             'unitQty':  [1, 1, 1, 1, 1]}
print(vegPrices)
#
# {'itemNo': [1, 3, 4, 8, 10],
# 'itemName': ['oranges', 'apples', 'bananas', 'peaches', 'grapes'],
# 'unitPrice': [3.5, 2.5, 4, 4, 2.5], 'unitQty': [1, 1, 1, 1, 1]}
#
len(vegPrices)
# 4
len(vegPrices['itemNo'])
# 5
```

Tutorial 2.3.6 Set operations

Note: - This section is optional for doing the tutorials/exercises in this book.

```
# Set is not ordered and cannot be updated
set1    = {'tomato', 'beans', 'carrot'}
set1[0]
# TypeError: 'set' object is not subscriptable
set1[1] = 'tomato'
# TypeError: 'set' object does not support item assignment

# Counting the number of elements in a set
len(set1)
# 3

# Duplicate elements are not stored in a set
set2 = {1, 5, 7, 9, 3, 1}
print(set2)
# {1, 3, 5, 7, 9} # note that item '1': duplicates are not stored

# Set items can be of different data types
set3 = {'tomato', 1, True, 40.4}

# Set operations
```

```
set1     = {'tomato', 'beans', 'carrot'}
set2     = {1, 5, 7, 9, 3, 1}
set2.union(set3)         # set union
# {1, 3, 40.4, 5, 7, 9, 'tomato'}

set4.intersection(set1)   # set intersection
# {'beans', 'carrot', 'tomato'}

set4.difference(set1)     # set difference
# {1, 3, 5, 7, 9, 40.4}
```

2.4   Control Flow

Control Flow Statements

Control flow statements work on the following principle—(a) make a logical decision and (b) based on the decision, decide which set of statements should be executed next. The following are the control flow statements in Python.

* if-else elif statement for decision-making

  if condition:

  > block of statementsto be executed if the condition is true

  elif condition:

  > block of statementsto be executed if the condition is true

  else:

  > block of statementsto be executed if the condition is true

* for loop for iteration

  for variable in range (start, stop, step):

  > block of statementsto be executed if the condition is true

* while condition:

  block of statementsto be executed if the condition is true

Indentation

Indentation refers to the space prefixing a line of code. Python uses indentation to identify a block of code under control structures (if, for, while), function definition, etc. Python will throw an error if indentation is not given correctly.

Breaking the Control Flow

Python supports the following statements for breaking the control flow statements:

- continue statement
- break statement
- pass statement.

Tutorial 2.4.1 Control Flow - introductory example

Let us get familiar with two distinct control flow statements by demonstration. Explanations are provided in the subsequent tutorials.

```
# if example
x, y = 20, 10
if x > y:
    print ('the bigger number is x')
# the bigger number is x

# for example
for x in range(4):
  print(x)
# 0   1 2 3
```

Tutorial 2.4.2 if

This is an important construct.
The logical expression x > y is evaluated in the if statement below. If the
outcome is True, the print statement is executed. The ':' after the condition
'x>y' signifies that the 'if block' statements follow. You can use several
statements, all of which must have the same indentation.

```
x, y = 20, 10
if x > y:
    print ('the bigger number is x')
# the bigger number is x
```

In the example below, the second print statement must be indented with the
same space as the first one. If the second statement is out of the if block,
then it must have the same indentation as if

```
#
if x > y:
    print ('the bigger number is x')
        print ("... indentation error!")
# *** Indentation Error: unexpected indent
```

In the example below, the second print statement must be indented with the
same space as the first one. If the second statement is out of the if block,
then it must have the same indentation as if

```
#
if x > y:
    print ('the bigger number is x')
   print ("... indentation error!")
# *** indentation Error: unindent does not match any outer indentation
```

The example below shows that the second print statement is outside the if block. It is fine.

```
#
if x > y:
    print ('the bigger number is x')
print ("... we are outside the if block!")
# the bigger number is x
# We are outside the if block!
```

Tutorial 2.4.3 if ... else ... example

The logical expression x > y is evaluated in the if statement below. If the outcome is True, the first print statement is executed. Otherwise, the second print statement is executed.

```
#
x, y = 10, 10
if x > y:
    print ('the bigger number is x')
else:
    print ('the bigger number is y / x and y are equal)
# the bigger number is y / x and y are equal
```

Tutorial 2.4.4 if ... elif ... else ... example

The logical expression x > y is evaluated in the if statement below. If the outcome is True, the first print statement is executed, followed by the print 'job over'. Otherwise, the 'elif' condition is checked. If the outcome is True, the second print statement is executed, followed by the print 'job over'. If the outcome of elif is False, the third print statement is executed, followed by the print 'job over'

```
#
x, y = 10, 20
if x > y:
    print ('the bigger number is x')
elif y > x:
    print ('the bigger number is y')
else:
    print ('x and y are equal')
print('job over')
# x and y are equal
# job over
```

Tutorial 2.4.5 Nested if

We can nest an if statement block within another if statement block and so on…, with progressive indentations.

```
#
x = 40
if x > 0:
  print('x is a positive number')
  if x % 7 == 0:
    print('and divisible by 7')
  else:
    print('but not divisible by 7')
# x is a positive number
# but not divisible by 7
```

Tutorial 2.4.6 pass (a verb used in if statements)

This is an important construct. Pass implies doing nothing. In for, and while loops the equivalent construct is 'continue.'

```
#
a = 33
b = 34
if b > a:
  pass
```

Tutorial 2.4.7 for loop with range

This is an important construct. A for loop repeatedly executes a code block while iterating over a sequence (range, string, list, tuple, set, or dictionary key/value). In the example below, print(x) is repeated 6 times.

```
#
for x in range(4):
  print(x)
# 0
# 1
# 2
# 3
```

Tutorial 2.4.8 for loop with range (start, stop, step)

The parameters of the range are (start, stop, step). It comes off the for loop when it reaches 'stop' or exceeds it.

```
#
for x in range(0, 12, 4):
  print(x)
# 0
# 4              <- (0+4)
# 8              <- (0+4+4)
```

The step can be negative, as shown below

```
#
for x in range(12, 4, -3):
    print(x)
# 12
# 9            <- 12-3
# 6            <- 12-3-3
```

Tutorial 2.4.9 for loop with a string

```
#
for x in 'mango':
  print(x)
# m
# a
# n
# g
# o
```

Tutorial 2.4.10 for loop with list, tuple, set

Note: - This section is optional for doing the tutorials / exercises in this book.

```
# Iterate through the items
list1 = ['books', 'pen', 'pencil']
for x in list1: print(x)
# books
# pen
# pencil

tuple1 = ('books', 'pen', 'pencil')
for x in tuple1: print(x)

set1   = {'books', 'pen', 'pencil'}
for x in tuple1: print(x)
```

Tutorial 2.4.11 break and continue

This is an important construct. When 'break' is encountered, the for loop terminates

```
#
list1 = ['books', 'pen', 'pencil', 'eraser']
for x in list1:
  if x == 'pencil':
    break
  print(x)
# books
# pen
```

When continue is encountered, the subsequent statements under the block are
not executed; the control jumps over to the next iteration. It is similar to
the 'pass' verb in the if statement. In the following example when the item
'pen' is reached, 'continue' is encountered. Control jumps over to the next
iteration, without printing 'pen'.

```
list1 = ['books', 'pen', 'pencil', 'eraser']
for x in list1:
  if x == 'pen':
    continue
  print(x)
# books
# pencil
# eraser
```

Tutorial 2.4.12 list comprehension

Data analytics professionals frequently use this method.

```
# Define list1
list1 = ['tomato', 'beans', 'carrot', 'beans']
print(list1)
# ['tomato', 'beans', 'carrot', 'beans']

# Get all items in the list that has the letter 'o' in them
list2 = [i for i in list1 if 'o' in i]
print(list2)
# ['tomato', 'carrot']

# Get all items in the list that are equal to 'beans'
list3 = [i for i in list1 if i == 'beans']
print(list3)
# ['beans', 'beans']

# List of all numbers divisible by 7
list4 = [i for i in range(0,100,1) if i%17 == 0]
print(list4)
# [0, 17, 34, 51, 68, 85]
```

Tutorial 2.4.13 iterations through a dictionary

Note: - This section is optional for doing the tutorials/exercises in this
book.

```
# Define Dictionary vegPrices, with 5 keys,
# itemNo, itemName, unitPrice, and uniQty
```

```
vegPrices={'itemNo':     [1, 3, 4, 8, 10],
           'itemName': ['oranges','apples','bananas','peaches','grapes'],
           'unitPrice': [3.5, 2.5, 4, 4, 2.5],
           'unitQty':    [1, 1, 1, 1, 1]}

#print the keys
for i in vegPrices.keys():
    print(i)
# itemNo
# itemName
# unitPrice
# unitQty

#print the values
for i in vegPrices.values():
    print(i)
# [1, 3, 4, 8, 10]
# ['oranges', 'apples', 'bananas', 'peaches', 'grapes']
# [3.5, 2.5, 4, 4, 2.5]
# [1, 1, 1, 1, 1]

#print the keys and values
for i in vegPrices.keys(), vegPrices.values():
  print(i)

# dict_keys(['itemNo', 'itemName', 'unitPrice', 'unitQty'])
# dict_values([[1, 3, 4, 8, 10], ['oranges', 'apples', 'bananas',
'peaches', 'grapes'], [3.5, 2.5, 4, 4, 2.5], [1, 1, 1, 1, 1]])
```

Tutorial 2.4.14 while loop

Note: - This section is optional for doing the tutorials/exercises in this book.

While loop is similar to for loop. The difference is that the condition check is done before the execution. In the following example, the value of 'i' is checked before entering the statement block below. The code block will be executed if i is less than 5; otherwise, the while loop terminates.

```
#
i = 0
while i < 5:
  print(i)
  i += 1
# 0
# 1
# 2
# 3
# 4
```

2.5   Functions and Libraries

Functions take a set of parameters separated by a ','. The parameters (or arguments) are enclosed within parathesis () immediately after the function name. All Python functions return one value. Numerous functions are available with Python libraries. Alternatively, we can define our functions.

Tutorial 2.5.1 Functions and Libraries - An introductory example

```
x = [7, 3, 3, 5, 3, 4, 6, 0, 6, 8, 3, 3, 5, 3, 4, 3, 3, 5, 3, 4]
```

Let us start with a commonly used library - 'statistics'. This library contains several functions available to all Python users. We need to import the library into our programming environment.

Let us try out the 'mode' function. It computes the statistics 'mode' of a distribution. The parameters of a function are provided in parenthesis (), immediately following the function name. A comma separates each parameter. The 'mode' function takes one parameter - a list of numbers.

```
#
from statistics import mode
mod = mode(x)
print ('mode is', mod)
# mode is 3
#
```

Tutorial 2.5.2 define your function

```
# Write a function to return the square of a given number
def square(x):
    return x * x
#
# Call 'square', with an argument with the appropriate data type
square(3)
# 9
#
# Write a function to return the volume of a solid
def volume(l,b,h):
    return l * b * h
```

Call the function volume, with the required parameters (arguments) matching in order and data type. Here, the parameters are length=2; breadth=3; and height=4; all integers. (Note that the size of the parameters must also match while using arrays or complex structures)

```
#
print ('volume =>', volume(2,3,4))
# 24
```

Tutorial 2.5.3 lambda - the function without a name

Lambda function is unique to Python. This function does not have a name. Like any other Python function, lambda takes several parameters (arguments) and returns one output. The difference is that the lambda function can evaluate only one expression.

```
# store the lambda function object in a variable 'volume'
volume = lambda l, b, h : l * b * h
# Use 'volume' as a function
print(volume(3, 4, 5))
# 60
```

Tutorial 2.5.4 commonly used in-built library functions

Let us take a look at standard functions such as abs, chr, format, int, input, len, max, min, pow, round, sum, type

```
#
abs(-100)          # 100
int(3.1416)        # 3
round(3.1416,2)    # 3.14    #rounds the result to 2 decimal places
pow(2,10)          # 1024    (power function: 2^10 = 1024)
The functions max, min, sum are self-explanatory
```

The function 'len' returns the number of items in a list, tuple, set, dictionary; the number of rows in a dataset, etc.

```
#
x = [7, 3, 3, 5, 3, 4, 6, 0, 6, 8, 3, 3, 5, 3, 4, 3, 3, 5, 3, 4]
len(x)   # 20
```

```
# format: alignment (justification), comma separator, signs etc
format(1380004385, ',')
# '1,380,004,385'
```

```
# input: prompt for an input string
x = input('Enter your name:'); print('Hello, ' + x)
# Enter your name:X
# Hello, X
```

Note: - The following tutorial on 'chr' is optional for doing the tutorials / exercises in this book.

```
chr: returns the character value of an integer in Unicode notation
for i in range(2309,2362): print(chr(i))   # अ, आ, इ, ई,...,श, ष, स, ह
for i in range(48,123): print(i,chr(i))   # 0..9, A..Z, a..z, special
chars
```

Tutorial 2.5.5 math library

The 'math' and 'cmath' libraries (complex numbers math) provide several mathematical functions.

The 'math' library has inbuilt functions for ceiling/floor, combination/ permutation, lcm/gcd, exponentiation (power); hyperbolic functions; logarithmic functions; trigonometric functions; special functions; and mathematical constants. Let us take a look at a few common 'math' functions: -

```
#
import math
math.sqrt(829921)    # 911.0
math.ceil(73.27)     # 74
math.floor(73.27)    # 73
math.pow(2,10)       # 1024
```

Tutorial 2.5.6 date time

```
import datetime
now = datetime.datetime.now()
print ("Current date and time : ")
print (now.strftime("%Y-%m-%d %H:%M:%S"))
# Current date and time :
# 2021-11-16 01:02:32
#
import time
s = time.time()
print(s)
# 1637004821.3311868
```

Tutorial 2.5.7 random numbers

See the Numpy section on generating random numbers

Tutorial 2.5.8 advanced string functions

Note: - This section is optional for doing the tutorials/exercises in this book.

```
#
s  = "on May 11 1997, in midtown Manhattan, IBM's Deep Blue \
beat the reigning world champion, Gary Kasparov, in a six-game match.\
Critics question the worth of research into computer chess. \
MIT linguist Noam Chomsky has said that \
a computer program's beating a grandmaster at chess in 1997 is about as\
interesting as a bulldozer winning an Olympic weight-lifting competition\
Deep Blue is indeed a bulldozer of sorts--its 256 parallel processors\
enable it to examine 200 million possible moves per second and to \
look ahead as many as fourteen turns of play."
```

```
#
s.find('computer')    #  167
s.find('www')         # -1 implies not found
s.count('as')         #  7
# parse into words separated by space
listk=list(set(s.split(' ')))
print(listk)
'''

['match.Critics', 'processorsenable', 'champion,', 'Kasparov,', ... ,
'Gary', 'per', 'possible', 'winning', 'indeed']
'''

#
listk.sort() # sort the words list
'''

['11', '1997', '1997,', '200', '256', ...,
'to', 'turns', 'weight-lifting', 'winning', 'world', 'worth']
'''

print(listk)
#
len(listk)   # how may words? 71
# word count
for i in listk: print(i, s.count(i)) # word count
'''

11 1
1997 2
...
the 2
to 4
turns 1
weight-lifting 1
winning 1
world 1
worth 1
'''

#
```

Tutorial 2.5.9 regular expressions library

Note: - This section is optional for doing the tutorials/exercises in this book.

Regular expressions allow us to match text strings, such as characters and words and extract string patterns from a given text. In this example, we consider the string 's' used in the Tutorial 'advanced string functions'.

```
#
import re
# locate the first occurrence of substring 'IBM'
x = re.search("IBM",s)
print(x) # <re.Match object; span=(38, 41), match='IBM'>
```

```
x.span() # (38, 41)

# locate the first occurrence of white space
re.search('\s',s) #<re.Match object; span=(2, 3), match=' '>

re.findall("the", s) # ['the', 'the'] - two occurrences listed

# list the items that are separated by 'space character'
re.split('\s',s)
# >> ['on', 'May', '11',..., 'turns', 'of', 'play.']

re.sub("\s", ",", s) #replace white space with comma
# on,May,11,... ,turns,of,play."

# replace the first two white spaces with #
re.sub("\s", "#", s,2) # "on#May#11 1997...
```

2.6 Pandas

Pandas is a powerful, flexible, and easy-to-use open-source data analysis and manipulation tool, built on top of the Python programming language. Pandas can access multiple file formats such as CSV (comma-separated values), excel, JSON, HTML, flat files with a fixed format, flat files with delimited values, clipboard, SQL, parquet (flat columns), pickle (to serialize/de-serialize objects for parallel processing), google big query, ORC, SAS, SPSS, and STATA. Pandas DataFrame is a 2D (two-dimensional) data structure, like a spreadsheet, with rows and columns. The Pandas project aims to become **the most powerful and flexible open-source data analysis/manipulation tool available in any language**. Pandas is actively supported by a vibrant community worldwide who contribute significantly to making open-source pandas possible. The Pandas' community experts provide immense support through Stack Overflow.

Having said the above, it may be noted that the section below merely aims to provide a simple introduction to the world of pandas to help us navigate through the book—nothing more and nothing less! You may refer to https://pandas.pydata.org/ for detailed information.

Tutorial 2.6.1 pandas - explore the 'iris' dataset

```
# Import seaborn library.
# refer: https://seaborn.pydata.org/
import seaborn as sb
```

```
# Read the iris data file using the load_dataset method of the seaborn library.
Save it in the pandas DataFrame 'pdf'.
    pdf = sb.load_dataset('iris')
```

```
# check the data type of 'pdf'.
   type(pdf)
   # pandas.core.frame.DataFrame

# check the dimensions of 'pdf' - #rows, #columns)
   pdf.shape
   # (150, 5)

   pdf.columns    # list the column names
   #'sepal_length', 'sepal_width', 'petal_length', 'petal_width','species'

   # Print a summary of the DataFrame - columns, data types, non-null counts
   pdf.info()
   #
   '''
<class 'pandas.core.frame.DataFrame'>
RangeIndex: 150 entries, 0 to 149          -> 150 rows
Data columns (total 5 columns):
 #   Column        Non-Null Count  Dtype
---  ------        --------------  -----
 0   sepal_length  150 non-null    float64
 1   sepal_width   150 non-null    float64
 2   petal_length  150 non-null    float64
 3   petal_width   150 non-null    float64
 4   species       150 non-null    object
dtypes: float64(4), object(1)
'''

   #
Descriptive statistics for numeric data - central tendencies, dispersion etc,
excluding 'NaN' (non-null) values

   pdf.describe()
   #
   '''
      sepal_length  sepal_width  petal_length  petal_width
count   150.000000   150.000000    150.000000   150.000000
mean      5.843333     3.057333      3.758000     1.199333
std       0.828066     0.435866      1.765298     0.762238
min       4.300000     2.000000      1.000000     0.100000
25%       5.100000     2.800000      1.600000     0.300000
50%       5.800000     3.000000      4.350000     1.300000
75%       6.400000     3.300000      5.100000     1.800000
max       7.900000     4.400000      6.900000     2.500000
   '''

   #
   # print the data type of each column
   pdf.dtypes
```

```
#
'''
sepal_length     float64
sepal_width      float64
petal_length     float64
petal_width      float64
species           object
dtype: object
'''

  #
  # print the first two rows
  pdf.head(2) #print the first 2 rows of the DataFrame
'''
   sepal_length  sepal_width  petal_length  petal_width species
0           5.1          3.5           1.4          0.2  setosa
1           4.9          3.0           1.4          0.2  setosa
'''

  # print the last two rows
  pdf.tail(2) #print the last  2 rows of the DataFrame
#
'''
     sepal_length  sepal_width  petal_length  petal_width    species
148           6.2          3.4           5.4          2.3  virginica
149           5.9          3.0           5.1          1.8  virginica
'''

  #
  pdf.species # access the column species; same as pdf['species']
  pdf['species'] # access the column species; same as pdf.species
#
'''
0 setosa
1 setosa
..
149 virginica
Name: species, Length: 150, dtype: object
'''
```

Tutorial 2.6.2 pandas - accessing a subset of rows/columns

'iloc' and 'loc' are available for subsetting pandas data frame. Both are discussed here. However, iloc is popular
The general method for subsetting of pandas data frame using 'iloc' is pdf.iloc[r1:r2,[c1,c2]], where rows {r1 .. (r2-1)} are selected from which columns c1 and c2 are selected.

```
  # load iris using load_dataset() method from seaborn library
  import seaborn as sb
  pdf=sb.load_dataset('iris')
  #
  pdf.iloc[0:2,[1,3]] # two rows = 0th, 1st; columns 1, and 3
```

```
#
'''
   sepal_width  petal_width
0          3.5          0.2
1          3.0          0.2
'''

#
# Other Methods of Subsetting (1) - using a column name in subsetting
pdf.species[0:2] # access the first two rows of the column 'species'
#
'''
0    setosa
1    setosa
'''

   #
# Other Methods of Subsetting (2) - no 'loc', no 'iloc'
   pdf[0:2]     # two rows = 0th, 1st
   #
'''
   sepal_length  sepal_width  petal_length  petal_width species
0           5.1          3.5           1.4          0.2 setosa
1           4.9          3.0           1.4          0.2 setosa
'''

   #
   # Other methods of Subsetting (3) - using 'loc' instead of iloc
   #
For selecting a subset of rows, 'loc' uses the index, whereas, iloc uses in-
teger position. Note that it is common to use iloc
   #
   pdf.loc[0:2]  # three rows = 0th, 1st, 2nd
   #
'''
   sepal_length  sepal_width  petal_length  petal_width species
0           5.1          3.5           1.4          0.2 setosa
1           4.9          3.0           1.4          0.2 setosa
2           4.7          3.2           1.3          0.2 setosa
'''
```

Tutorial 2.6.3 pandas - filter, to access a subset of data

```
   # load iris using load_dataset() method from seaborn library
   import seaborn as sb
   pdf=sb.load_dataset('iris')

   # filter-1: bool
   pdf.species == 'setosa'
   # returns True or False for each of the 150 rows
'''
0        True
1        True
...
148     False
149     False
Name: species, Length: 150, dtype: bool
'''
```

```
  #
  # filter-2: selection of a set of rows based on a logical condition
  pdf[pdf.species == 'setosa']
  #
  # filter-3: selection of a set of rows based on a logical condition
  pdf[pdf['species'] == 'setosa'] # filter, same as above
  #
'''
```

	sepal_length	sepal_width	petal_length	petal_width	species
0	5.1	3.5	1.4	0.2	setosa
1	4.9	3.0	1.4	0.2	setosa
2	4.7	3.2	1.3	0.2	setosa
..	...	...	...	...	...
47	4.6	3.2	1.4	0.2	setosa
48	5.3	3.7	1.5	0.2	setosa
49	5.0	3.3	1.4	0.2	setosa

```
'''

  #
  # filter-4: selection of a set of rows based on a multiple conditions
  # get all rows for species setosa or versicolor
  pdf[(pdf['species']=='setosa')   | (pdf['species']=='versicolor')]
  #
'''
```

	sepal_length	sepal_width	petal_length	petal_width	species
0	5.1	3.5	1.4	0.2	setosa
1	4.9	3.0	1.4	0.2	setosa
2	4.7	3.2	1.3	0.2	setosa
..	...	...	...	...	...
97	6.2	2.9	4.3	1.3	versicolor
98	5.1	2.5	3.0	1.1	versicolor
99	5.7	2.8	4.1	1.3	versicolor

```
[100 rows x 5 columns]
'''

  #
  # filter-5: selecting a set of rows based on an inequality condition
pdf[pdf.sepal_length > 7.5]
  #
'''
```

	sepal_length	sepal_width	petal_length	petal_width	species
105	7.6	3.0	6.6	2.1	virginica
117	7.7	3.8	6.7	2.2	virginica
118	7.7	2.6	6.9	2.3	virginica
122	7.7	2.8	6.7	2.0	virginica
131	7.9	3.8	6.4	2.0	virginica
135	7.7	3.0	6.1	2.3	virginica

```
'''

  #
```

Tutorial 2.6.4 pandas – sort

```
import seaborn as sb
pdf=sb.load_dataset('iris')
pdf.columns
#

...

'sepal_length', 'sepal_width', 'petal_length', 'petal_width', 'species'
...

# sorting can be ascending or descending, with multiple sort keys
pdf.sort_values('species',ascending=False)
pdf.sort_values(['species','sepal_length'],ascending=False)
pdf.sort_values(['species','sepal_length','sepal_width'],ascending=True)
#
     sepal_length  sepal_width  petal_length  petal_width    species
13            4.3          3.0           1.1          0.1     setosa
8             4.4          2.9           1.4          0.2     setosa
38            4.4          3.0           1.3          0.2     setosa
..            ...          ...           ...          ...        ...
135           7.7          3.0           6.1          2.3  virginica
117           7.7          3.8           6.7          2.2  virginica
131           7.9          3.8           6.4          2.0  virginica
```

Tutorial 2.6.5 Correlation between the variables (using pandas function)

```
# pandas correlation, rounded to 2 decimals
round(pdf.corr(),2)
#
              sepal_length  sepal_width  petal_length  petal_width
sepal_length          1.00        -0.12          0.87         0.82
sepal_width          -0.12         1.00         -0.43        -0.37
petal_length          0.87        -0.43          1.00         0.96
petal_width           0.82        -0.37          0.96         1.00
```

2.7 Numpy

NumPy Basics

NumPy is short for Numerical Python. NumPy serves as the foundational package for scientific computing in Python. Functioning as a Python library, it offers a multidimensional array object, diverse derived objects (like masked arrays and matrices), and a collection of routines designed for swift operations on arrays. These operations encompass mathematical, logical, shape manipulation, sorting, selecting, I/O (input/output), discrete Fourier transforms, introductory linear algebra, basic statistical operations, random simulation, and various other functionalities. For details, please refer to https://numpy.org/doc/stable/user/.

NumPy is a community-driven open-source project developed by a diverse group of contributors. The objective of the section below is limited to a window to the world of NumPy that is required to help us navigate through the rest of the book.

Tutorial 2.7.1 NumPy - generating random numbers

Let us start with a popular but simple application of NumPy – generation of random numbers using numpy library functions

```
#
# get one random decimal in the range 0 to 1
import numpy as np
np.random.rand()    #  0.44117932776647417

# get one random integer in the range 0 to 100
np.random.randint(100)    # 62

# create a random integer array of 3*5
x = np.random.randint(100, size=(3, 5))
print(x)
#
'''
[[98 62 99 42 23]
 [31 94 48 71 69]
 [ 2 10 46  9  5]]
'''
```

Tutorial 2.7.2 NumPy - creating ndarray using data enumeration

A ndarray array is an object. In a ndarray, every column must be of the same data type. The array has strings, integer, float, and bool in the case shown below. All these columns will be stored as 'string' type

```
#
import numpy as np
a1 = np.array(['a','b','c',1,2.5,True])
a1.dtype        # dtype('U32')
# items in 'a1' are stored as unicode strings with size < = 32 characters
a1.ndim       # 1                # 1-dimension
a1.shape      # (6,)             # 6 rows
```

Tutorial 2.7.3 NumPy - creating ndarray using zeros, ones, arrange

This is a popular way of creating a NumPy array of specified dimensions

```
import numpy as np
np.ones(5)    #([1., 1., 1., 1., 1.])
```

```
np.zeros((2, 9))
# array([[0., 0., 0., 0., 0., 0., 0., 0., 0.],
#        [0., 0., 0., 0., 0., 0., 0., 0., 0.]])

x = np.arange(10)
x       # ([0, 1, 2, 3, 4, 5, 6, 7, 8, 9])
```

Tutorial 2.7.4 NumPy - accessing array elements / subsetting

The general method for subsetting in numpy is: Xarray [r1:r2, c1:c2], where
r1.. (r2-1) rows; c1..(c2-1) columns will be selected.
(Note that NumPy subsetting is similar to the iloc command syntax of pandas
DataFrame subsetting. The general method for subsetting of pandas data frame
using 'iloc' is pdf.iloc[r1:r2,[c1,c2]], where rows {r1 .. (r2-1)} are selected from which columns c1 and c2 are selected. The numpy method allows a 'range'
of columns to be selected. Note the difference and try to avoid confusion)

```
#
import numpy as np
a2 = np.array([[1,2,3,4,5], [6,7,8,9,10],['a','b','c','d','e']])

a2[0, 4]    # 5 (element in the cell row:0, column 4)
a2[2, 0]    # a (element in the cell row:3, column 1)
a2[1:3,]    # select all the elements of row-1 and row-2
#
'''
array([['6', '7', '8', '9', '10'],
       ['a', 'b', 'c', 'd', 'e']], dtype='<U11')
'''

#
# subset all the elements falling under (row-1 to 2 and columns 2 to 3)
a2[0:2,2:4]
#
'''
array([['3', '4'],
       ['8', '9']], dtype='<U11')
'''
```

Tutorial 2.7.5 NumPy - rules for setting indices in subsetting

```
import numpy as np
a3 =             np.array(['a','b','c','d','e'])
# array index positions    0   1   2   3   4
# array index positions   -5  -4  -3  -2  -1

# select elements at positions 2, 3; do not select the 4th element
a3[2:4]   # ['c' 'd']

# select elements at positions -4, -3, -2
print(a3[-4:-1])   # ['b' 'c' 'd']
```

```
# select elements starting at position '0', up to but not including '6'
# traversing by steps of 2  (ie., 0, 2, 4)
print(a3[0:6:2])   # ['a' 'c' 'e']

# select elements below position '5'
print(a3[:5])   # ['a' 'b' 'c' 'd' 'e']
print(a3[:99])  # ['a' 'b' 'c' 'd' 'e']  - as explained above

# select elements from position '3' onwards
print(a3[3:])   # ['d' 'e']
```

Tutorial 2.7.6 NumPy - how to copy an array

```
a4 = np.array([1, 2, 3, 4, 5])
x = a4
```
This method is not advisable. Changes to a4 will affect x. However, changes in x will not affect a4

```
a4[0] = 666   # a4 changed
x             # [666, 2, 3, 4, 5] -> changes in a4 affects 'x'
#
```
Use copy() method to make a copy of an array, e.g., y = a4.copy(). Then y owns the data. The changes in a4 will not affect y

```
#
a4 = np.array([1, 2, 3, 4, 5])
y = a4.copy()
a4[0] = 666
y                       # [1, 2, 3, 4, 5] -> changes in a4 does affect 'y'
```

Tutorial 2.7.7 NumPy - Filter

```
import numpy as np
a8 = np.array([10,20,30,40,50,60,70,80,90,100])

filter1 = a8 > 40
a8[filter1]                # [ 50,  60,  70,  80,  90, 100]

filter2 = (a8 > 40) & (a8 < 70)
a8[filter2]                # [50, 60]

filter3 = (a8 < 20) | (a8 > 70)
a8[filter3]                # [ 10,  80,  90, 100]
```

Assume we have items with multiple data types - bool, int, float, str. They will be all designated as the highest data type present - 'str'

```
a9 = np.array([1,2,3,4,5,6,7,8,9,10,'a','b','c','d','e',
        11.0,12.0,13.0,14.0,15.0,True,False,True,False])
```

```
filter4 = a9 > 'a'
a9[filter4]                # ['b', 'c', 'd', 'e'], dtype='<U32'
```

The following example gives seemingly strange results. The sorting results in string order are different from that of number order. For example, string '2', '3' etc. are > string '15.0'.

```
#
filter5 = (a9 > '15.0')
a9[filter5]
#
['2', '3', '4', '5', '6', '7', '8', '9', 'a', 'b', 'c', 'd', 'e', 'True',
'False', 'True', 'False']
```

Tutorial 2.7.8 NumPy Functions - Set, Mean, Median, SD, Round

Note: - This section is optional for doing the tutorials/exercises in this book.

```
import numpy as np
x = [7, 3, 3, 5, 3, 4, 6, 0, 6, 8, 3, 3, 5, 3, 4, 3, 3, 5, 3, 4]
mea = np.mean(x)
med = np.median(x)
print ('mean is', mea, 'median is',med)
# mean is 4.05 median is 3.5
```

The mean and median functions are available as NumPy functions; Mode is un-available in NumPy. For computing mode, see 'Functions and Libraries'

```
# np.max, np.std(1) ...
# np.round, np.fix, np.ceil, np.floor, np.trunc ...

# Trignometric / Hyperbolic functions
import numpy as np
h = np.array([1, 2, 3, 4, 5, 6, 7, 8])
np.round(np.tanh(h),3)  # hyperbolic tangent, rounded to 3 decimals
# ([0.762, 0.964, 0.995, 0.999, 1.   , 1.   , 1.   , 1.   ])

# set operations on 1D array
s1 = np.array([1, 2, 3, 4, 5, 6, 7, 8])
s2 = np.array([2, 4, 6, 8])
np.union1d(s1,s2)             # [1, 2, 3, 4, 5, 6, 7, 8]

np.intersect1d(s1,s2) # [2, 4, 6, 8]
```

Tutorial 2.7.9 NumPy - Flatten (reshape to 1D)

```
import numpy as np
a6 = np.array([[1,2,3,4], [5,6,7,8],['a','b','c','d']])
a6
#
array([['1', '2', '3', '4'],
       ['5', '6', '7', '8'],
       ['a', 'b', 'c', 'd']], dtype='<U11')
#
a6.reshape(-1)      #reshape(-1) returns 1D array
#
array(['1', '2', '3', '4', '5', '6', '7', '8', 'a', 'b', 'c', 'd'],
      dtype='<U11')
#
a6.flatten()        #flatten returns 1D array; same as reshape(-1)
a6.flatten('C')     #column wise, the default option
a6.flatten('F')     #flatten row wise
#
array(['1', '5', 'a', '2', '6', 'b', '3', '7', 'c', '4', '8', 'd'],
      dtype='<U11')
```

NumPy—Advanced

The learning from this section is optional for doing the tutorials/exercises in this book.

Tutorial 2.7.10 NumPy Advanced - ndarray Reshape; dimensions MN to PQ

We can reshape an array M x N to any shape P x Q, so long as all elements are fully covered (MN = PQ). Note that the rows are filled up first

```
#
import numpy as np
a2 = np.array([[1,2,3,4,5], [6,7,8,9,10],['a','b','c','d','e']])
a2
#
array([['1', '2', '3', '4', '5'],
       ['6', '7', '8', '9', '10'],
       ['a', 'b', 'c', 'd', 'e']], dtype='<U11')
# reshape from 3rows x 5 cols to 5 rows x 3 cols
a5 = a2.reshape(5,3)
a5
#
array([['1', '2', '3'],
       ['4', '5', '6'],
       ['7', '8', '9'],
       ['10', 'a', 'b'],
       ['c', 'd', 'e']], dtype='<U11')
#
# Note that items are filled up row-wise in the above example
#
```

Tutorial 2.7.11 NumPy Advanced - ndarray Reshape; dimensions MN to PQR

We can reshape an array M x N to any shape P x Q x R, so long as all
elements are fully covered (MN = PQR). Note that the rows are filled up
first

```
#
import numpy as np
a6 = np.array([[1,2,3,4], [5,6,7,8],['a','b','c','d']])
a6
#
'''
array([['1', '2', '3', '4'],
       ['5', '6', '7', '8'],
       ['a', 'b', 'c', 'd']], dtype='<U11')
'''

# reshape from 3rows x 4 cols to 2 x 2 x 3
a7 = a6.reshape(2,2,3)
# Note that the inner dimensions are filled up first
a7
#
array([[['1', '2', '3'],
        ['4', '5', '6']],

       [['7', '8', 'a'],
        ['b', 'c', 'd']]], dtype='<U11')

#
a8 = np.array([[1,2,3,4,5], [6,7,8,9,10],
     ['a','b','c','d','e'], [11.0,12.0,13.0,14.0,15.0]])
a9 = a8.reshape(5,2,2)
a9
#
array([[['1', '2'],
        ['3', '4']],
       [['5', '6'],
        ['7', '8']],
       [['9', '10'],
        ['a', 'b']],
       [['c', 'd'],
        ['e', '11.0']],
       [['12.0', '13.0'],
        ['14.0', '15.0']]], dtype='<U32')

#
```

Tutorial 2.7.12 NumPy Advanced - ndarray search

```
import numpy as np
sa = np.array([1,4,5,6,7,2,3,10,8,9,11.5,12.8,13.1,14.6,15.7])

np.where(sa == 6)      # where is '6'? in the array 'sa'?
# array([3]

np.where(sa == 11.5)  # where is '11.5'? in the array 'sa'?
# array([10]
```

Tutorial 2.7.13 NumPy Advanced - simple sort

The numpy sort function provides many options. For performance, you can even
specify the sort algorithms - 'quicksort', 'heapsort', 'mergesort', 'tim-
sort'. However, we are exploring only a few options here.

```
#
import numpy as np

# example-1
sa = np.array([1,4,5,6,7,2,3,10,8,9,11.5,12.8,13.1,14.6,15.7])
sa.dtype      # dtype('float64')
np.sort(sa)   # number order
#
array([ 1. ,   2. ,   3. ,   4. ,   5. ,   6. ,   7. ,   8. ,   9. ,  10. ,  11.5,
       12.8, 13.1, 14.6, 15.7])
#
# example-2
sb = np.array([1,4,5,6,7,2,3,10,8,9,'c','d','e','a','b',
          11.0,12.0,13.0,14.0,15.0])
sb.dtype        # dtype('<U32')
np.sort(sb)   # string order
#
array(['1', '10', '11.0', '12.0', '13.0', '14.0', '15.0', '2', '3', '4',
       '5', '6', '7', '8', '9', 'a', 'b', 'c', 'd', 'e'], dtype='<U32')
```

Tutorial 2.7.14 NumPy Advanced - sort dimensions

```
import numpy as np
sd  = np.array([1,4,5,6,7,2,3,10,8,9,11.5,12.8,13.1,14.6])
len(sd) #14
sd1 = sd.reshape(2,7)  # sd1 is an array with 2 rows, 7 columns
sd1
#
array([[ 1. ,   4. ,   5. ,   6. ,   7. ,   2. ,   3. ],
       [10. ,   8. ,   9. ,  11.5, 12.8, 13.1, 14.6]])
```

```
  #
  # sort example-1: row wise sort {this is the default method}
  np.sort(sd1, axis=0)
array([[ 1. ,   4. ,   5. ,   6. ,   7. ,   2. ,   3. ],
       [10. ,   8. ,   9. ,  11.5, 12.8, 13.1, 14.6]])
  #
  # sort example-1: column wise sort
  np.sort(sd1, axis=1)   # column wise sort
  #
array([[ 1. ,   2. ,   3. ,   4. ,   5. ,   6. ,   7. ],
       [ 8. ,   9. ,  10. ,  11.5, 12.8, 13.1, 14.6]])
  #
  # sort example-3: sorts along the last axis
  np.sort(sd1, axis=-1)
  #
array([[ 1. ,   2. ,   3. ,   4. ,   5. ,   6. ,   7. ],
       [ 8. ,   9. ,  10. ,  11.5, 12.8, 13.1, 14.6]])
  #
  # sort example-3: flatten the data; then sort
  np.sort(sd1, axis=None)
array([ 1. ,   2. ,   3. ,   4. ,   5. ,   6. ,   7. ,   8. ,   9. ,  10. ,  11.5,
       12.8, 13.1, 14.6])
```

Tutorial 2.7.15 NumPy Advanced - argsort

```
  import numpy as np
  chickweight = np.array([['D1', '102.65'],
                          ['D2', '122.62'],
                          ['D3', '142.95'],
                          ['D4', '135.26']])

  # sort and print column 1
  np.sort(chickweight[:,1])
  ['102.65', '122.62', '135.26', '142.95']

  # sort index of column 1
  chickweight[:,1].argsort()
  [0, 1, 3, 2]

  # sort index of column 1, apply sorted index
  chickweight[chickweight[:,1].argsort()]
  #
  '''
array([['D1', '102.65'],
       ['D2', '122.62'],
       ['D4', '135.26'],
       ['D3', '142.95']], dtype='<U6')
  '''
```

```
#
# sort index of column 0, apply sorted index
chickweight[:, chickweight[0, :].argsort()]
#
array([['102.65', 'D1'],
       ['122.62', 'D2'],
       ['142.95', 'D3'],
       ['135.26', 'D4']], dtype='<U6')
```

Tutorial 2.7.16 NumPy - join and split arrays

```
#join arrays
j1 = np.array([1, 2, 3, 4, 5, 6])
j2 = np.array([7, 8, 9, 10, 11, 12])
# the default axis is '0' for an nD array
j3 = np.concatenate((j1,j2),axis = 0)
j3    # array([ 1,   2,   3,   4,   5,   6,   7,   8,   9, 10, 11, 12])
#
# split an array into three arrays
k1, k2, k3 = np.split(j3,3)
k1    #   array([1, 2, 3, 4])
k2    # array([5, 6, 7, 8])
k3    # array([ 9, 10, 11, 12])
```

2.8   Data Processing

The cardinal principles of data mining include—(a) identifying and defining the problem to be solved in clear terms and (b) gaining a good understanding of the data to be processed. This section introduces the basic concepts of data processing and tutorials that will help us navigate the rest of the book.

2.8.1   *Measurements, Features, and Targets*

Figure 2.1 shows the features that determine customer satisfaction. We have a set of features that include functional, financial, psychological, service, etc. (Sorger 2013). These features determine the target, customer satisfaction. The features are also called feature variables, independent variables, predictors, inputs, or attributes. The target goes by different names, such as the target variable, the dependent variable, the predicted variable, or the outcome. In our discussions, we denote features as vector X of k-dimensions and the target by 'y'. The features and targets are represented as columns in a dataset. The rows in a dataset represent observations, cases, data instances, or data objects, as they are variously called. These are gathered from various data sources.

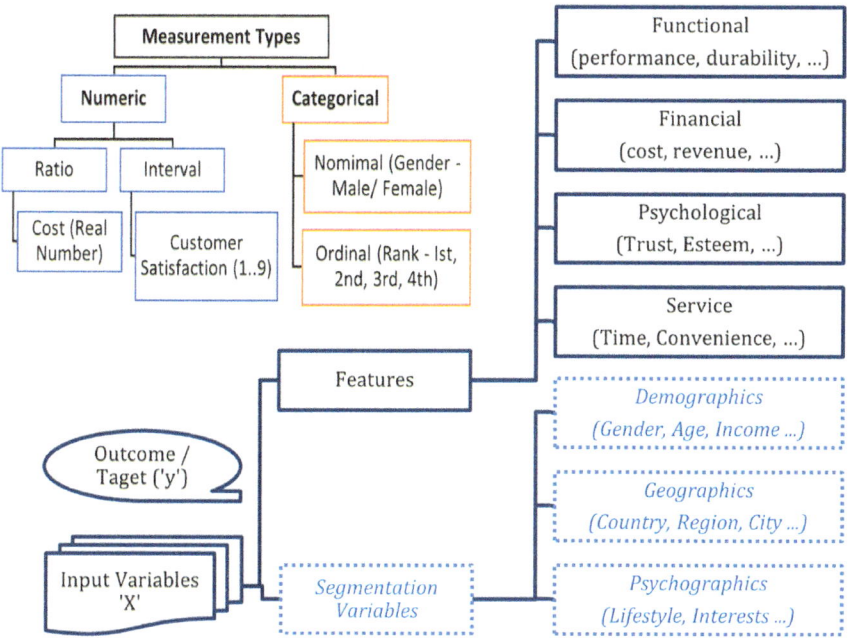

Fig. 2.1   Measurement types, input variables, and outcome/target variables

In Fig. 2.1, the demographic, geographic, and psychographic variables are grouped under the umbrella 'segmentation' variables. They help group the data rows into meaningful groups. There are various types of measurements. Cost is a decimal. Technically it is a ratio scaled or continuous-valued variable. Customer Satisfaction could be measured on an interval scale, e.g., 1–9, where 1 represents least satisfied and 9 highly satisfied. All these variables (ratio scaled and interval scaled) are called numeric variables.

Gender is a categorical variable with two values. Consider the ranks obtained by students in an entrance examination. The rank is a categorical variable. Technically, gender is a categorical variable of subtype nominal; rank is a categorical variable of subtype ordinal.

2.8.2   Data Preprocessing Concepts

Data preprocessing is essential before data mining and model building (Han and Kamber 2014). Missing values, noise, outliers, inconsistencies, multiple data formats, and the like make data preprocessing challenging. Data preprocessing encompasses critical steps, including cleaning, integration, reduction (both in terms of features and data volume), transformation, discretization, standardization, and summarization. This intricate process constitutes a substantial portion, approximately 70%, of a typical data mining project. Given its magnitude, it requires specific attention, but delving into its intricacies goes beyond the scope of this book.

Missing data impedes numeric computations. We may discard all the rows with missing values for features of interest. Alternatively, the missing values can be imputed with the mean (or median or mode), zeros, or other values. However, it may be noted that any imputation will affect the original characteristics of the feature.

Outliers are the data points that deviate considerably from the rest of the data in a dataset. They can substantially impact statistical analyses, building machine learning models, and visualizing data. Outliers happen for various reasons, such as measurement errors, errors in data capturing, possible natural variations, etc. For outlier detection, we may use visual inspection, interquartile range (see Sect. 3.1.3 visual description), Mahalanobis distance, the z-score of the data points, etc. Care must be taken in removing outliers, as it may lead to information loss and bias in the analysis. It demands domain knowledge and depends on the goals of the analysis. The commonly used methods are data transformation (e.g., log transformations), imputation (e.g., replacement of data points by representative values such as the mean), winsorization (replacing extreme values with the nearest values), etc.

Data standardization is a process that involves transformation to a common format, discretization, normalization, etc. Data transformation to a common format may include assigning labels or codes, unit conversion, unified date/time format, etc. A numeric variable can be grouped into categories—for example, the price can be categorized as low, medium, or high. This method is called **discretization**. Normalization techniques bring data within a specific range or scale. There are many methods for data **normalization**: min–max, z-score, decimal scaling, etc. Z-score standardization is popularly used in statistical data mining. A variable is standardized by subtracting from its sample mean and by dividing it by its standard deviation. An observation x_i can be transformed into z-score as follows:

$$z_i = (x_i - x_{mean})/sd, \text{ where sd is the standard deviation of the sample.}$$

Data standardization helps improve the performance of algorithms like gradient descent (a commonly used optimization technique in machine learning).

2.8.3 Data Preprocessing Examples

Tutorial 2.8.1 Preprocessing: recode category -> integer

Recode category labels as integers, so that they may be used in computations. The integer codes are saved in a new column 'species_coded'

```
#
# list the category labels under the 'species' column
pdf.species.unique()
# ['setosa', 'versicolor', 'virginica']
```

```
    #
    # list the number of entries (rows) in each category
    pdf.species.value_counts()
    #
setosa        50
versicolor    50
virginica     50
    #
    # recode category labels as integers
    pdf['species_coded'] = pdf.species.astype("category").cat.codes
    Pdf

      sepal_length  sepal_width  ...     species  species_coded
0              5.1          3.5  ...     setosa              0
1              4.9          3.0  ...     setosa              0
..             ...          ...  ...         ...            ...
148            6.2          3.4  ...   virginica              2
149            5.9          3.0  ...   virginica              2

[150 rows x 6 columns]
```

Tutorial 2.8.2 Preprocessing: handling missing values

```
    # Missing values must be removed or recoded.
    #
    import seaborn as sb
    pdf  = sb.load_dataset('penguins')
    pdf.info()
    #
<class 'pandas.core.frame.DataFrame'>
RangeIndex: 344 entries, 0 to 343
Data columns (total 7 columns):
 #   Column          Non-Null Count  Dtype
---  ------          --------------  -----
 0   species         344 non-null    object
 1   island          344 non-null    object
 2   bill_length_mm  342 non-null    float64  →  2 null values
 3   bill_depth_mm   342 non-null    float64  →  2 null values
 4   flipper_length_mm 342 non-null  float64  →  2 null values
 5   body_mass_g     342 non-null    float64  →  2 null values
 6   sex             333 non-null    object   → 11 null values
dtypes: float64(4), object(3)
```

There are a total of 344 rows. However, some columns have missing values as indicated above. For example, in 11 rows, sex is missing

```
    #
    # Tutorial 2.8.2.1 Preprocessing:  drop rows with missing values
    pdf1 = pdf.dropna()  # 9 rows dropped (344 rows to 333 rows)
    #
```

```
# Tutorial 2.8.2.2 Preprocessing:  dropna inplace
# deletes rows from the original dataset, with 'inplace' parameter
pdf2 = pdf.copy()
pdf2.shape     # (344, 7)
pdf2.dropna(inplace=True)
pdf2.shape     # (333, 7)
#
# Tutorial 2.8.2.2 Preprocessing:  data imputation
# copy pdf to pdf4, so that changes to 'pdf' will not affect 'pdf4'
pdf4  = pdf.copy()
#
# replace missing values with mean
blMean = pdf.bill_length_mm.mean()    # compute mean
print(blMean) # 43.92
pdf4.bill_length_mm.fillna(blMean,inplace=True) # impute with mean
#
# replace missing values with mode
SMode = pdf.sex.mode()    # compute mode
print(SMode)  # Male
# replace missing values with mode. The sample is predominantly male.
# If sex is missing, enter it as 'male'
pdf4.sex.fillna(SMode,inplace=True)    # impute with mode
```

Tutorial 2.8.3 Preprocessing: z-score standardization

Data standardization (see the discussion in this section) improves the performance of many algorithms and helps us evaluate the comparative influence
of features

```
#
import seaborn as sb
from   scipy import stats     # for z-score standardization

pdf = sb.load_dataset('iris')
X =   pdf[['sepal_length', 'sepal_width', 'petal_length', 'petal_width']]
Xz = stats.zscore(X)
X.head(2)
#
  sepal_length  sepal_width  petal_length  petal_width
0          5.1          3.5           1.4          0.2
1          4.9          3.0           1.4          0.2
#
Xz.head(2)  # z score standardised; mean =0; sd = 1
#
  sepal_length  sepal_width  petal_length  petal_width
0    -0.900681     1.019004     -1.340227    -1.315444
1    -1.143017    -0.131979     -1.340227    -1.315444
```

Tutorial 2.8.4 Preprocessing: categorizing numeric values

```
pdf4.body_mass_g.describe()
  #
mean      4201.754386
std        801.954536
min       2700.000000
25%       3550.000000
50%       4050.000000
75%       4750.000000
max       6300.000000
  #
We are categorizing penguins by mass. We will divide the penguins into
four categories based on the mass quartiles.
  #
pdf4['WeightCat'] = 0
for i in pdf4.index:
    if   pdf4.loc[i, 'body_mass_g'] > 4750:
         pdf4.loc[i, 'WeightCat'] = 3
    elif pdf4.loc[i, 'body_mass_g'] > 4050:
         pdf4.loc[i, 'WeightCat'] = 2
    elif pdf4.loc[i, 'body_mass_g'] > 3550:
         pdf4.loc[i, 'WeightCat'] = 1
    else:
         pdf4.loc[i, 'WeightCat'] = 0
  #
pdf4
  #
     species     island bill_length_mm ... body_mass_g     sex WeightCat
0     Adelie  Torgersen        39.10000 ...      3750.0    Male       1.0
1     Adelie  Torgersen        39.50000 ...      3800.0  Female       1.0
2     Adelie  Torgersen        40.30000 ...      3250.0  Female       0.0
3     Adelie  Torgersen        43.92193 ...         NaN     NaN       0.0
..       ...        ...             ... ...         ...     ...       ...
339   Gentoo     Biscoe        43.92193 ...         NaN     NaN       0.0
340   Gentoo     Biscoe        46.80000 ...      4850.0  Female       3.0
341   Gentoo     Biscoe        50.40000 ...      5750.0    Male       3.0
342   Gentoo     Biscoe        45.20000 ...      5200.0  Female       3.0
343   Gentoo     Biscoe        49.90000 ...      5400.0    Male       3.0

[344 rows x 8 columns]
  '''
```

Tutorial 2.8.5 Preprocessing: Creating Pandas DataFrame from arrays

```
# Tutorial 2.8.5.1 pandas - Creating Pandas DataFrame from a 2D array
import pandas as pd
a = [[3,2],[5,2]]                    #data
column_names = ['project_perf', 'project_plan']  #columns
# create pdf
pdf = pd.DataFrame(columns = column_names, data=a)
pdf    # print the contents of pdf
#
  project_perf  project_plan
0            3             2
1            5             2
  #
# Tutorial 2.8.5.2 pandas - Creating Pandas DataFrame from 1D arrays
import pandas as pd
# create an empty pandas DataFrame
pdf = pd.DataFrame()
# load pdf DataFrame from lists, into two newly created columns
pdf['project_perf'] = [3,2,5,3,2,6,1,2,5,2,3,4,3,3,4]
pdf['project_plan'] = [5,2,4,4,3,5,2,2,4,2,4,4,4,4,3]
```

Tutorial 2.8.6 Preprocessing: Exploring Pandas DataFrame Data structure

```
# Tutorial 2.8.6.1 create pandas DataFrame and Explore it
pdf = pd.DataFrame()
pdf['project_perf'] = [3,2,5,3,2,6,1,2,5,2,3,4,3,3,4]
pdf['project_plan'] = [5,2,4,4,3,5,2,2,4,2,4,4,4,4,3]
#
pdf    # print the contents of pdf
#
   project_perf  project_plan
0             3             5
1             2             2
..
13            3             4
14            4             3
  #
pdf.project_perf    # print the first column
pdf.project_plan    # print the second column
pdf.iloc[0]         # print the first row
  #
project_perf    3
project_plan    5
Name: 0, dtype: int64
  #
```

```
   # Tutorial 2.8.6.2 column names and data values
   #
   print (pdf.columns)    # print the column names
   # Index(['project_perf', 'project_plan'], dtype='object')
   #
   pdf.values     # print the data
   #
array([[3, 5],
       [2, 2],
       [5, 4],
..
<more data>
..
   #
   # Save the column names and print them iteratively
   col = list(pdf.columns)    # save the column names in a list
   for i in col: print(i)     # print the column names from the list
   #
   # print the column values column by column
   for i in col:
       print('contents of column',i,'\n',pdf[i])
   #
   # Tutorial 2.8.6.3 using row index
   # print the first two rows programmatically
   for i in pdf.index:
       print(i,pdf.iloc[i])
       if i == 1: break
   #
0 project_perf     3
  project_plan     5
Name: 0, dtype: int64
1 project_perf     2
  project_plan     2
Name: 1, dtype: int64
   #
   pdf.shape      # (15, 2)
   #
   # Tutorial 2.8.6.4 remove duplicates
   # remove the duplicates; save the result in another pdf
   pdf2 = pdf.drop_duplicates()
   pdf2.shape     # (9, 2);  15 rows to 9 rows (6 rows deleted)
   #
   # delete the duplicates "inplace"
   pdf3 = pdf.copy()
   # physically delete the duplicates of pdf3 dataframe
   pdf3.drop_duplicates(inplace=True)
   pdf3.shape     # (9, 2)
```

Tutorial 2.8.7 Preprocessing: Pandas to/from NumPy

```
import numpy as np
import pandas as pd
#
# Creating and loading a pandas data frame from ndarray
array1 = np.zeros([15,2])
array1[:,0] = np.array([5,2,4,4,3,5,2,2,4,2,4,4,4,4,3])
array1[:,1] = np.array([3,2,5,3,2,6,1,2,5,2,3,4,3,3,4])
array1[0:2]
array([[5., 3.],
       [2., 2.]])
# creating column names
column_names = ['performance','plan']
# creating the DataFrame
pdf = pd.DataFrame(data = array1, columns = column_names)
pdf.head(2)
#
  performance   plan
0          5.0    3.0
1          2.0    2.0
#
array2 = pdf.to_numpy() # convert back pandas DataFrame to ndarray
array2[0:2]
array([[5., 3.],
       [2., 2.]])
```

Tutorial 2.8.8 Preprocessing: NumPy to PDF, Write to local disk

```
import pandas as pd
import numpy as np
from sklearn.datasets import load_iris
#
ndaX, nday = load_iris(return_X_y=True)
type(ndaX) #numpy.ndarray
ndaX[0:2]
#
array([[5.1, 3.5, 1.4, 0.2],
       [4.9, 3. , 1.4, 0.2]])
#
pdfX=pd.DataFrame(ndaX)   # Convert ndarray to pandas DataFrame
pdfX.iloc[0:2]
#
```

```
    0    1    2    3
0  5.1  3.5  1.4  0.2
1  4.9  3.0  1.4  0.2
   # Write the PDF out to the local disk as CSV file
   # Do not save 'row indices'
   pdfX.to_csv('pdfX.csv', index=False)
   #
   # read from local disk.
   pdfXread = pd.read_csv('pdfX.csv')
   pdfXread.head(2)


    0    1    2    3
0  5.1  3.5  1.4  0.2
1  4.9  3.0  1.4  0.2
```

Data Analytics in Action

Python Interface to Apache Spark for Big Data Analysis

Apache Spark is an open-source data processing engine for big datasets, like Hadoop. Spark can execute tasks 100 times faster than Hadoop for smaller workloads, using random access memory (RAM) to cache and process data instead of a file system. You may refer to Apache Spark (2023) for detailed information.

Spark's SQL analytics engine processes distributed SQL queries faster than most data warehouses. Spark supports big data analysis on a petabyte scale. Spark platform helps to develop machine learning models and scale them up for deployment in fault-tolerant clusters with thousands of cores. Spark is written in Scala programming language. Spark provides a powerful platform for real-time processing with streaming data and batch data processing, using Python, SQL, Scala, Java, or R. Spark features include Spark Streaming, MLlib (Machine Learning), Spark SQL, and GraphX.

PySpark is a Python API for Spark. In addition, PySpark provides a Python interface with resilient distributed datasets (RDDs), the fundamental data structure of Spark. PySpark allows us to write Spark applications using Python APIs. PySpark shell allows interactive analysis of data in a distributed computing environment. PySpark supports most of Spark's features, such as Spark SQL, DataFrame, Streaming, MLlib (Machine Learning), and Spark Core. Please refer to Apache Spark (2023) for details.

Summary

We did a whirlwind tour of programming in Python and covered necessary programming concepts to do the tutorials and exercises for applying statistical, computational, and machine learning techniques.

Python is one of the top two popular programming languages. Python's standard libraries and ever-growing community contributions allow for endless programming possibilities. Python is supported by popular packages such as Pandas for data analysis and manipulation, NumPy for matrix algebra and numerical operations, NetworkX for Graphs and Network Programming, Matplotlib, and Seaborn for data visualization, SciPy and Statsmodels for statistics, and optimization, Scikit-learn for statistics and machine learning, TensorFlow and Keras framework for deep learning, etc.

An integrated development environment (IDE) is a set of software tools that helps a programmer develop and test computer programs. There are numerous IDEs. Spyder IDE or Jupyter Notebook IDE can be installed in our local system. Google Colab IDE needs internet access. The code snippets in this book are tested in Google Colab IDE and Spyder IDE. Google Colab makes code sharing easy and allows parallelism through GPUs.

Python is an object-oriented programming language. Python is also a structured language and a procedural language. The Python program is first compiled into byte code. The byte code is platform-independent. The interpreter converts byte code to instructions specific to an operating system (Windows, Mac-OS, etc.). To the programmer, the above two steps are transparent. In Python, variables are not explicitly declared. They are created from assignment statements. The primary data types include Integer (int), Decimal Numbers (float), Character String (str), True/False (bool), etc. List, Tuple, Set, and Dictionary are advanced data structures.

Pandas is a powerful, flexible, and easy-to-use data analysis and manipulation tool, built on top of the Python programming language. Pandas can access multiple file formats such as CSV (comma-separated values), excel, JSON, HTML, SQL, parquet, pickle, and Google Big Query. Pandas DataFrame is a 2D (two-dimensional) data structure, like a spreadsheet, with rows and columns. NumPy is short for Numerical Python. It is highly effective in array maths like linear algebra, Fourier transforms, matrix manipulation, etc.

In marketing research, data may typically consist of three groups—a set of feature variables, a target variable, and a set of segmentation variables. In our discussions, we denote features as vector 'X' of k-dimensions and target by 'y'. The features and targets are represented as columns in a dataset. The rows in a dataset represent observations, cases, data instances, or data objects, as they are variously called. These are gathered from various data sources.

There are various types of measurements—numeric (both ratio-scaled and interval-scaled), and categorical (nominal or ordinal).

Data preprocessing is an essential step before data mining and model building. Missing values, noise, outliers, inconsistencies, multiple data formats, and the like make data preprocessing challenging. Data preprocessing includes cleaning, integration, reduction (of features and data volume), transformation, discretization, standardization, and summarization.

We may discard the data instances with missing values or impute them with a value like the mean. Data standardization helps us compare variables measured on different scales, e.g., pounds and kilograms. There are many methods for data standardization—min–max, z-score, decimal scaling, etc. Data standardization helps in improving the performance of certain algorithms. A numeric variable can be grouped into categories—this method is called discretization.

Questions

Comprehension

1. How do you define Python Programming Language? Is it an interpreted language or a compiled language?
2. Is Python a procedural language or an object-oriented programming language?
3. Describe the Python environment (associated packages/libraries).
4. How are strings and arrays related in Python?
5. Describe the basics of data preprocessing.

Write short notes on the following, with suitable examples:

6. Python variable naming
7. Python basic data types
8. Python data structure—List, Tuple, Set, and Dictionary
9. Control flow in the Python programming language
10. Python 'function'
11. Python lambda function
12. missing value analysis and processing.
13. the difference between the Python operators '=' and '=='
14. range data type
15. list—subsetting, lookup for an item
16. break and continue in for loop
17. pandas—filter, sort
18. ndarray—accessing array elements, subsetting, reshape, filter, search, sort, argsort

Application

19. Model a business application, e.g., understanding customer satisfaction and relating major features with outcomes.
20. Write code to generate 10 random numbers between 0 and 1.
21. Write a note on list comprehension with examples.

22. Write a note on regular expression functions with examples.
23. Write a note on missing value handling in Pandas, with code examples.
24. Write coding examples to show how to transform a 2D NumPy array to Pandas DataFrame and vice versa.

Exercises

The questions in this section are based on two datasets available with the seaborn package—the diamonds and planets datasets. Use the diamonds dataset for exercises 2.1 to 2.7. Use the planets dataset discussed in this Chapter for exercise 2.8.

Exercise 2.1 Variables Tryout the learnings from tutorials under Variables.

Exercise 2.2 Operators Tryout the learnings from tutorials under Operators.

Exercise 2.3 Advanced Data Types Tryout the learnings from tutorials under Advanced Data Types.

Exercise 2.4 Control Flow Tryout the learnings from tutorials under Control Flow.

Exercise 2.5 Functions and Libraries Tryout the learnings from tutorials under Functions And Libraries.

Exercise 2.6 Pandas Tryout the learnings from tutorials under Pandas.

Exercise 2.7 NumPy Tryout the learnings from tutorials under NumPy.

Exercise 2.8 Data Preprocessing Using the dataset planets from Seaborn, try the learnings from tutorials in 'Data Preprocessing'.

References

Apache Spark (2023) Apache Spark. Apache Spark. https://spark.apache.org/
Han J, Kamber M, Pei J (2014) Data mining. Concepts and techniques (The Morgan Kaufmann Series in Data Management Systems). In: Proceedings—2013 international conference on machine intelligence research and advancement, ICMIRA 2013, 3rd ed
Sorger S (2013) Marketing analytics: strategic models and metrics

Chapter 3
Introduction to Probability and Statistics

Learning Objectives

- Describe random numbers and the theory of probability.
- Obtain an overview of descriptive statistics.
- Examine discrete probability distributions.
- Examine continuous probability distributions.
- Illustrate sampling and sampling methods.
- Illustrate sampling distributions and central limit theorem.
- Discuss point and interval estimates.

Overview

This chapter provides an overview of the basic concepts of probability, inferential, and descriptive statistics. The chapter explores discrete and continuous probability distribution functions. An examination of sampling methods, sampling distributions, and central limit theorem follows this. Finally, point and interval estimates are introduced.

Definitions

Binomial Distribution: The binomial distribution represents the discrete probability distribution of the count of successes in a series of 'n' independent

Supplementary Information The online version contains supplementary material available at https://doi.org/10.1007/978-981-99-0353-5_3.

experiments, where each experiment is a success/failure (dichotomous) event with a probability of success denoted by 'p'.

Box Plot: Box plots help us to visualize the measures of location and dispersion using simple plots.

Central Limit Theorem: CLT states that, as the sample size increases, the sampling distributions closely approximate the normal distribution and become clustered around the population mean for all distributions of independent, identically distributed variables with finite variance.

Chi-Square Distribution: Chi-Square Distribution is a continuous probability distribution used to compare the distribution of a categorical variable in a sample with the distribution of a categorical variable in another sample. The curve starts with a zero on the left side and extends to infinity on the right.

Descriptive statistics: There are two approaches to descriptive statistics—Visualization and Summarization using numbers. Visual techniques include box plots, histograms, frequency distributions, scatter plots, etc. Summary statistics include location or central tendencies, dispersion, and shape.

Event: A random experiment is a phenomenon whose outcome cannot be predicted with certainty. The set of possible outcomes is called the sample space. An event (a probabilistic event) is one or more possible outcomes of a random experiment.

Histogram: A histogram shows the frequency distribution of a numeric variable.

Inferential Statistics: The process of estimating the population parameters (for example, mean and standard deviation) from sample data is called statistical inference.

Interval estimate: A statistic (e.g., mean) computed from a data sample gives a point estimate of the population from which the sample is drawn. The computed mean will vary from sample to sample. Therefore, we state that the population means lie within a specific interval, with a certain confidence. This is called an interval estimate.

Kurtosis: Kurtosis indicates the relative peakedness or flatness of the frequency distribution. Kurtosis is the fourth moment of the standardized score of X.

Normal distribution: The normal (or Gaussian) distribution is a continuous probability distribution with a bell-shaped probability density function.

Probability: Probability is the likelihood of an event occurring. Probabilities are expressed as fractions/decimals between zero and one.

Sampling distribution: The probability distribution of all possible values of the sample statistic is known as the sampling distribution.

Sampling: In practice, investigating the entire population may not be possible. Therefore, information is collected from a part of the population, known as a sample. We call this method 'sampling'.

Scatter Plot: A scatter plot is a joint distribution of the two variables.

Skewness: Skewness is the degree of distortion from a normal distribution. Skewness is the third moment of the standardized score of X.

Standard normal distribution: The standard normal distribution is a special case of the normal distribution with the mean equal to zero and the standard deviation equal to 1.

Statistical independence: If the outcome of an event does not affect the outcome of another, then the events are called independent. There are three types of probabilities under statistical independence—marginal probability, joint probability, and conditional probability.

Stem and leaf plot: A stem and leaf plot shows the rank order of all the items in a dataset and the shape of the distribution.

T-distribution: The 't-distribution' is a continuous probability distribution that assumes a bell-shaped symmetric curve. As the sample size, n increases, the t-distribution approaches the standard normal distribution.

3.1 Introduction

The official government statistics are as old as recorded history. However, the word statistics was first used by Gottfried Achenwall (Year 1719–72). He used the term to refer to a comprehensive summary of a nation's social, political, and economic aspects.

While we witness the breathtaking pace at which machine learning and deep learning bring disruptive innovations in all spheres of life, it is incredible that this domain has humble origins in twentieth-century statistics. Karl Pearson, who is known for the chi-square test (The year 1900); William Sealy Gosset who is known for the t-test (The year 1908); and Ronald A Fisher, who is known for ANOVA and F-test (The year 1918), are considered to be the pioneers of modern statistics. Today, statistics find application in almost all branches of science. Machine learning and deep learning rely heavily on statistics.

This chapter introduces the fundamentals of probability and statistics that form the basis of the rest of the book. Under probability, we will learn the basics of random events and the rules that govern them. Statistical methods can be broadly divided into two categories, as shown below.

- Descriptive statistics
- Inferential Statistics

As we know, graphs, charts, tables, and aggregations make it easier to understand data. All these are examples of descriptive statistics. While we can compute the mean or variance of a data sample, we may need to be certain that these 'point' estimates will hold good for the entire population. Statistical inference is estimating the population parameters (for example, mean 'μ' and standard deviation 'σ') from sample data. The method and techniques of statistical inference can be used for making decisions under conditions of uncertainty.

3.2   Descriptive Statistics

There are two approaches to descriptive statistics—Visualization and Summarization using numbers. Visual techniques such as box plots, histograms, frequency distributions, and scatter plots, give a visual description of the data. Summarization using numbers, or summary statistics as they are called, are numbers that emerge from a broad category of measurements such as (Malhotra, 2020):

1. Location or Central Tendencies (Mean, Median, Mode, Quartiles ...)
2. Dispersion (Variance, Standard Deviation, Range, Outliers ...)
3. Shape (Skewness, Kurtosis ...)

3.2.1   Measures of Location

Mean (or arithmetic mean) is highly influenced by extreme values. While dealing with variables whose values change over a period of time, we observe the average rate of change. In such cases, the geometric mean is preferred over the arithmetic mean. For example, the geometric mean shows multiplicative effects over time in compound interest and inflation calculations.

The median is a single value from the dataset that measures the central item in the data. We must sort the data before we can calculate the median. Consider the penguin dataset shown in Table 3.1/Chapter 1.6. The median mass is 4050 gm. This implies that about 166 penguins have a mass above 4050 gm, and 166 have a mass below 4050 gm. We need to sort the data before we calculate the median. Extreme values do not influence the median. For example, there may be a few obese penguins, and they will not sway our calculations.

Mode differs from the mean but is similar to the median because the arithmetic process does not calculate it. The mode can be used as a measure of location for

Table 3.1   Penguin DataFrame

Features measured	Data type	Statistics	bill_length_mm	bill_depth_mm	flipper_length_mm	body_mass_g
Species	object	count	333	333	333	333
Island	object	Mean	**43.99**	**17.16**	**200.97**	**4207.06**
Bill_length_mm	float64	SD	**5.47**	**1.97**	**14.02**	**805.22**
Bill_depth_mm	float64	Min	32.1	13.1	172	2700
Flipper_length	float64	Q1–25%	39.5	15.6	190	3550
Body_mass_g	float64	Q2–50%	**44.5**	**17.3**	**197**	**4050**
Sex	object	Q3–75%	48.6	18.7	213	4775
Total rows	333	Max	59.6	21.5	231	6300

quantitative and qualitative data. Mode is unaffected by extreme values and can be used for open-ended data. If the frequency of penguins is the highest in the range of 3700–3800 gm (we have 25 penguins in this category), the mode is roughly in the range of 3700–3800 gm. Mode is not affected by extreme values, just like the median. Note that data can be multi-modal—having two or more modes.

3.2.2   Measures of Dispersion

The range is the difference between the highest and lowest observed values. The variance of a random variable or distribution is the average or mean of the squared difference between each value of the variable and its expected value or mean. Standard deviation is the square root of the variance. It is a widely used measure of dispersion. Extreme values in distribution affect the value of standard deviation, as well as the mean.

3.2.3   Visual Description

Box plots help us to visualize the measures of location and dispersion using simple plots. See Figs. 3.1a andb. The boxes show quartiles (Q1, Q2=Median, Q3). The whiskers are at a distance of 1.5 * (Q3–Q1) from Q2. The points outside the whiskers are marked as outliers (see Sect. 2.8.2). For example, check Fig. 3.2: Chinstrap species of dream island. Note that there are no clear-cut rules for determining the distance criteria for outliers; it rests at the researcher's discretion based on the research problem.

A histogram shows the frequency distribution of a numeric variable. It is a series of rectangles, each proportional in width to the range of values within a class and proportional in height to the number of items falling in the class. The

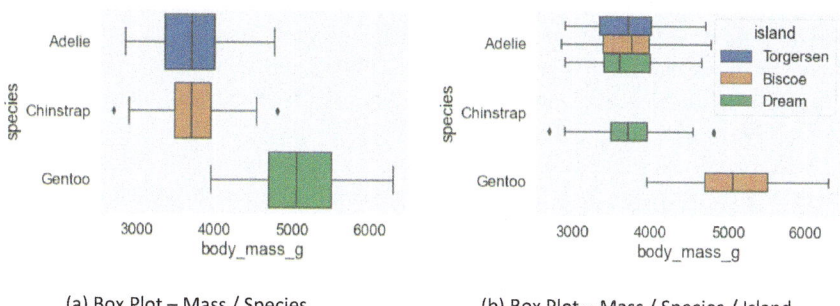

(a) Box Plot – Mass / Species   (b) Box Plot – Mass / Species / Island

Fig. 3.1 Box Plot—mass/species/island

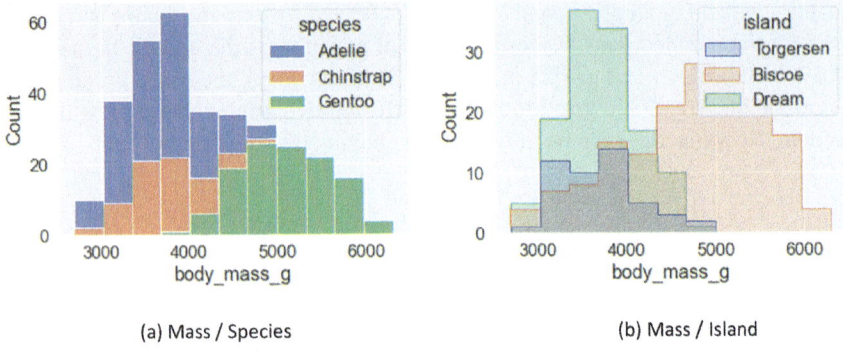

(a) Mass / Species   (b) Mass / Island

Fig. 3.2 Histogram—mass/species/island

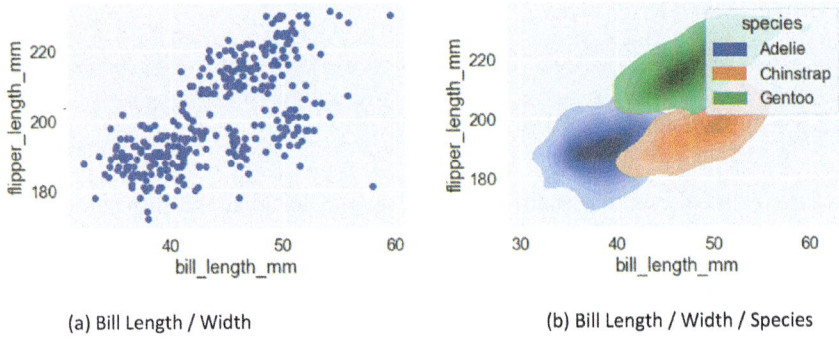

(a) Bill Length / Width   (b) Bill Length / Width / Species

Fig. 3.3 Scatter plot and KDE plot

height of the bar for each class corresponds to the number of items in the class. See Figs. 3.2a and b. A frequency curve sketches an outline of the data pattern more evident than a histogram.

Figure 3.3a shows a scatter plot of two variables—Bill Length by Bill Width. This is a joint distribution of the two variables. Figure 3.3b shows a two-dimensional KDE (kernel density estimation) plot. Since two variables are involved, it is equivalent to the scatter plot. The distribution is further subdivided into three groups, by species, using color coding (Figs. 3.4 and 3.5).

A stem and leaf plot shows the rank order of all the items in a dataset and the shape of the distribution. It is in a way similar to a histogram. It is used in exploratory data analysis.

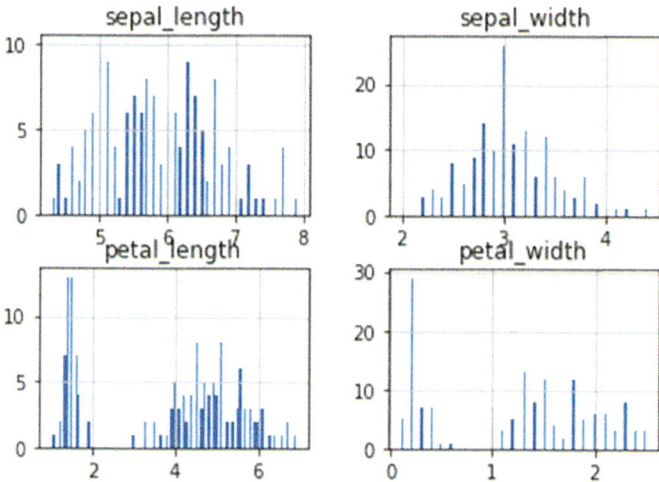

Fig. 3.4   Pandas plot (Histogram of all Numeric Variables)

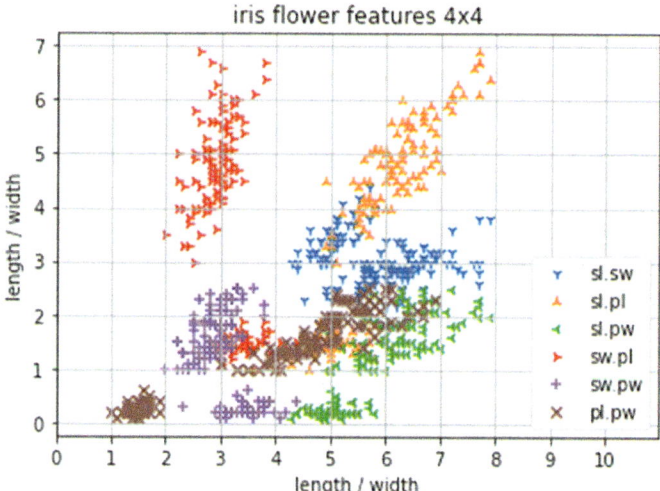

Fig. 3.5   matplotlib.pyplot—scatter plot of all pairs of numeric variables

Tutorial 3.2.1 Data Description by Visual Plots

Matplotlib is a Python library for plots. Seaborn gives rich graphics, and it uses matplotlib functions - refer https://seaborn.pydata.org/

```
import matplotlib.pyplot as plt
import seaborn as sb
sb.set(font_scale=1.5)  #set font size to 150%

d=sb.load_dataset('penguins')
d.dropna(inplace=True) # drop rows having null valued cells
d.describe()    # descriptive statistics of all numeric variables
```

Figure 3-1(a): boxplot of mass by species
```
sb.boxplot (data=d, x='body_mass_g', y='species')
```

Figure 3-1(b): boxplot of mass by species and island
```
sb.boxplot (data=d, x='body_mass_g', y='species', hue='island')
```

Figure 3-2(a): histogram of mass by species
```
sb.histplot(data=d, x='body_mass_g', hue='species',multiple='stack')
```

Figure 3-2(b): histogram of mass by island
```
sb.histplot(data=d, x='body_mass_g', hue='island',element='step' )
```

Figure 3-3(a) Scatter Plot
```
plt.scatter(data=d, x='bill_length_mm', y ='flipper_length_mm')
plt.xlabel('bill_length_mm')
plt.ylabel('flipper_length_mm')
```

Figure 3-3(b) KDE Plot
```
sb.kdeplot (data=d, x='bill_length_mm', y ='flipper_length_mm',
            hue='species', shade=True)
```

Strip Plot
```
sb.stripplot(hue='sex', y='body_mass_g', x='island', data=d)
```

Tutorial 3.2.2 Pandas Plot

```
import seaborn as sb
pdf = sb.load_dataset('iris')
```

See Figure 3-4: Histograms of all numeric variables
```
pdf.hist(bins=100)
#pdf.hist(bins=100, by='species')
```

Tutorial 3.2.3 Basic Matplotlib Plot

```
import seaborn as sb                      # for loading the iris dataset
import matplotlib.pyplot as plt           # for graphics

pdf  = sb.load_dataset('iris')
leg  = ['sl.sw', 'sl.pl', 'sl.pw', 'sw.pl', 'sw.pw', 'pl.pw' ]
```

See Figure 3-5: Scatter Plot of all pairs of Numeric Variables
```
    plt.scatter(pdf['sepal_length'], pdf['sepal_width'],  marker='1')
    plt.scatter(pdf['sepal_length'], pdf['petal_length'], marker='2')
    plt.scatter(pdf['sepal_length'], pdf['petal_width'],  marker='3')
    plt.scatter(pdf['sepal_width'],  pdf['petal_length'], marker='4')
    plt.scatter(pdf['sepal_width'],  pdf['petal_width'],  marker='+')
    plt.scatter(pdf['petal_length'], pdf['petal_width'],  marker='x')

    plt.title('iris flower features 4x4', fontsize=14)
    plt.xlabel('length / width', fontsize=14)
    plt.ylabel('length / width', fontsize=14)
    plt.legend(leg, loc='lower right')
    plt.xlim(0,11)
    plt.xticks(range(0,11,1))
    plt.grid()
    plt.show()
```

3.3 Probability

Probability is the likelihood of an event occurring. Many events cannot be predicted with total certainty. How likely they are to happen, using the concepts of probability.

Jacob Bernoulli (1654–1705), Abraham de Moivre (1667–1754), Thomas Bayes (1702–61), and Joseph Lagrange (1736–1813) are credited with the development of probability formulae and techniques. The theory of probability forms the basis for statistical applications. Probability is the chance that something will happen. Probabilities are expressed as fractions/decimals between zero and one. Assigning a probability of zero means that something will never happen, a probability of one indicates something will happen.

A random experiment is a phenomenon whose outcome cannot be predicted with certainty, such as flipping a coin or rolling a die. See Fig. 3.6.

Assume a random experiment. The set of possible outcomes is called the sample space. In probability theory, an event is one or more of the possible outcomes of a random experiment. Consider a coin-tossing experiment. Getting a tail when tossing a coin is an event. The probability of this event is 0.5. Getting two tails when tossing a coin two times is an event. The probability of this event is $0.5 \times 0.5 = 0.25$. Consider choosing a card from a deck of playing cards—choosing a King from a deck of cards (any of the 4 Kings out of 52 cards) is an event (with a probability of 4/52). Consider rolling two dice. There are 6×6 possible

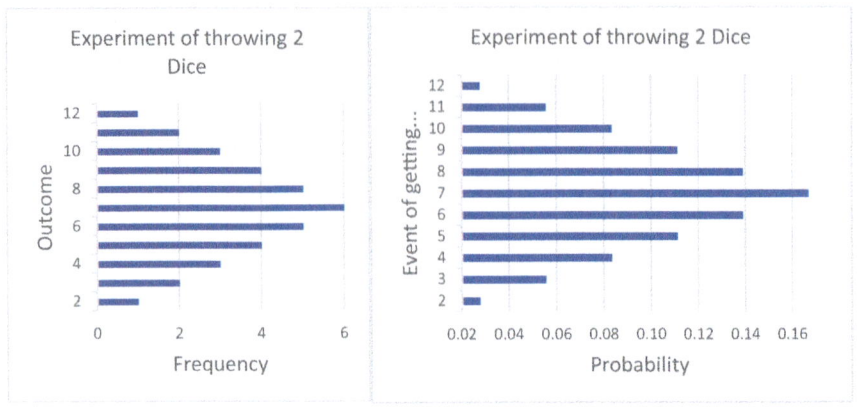

Fig. 3.6 a Experiment and outcome space. **b** probability of events

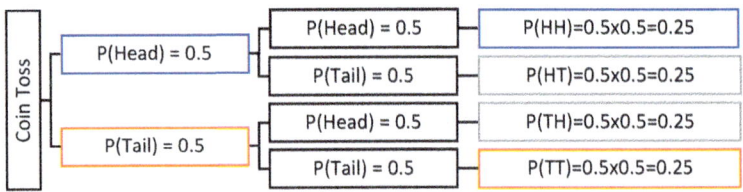

Fig. 3.7 Fare coin toss

outcomes from throwing two dice, which we call the outcome space. There are six ways of getting the outcome '7'—(1,6), (2,5), (3,4), (4,3), (5,2), (6,1). So, the probability of the event (of getting '7') is 6/36=0.167. See Fig. 3.6a, and b.

3.3.1 Rules of Probability

For any event, A, the probability of occurrence of A, is represented as P(A). P(A) is called marginal or unconditional probability (Figs. 3.7 and 3.8).

Axioms of probability
1. For any event A, P(A)> =0.
2. The probability of sample space P(S)=1.
3. For disjoint events (mutually exclusive events) P (A U B)=P (A)+P(B).

Addition rule for probability

$$P \ (A \ U \ B) = P \ (A) + P(B) - P \ (A \ and \ B).$$

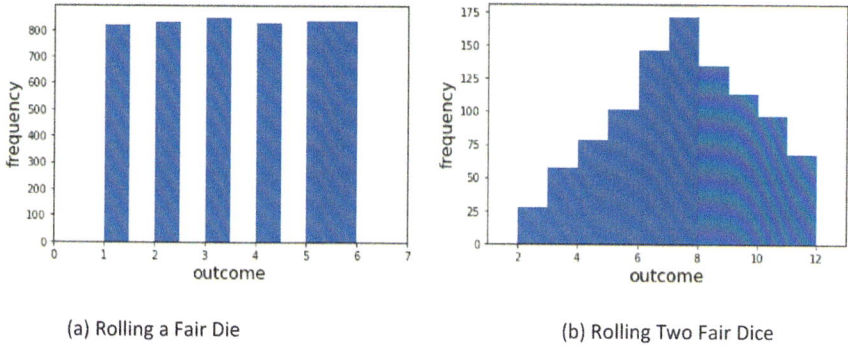

(a) Rolling a Fair Die   (b) Rolling Two Fair Dice

Fig. 3.8 Rolling fair die(s)

3.3.2 Independent Events

If the outcome of an event does not affect the outcome of another, then the events are called independent. There are three types of probabilities under statistical independence:

A. Marginal Probability
Marginal probability is the simple probability of occurrence of an event. In a fair coin toss, marginal probability $P(Head) = 0.5$, and $P(Tail) = 0.5$.

B. Joint Probabilitys
The joint probability of two independent events occurring together (or in succession) is the product of their marginal probabilities. $P(AB) = P(A) \times P(B)$
In the case of a fair coin toss, the probability of
two heads appearing on two successive tosses:

$$P(H1\ H2) = P(H1) \times P(H2) = 0.5 \times 0.5 = 0.25 \text{(see Figure 3.7)}$$

Similarly, $P(HHH) = P(H) \times P(H) \times P(H) = 0.5 \times 0.5 \times 0.5 = 0.125$.

C. Conditional Probability of Independent Events
For statistically independent events, the conditional probability of event B given that event A has occurred is the probability of event B
$P(A|B) = P(A)$,
where A and B are independent events
Similarly, $P(B|A) = P(B)$

3.3.3 Statistical Dependence and Bayes Theorem

If an event's outcome is affected by another's outcome, then the events are dependent. In this section, we discuss the rules for (a) Conditional Probability for Statistically Dependent Events and (b) Joint Probability for Statistically Dependent Events.

If an event's outcome is affected by another's outcome, then the events are dependent. Assume that event B is statistically dependent on event A.

Let P(A) be the probability that event A will happen.

Let P(B|A) be the probability of event B to happen, given that event A has happened.

Let P(BA) be the joint probability for statistically dependent events B and A happening together or in succession.

The joint probability of events B and A happening together or in succession

=Probability of event B given that event A has happened *

The probability that event A will happen.   (3.1)

$$P(BA) = P(B|A) \times P(A)$$

The conditional probability for the statistically dependent event B

to happen given that event A has happened

= The joint probability of events B and A happening together or in succession /

The probability that event A will happen

$$\text{Posterior}(B|A) = P(BA)/P(A)$$

(3.2)

Note that, P(B|A) and P(A|B) represent different conditional probabilities.

For any two events A and B, *where* P(A)<>0, Bayes Theorem states that 'The posterior probability equals the prior probability times the likelihood ratio' (Hutten, 1958).

Posteriori probability P(B|A) =

$$\text{prior-probability P(B)} \frac{\text{likelihood P(A|B)}}{\text{evidence P(A)}}$$

$$P(B|A) = P(B) \frac{P(A|B)}{P(A)}$$ (3.3)

3.4   Discrete Probability Distributions

A random experiment is a phenomenon whose outcome cannot be predicted with certainty, such as flipping a coin or rolling a die. Assume a random experiment. Random experiments may have numerical outputs, such as the lifetime of an electric bulb. In the absence of numerical outputs, they can be assumed. A variable is random if it takes on different values due to the outcomes of a random

experiment. The values of a random variable are the numerical values corresponding to each possible outcome of the random experiment.

The discrete random variable takes a limited number of values, which can be
enumerated. A continuous random variable can take any value within a given
range, which cannot be enumerated, e.g., $f(x) = e^{-x}$. If the probability distribution
of the discrete random variable is called PMF (probability mass function). The
probability distribution is called PDF (probability density function) for a continuous random variable. This section (Sect. 3.4) will discuss the discrete probability
distribution functions—Binomial distribution and Poisson distribution.

Tutorial 3.4.1 Random Numbers

```
import numpy as np
from numpy import random

random.randint(10)    # 5: a random integer between 0 and 1
random.randint(1,6,3) # [2,3,4] an array of 3 random integers between
  1...6
```
2D array of 2 rows by 3 columns, of random integers between 1 and 6
```
  random.randint(1,6,size=(2,3))
array([[4, 5, 2],
       [3, 1, 5]])

  rf = random.random(100)   #returns 100 random floats between 0 and 1
  np.round(rf,2)            #rounded to 2 decimals
[0.98, 0.82, 0.38, 0.06, 0.23, 0.67,...,0.66, 0.67, 0.33,0.23]
```

Array of 2 rows x 3 columns, filled with random floats between 0 and 1
```
  random.rand(2,3)
array([[0.38999024, 0.30542701, 0.1800722 ],
       [0.1303612 , 0.94954896, 0.40937711]])
```

Tutorial 3.4.2 Rolling a Die

```
See Figure 3-8(a): Rolling a Fair Die
  d1 = random.randint(1,7,5000)  #random values 1,2,3,4,5,6
  plt.hist(d1)
  plt.xlabel('outcome', fontsize=16)
  plt.ylabel('frequency', fontsize=16)
  plt.xlim(0, 7)
  plt.show()
```

Tutorial 3.4.3 Rolling Two Die

```
  See Figure 3-8(b): Rolling Two Fair Dice
  d1 = random.randint(1,7,1000)
  d2 = random.randint(1,7,1000)
  plt.hist(d1+d2)
  plt.xlabel('outcome', fontsize=16)
  plt.ylabel('frequency', fontsize=16)
  plt.xlim(1, 13)
  plt.show()
```

Tutorial 3.4.4 Binomial Distribution

A researcher conducts a study on the fair coin toss. Four sets of experiments are designed. Each set of experiments has 10 experiments each. Each experiment consists of 10 Bernoulli trials. The tosses per trial varies across the experiment sets—10, 20, 50, 100. Every trial throws an outcome, e.g., the count of 'heads'. The binominal probability distribution for the above experimental design, assuming a fair coin toss is illustrated below.

```
import numpy as np
from numpy import random
import seaborn as sb
import matplotlib.pyplot as plt
nSuccess    = 10

nTossesPerExpt = [10,20,50,100]   # number of tosses in an experiment
pSuccess = 0.5 # probability of getting a head in Fair coin toss
```

mean value of success (count of heads in an experiment)
= np = nTossesPerExpt x pSuccess

```
ls=['dashdot','dashed','dotted','solid']
i=0
for nTosses in nTossesPerExpt:
    x = random.binomial(size=nSuccess, n=nTosses, p=pSuccess)
    print(x)
    sb.kdeplot(x,linestyle=ls[i])
    i+=1
plt.title ('smoothened binomial p=0.5, n=10..100', fontsize=16)
plt.xlabel('Number of Successes' ,fontsize=16)
plt.ylabel('Probability Density', fontsize=16)
plt.xticks( list(range(0,80,10)) ,fontsize=16)
plt.yticks( list(np.arange(0,0.35,0.1)), fontsize=16)
plt.legend( nTossesPerExpt, fontsize=16)
```
See Figure 3-9(a): Binomial and Poisson Distributions; Table 3.3

Tutorial 3.4.5 Poisson Distribution

```
from numpy import random
import matplotlib.pyplot as plt
import seaborn as sb
import numpy as np
leg  = ['binomial', 'poisson' ]
plt.title('binomial with large n, small p ~ poisson', fontsize=16)
b = random.binomial(n = 100, p = 0.05, size = 50)
sb.kdeplot(b,linestyle='solid')
p = random.poisson(lam = 5, size = 50)
```

```
sb.kdeplot(p,linestyle='dashed')
plt.xlabel('Number of Successes (np)', fontsize=16)
plt.ylabel('Probability Density',      fontsize=16)
plt.legend(leg, loc='upper right',     fontsize=16)
plt.ylim(0,0.21)
plt.yticks(np.arange(0,0.21,0.05))
See Figure 3-9 (b): Binomial and Poisson Distributions
```

3.4.1   Mean and Standard Deviation of a Discrete Random Variable

The mean of a discrete random variable is denoted by the expected value E(X), which can be expressed as

$$E(X) = \frac{1}{N} \sum_{i=1}^{N} x_i P(X = x_i)$$

E(X) approximates to μ, the population mean, for large N.

Assume that 10 experiments were conducted. In each experiment, a fair coin is tossed ten times (N = 10). The counts of heads observed from the experiments are {5 5 4 6 7 4 7 4 8 4}. From the above set of outcomes, we observe five events—five heads occurred twice, four heads four times, six heads once, seven heads twice, and eight heads once. This is shown in Table 3.2. From that, E(X) is calculated to be 5.4. Given the total number of trials N = 10, the probability of success was observed to be (5.4/10=) 0.54. This is close to the expected probability of headcount in a fair coin toss, P = 0.5. As stated earlier, E(X) approximates to μ, the population mean, for large N. This implies that, given the number of trials is large, and the coin toss is fair, the probability of success is expected to be near to 0.5 (Table 3.3).

Table 3.2 Calculation of expected value E(X)

	Event-1 (5 heads)	Event-2 (4 heads)	Event-3 (6 heads)	Event-4 (7 heads)	Event-5 (8 heads)	Sum
Number of heads (x_i) =	5	4	6	7	8	54
Observed frequency	2	4	1	2	1	10
Observed P $(X = x_i)$ = Observed frequency/ Number of trials (N = 10)	0.2	0.4	0.1	0.2	0.1	1
x_i * Observed P $(X = x_i)$	1	1.6	0.6	1.4	0.8	5.4

Table 3.3 The count of heads in a fair coin toss

Experiment design	Number of coin tosses per experiment	Number of experiments	Headcount observed in each experiment
1	10	10	[5 5 4 6 7 4 7 4 8 4]
2	20	10	[10 10 10 11 10 12 9 12 10 7]
3	50	10	[28 24 26 24 20 31 26 30 27 24]
4	100	10	[57 46 50 47 43 56 54 51 40 48]

3.4.2 Binomial Distribution

The binomial distribution is the basis for the popular binomial test of statistical significance. The binomial distribution describes discrete data, resulting from Bernoulli trials. Assume that we conduct **several Bernoulli trials**. Note that.

1. Each trial has only two possible outcomes: (e.g., head or tail for each coin toss; yes or no; success or failure).
2. The probability of the outcome of any trial remains fixed over time.
3. The trials are statistically independent. That is, one toss's outcome does not affect another toss's outcome.

Let p be the probability of success in one trial. **The probability of r successes in n trials, denoted by P (X=r), is known as a binomial distribution**. The probability mass function of binomial distribution can be expressed as.

$$P (X = r) = {}^nC_r p^r (1 - p)^{(n-r)}$$

The parameters of the binomial distribution are:

- Mean of a: $\mu = n.p$
- Standard deviation of a binomial distribution: $\sigma = \sqrt{npq}$.

The binomial distribution is the discrete probability distribution of the number of successes in a sequence of 'n' independent trials, where each trial gives a dichotomous outcome (success/failure). The binomial distribution is usually used to model the number of successes in a sample of size n drawn with replacement from a population of size N.

The binomial distribution is symmetrical when the probability of success p is close to 0.5. When p moves towards 1 or 0, skewness increases. For a binomial distribution, with large n (>20), n.p> =5, and n.q> =5, normal distribution can be used as an approximation. However, the binomial distribution becomes increasingly skewed as p moves towards 0 or 1. When p is close to 1 or 0, the distribution becomes highly skewed and more asymmetrical.

3.4.3 Poisson Distribution

Poisson distribution is a discrete probability distribution that expresses the probability of a given number of events occurring in a fixed interval of time and/ or space if these events occur with a known average rate and are independent of the time since the last event (Levin, 2011), (Zaki & Meira, 2014). The probability mass function (PMF) of the Poisson distribution can be expressed as

$$P(X = k) = e^{-\lambda}\lambda^{k}/k!$$

where

P(X = k) is the probability of observing k events in the interval,
λ is the average rate of events in the interval,
e ~ 2.718.

The Poisson distribution is also used for events in other intervals, such as distance, area, or volume. It has some properties that make it applicable to many situations that require making inferences by taking samples. The parameters of Poisson distribution are:

λ: The mean number of occurrences of an event per interval of time.
σ: The standard deviation $= \sqrt{\lambda}$.

Note: When n is large and p is very small, the Poisson distribution can be used to approximate the binomial distribution. For example, $n >= 20$, and $p <= 0.05$. In this case, we can substitute the Poisson distribution's mean (λ) with the binomial distribution's mean (np). See Fig. 3.9(b)/Tutorial 3.4.5. Poisson distribution is right-skewed.

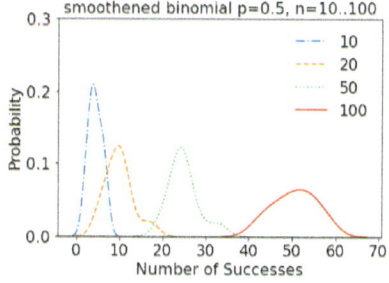

(a) Binomial Dist. Coin Toss, BD, p=0.5, n = 10..100

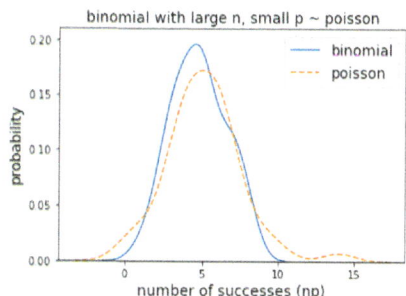

(b) Binomial and Poisson, p=0.05, n=100

Fig. 3.9 Binomial and Poisson Distributions

3.5 Continuous Probability Distributions

As discussed earlier, a probability distribution function can be discrete, or continuous. In this section, we will discuss three continuous-valued probability distribution functions—normal, t, and chi-square (Levin, 2011), (Zaki & Meira, 2014).

3.5.1 Normal Distribution

The normal (or Gaussian) distribution is a continuous probability distribution with a bell-shaped probability density function, known as the Gaussian function or the bell curve's informally. Carl Friedrich Gauss (1777–1855) was one of the earliest researchers to explore normal distribution. He applied it to analyze errors in astronomical observations and advanced the least squares approximation method.

The mean of a normal probability distribution is denoted by μ and its standard deviation by σ. A continuous random variable x with a normal distribution is called a normal random variable. The normal distribution comes close to fitting the observed frequency distributions of many phenomena (Central limit theorem).

The probability density function (PDF) of a normal distribution is expressed as

$$f(x) = \frac{1}{\sqrt{2\pi}} e^{-(1/2)x^2}$$

The cumulative distribution function (CDF) of a normal distribution is expressed as

$$F(x < z) = \frac{1}{\sqrt{2\pi}} \int_{-\infty}^{z} e^{-(1/2)x^2} dx$$

The standard normal distribution represents a specific instance of the normal distribution, where the mean (μ) is zero and the standard deviation (σ) is one. The random variable associated with the standard normal distribution is typically denoted as 'z.' The unit of measurement for this distribution is expressed in terms of z-scores (refer to Fig. 3.10). Tables illustrating standard normal probability distribution computations are widely accessible, providing probabilities for various ranges (areas under any normal curve). Utilizing these tables, we can ascertain the probability of a randomly distributed variable falling within specific distances from the mean, defined in terms of standard deviations.

Key characteristics of a normal probability distribution (normal curve) include being unimodal with a single peak, exhibiting a bell-shaped symmetrical curve, having the mean located at the center, and featuring overlapping mean, median, and mode. Additionally, the two tails of the normal curve extend indefinitely without touching the horizontal axis.

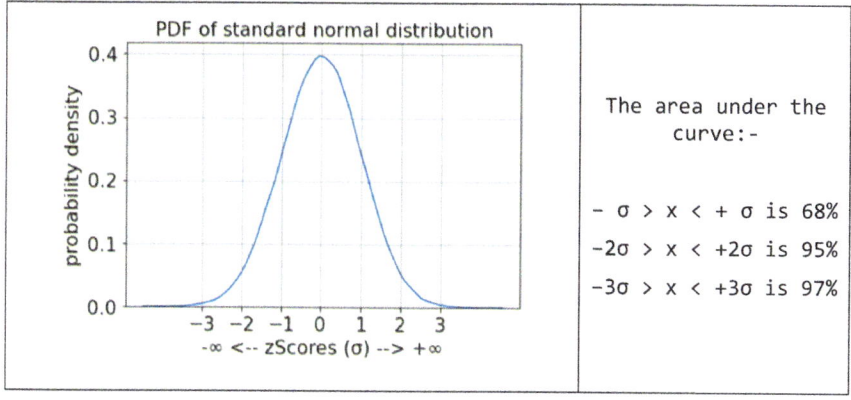

Fig. 3.10 The probability density function (PDF) of a standard normal distribution

Parameters of Normal Distribution

- The mean of the normal distribution is (μ) and the standard deviation (σ).
- A std normal distribution has $\mu = 0$ and $\sigma = 1$.
- Standardizing a normal random variable: $Z = (X - \mu)/\sigma$.

Measures of Shape

In statistical theory, location and variability are referred to as a distribution's first and second moments. The third and fourth moments are called skewness and kurtosis. Skewness is the degree of distortion from a normal distribution. Skewness is the third moment of the standardized score of X (Fernandez-Granda, 2017).

$$\text{Skewness}(X) = E\left[\left(\frac{(X - \mu)}{\sigma}\right)^3\right]$$

A distribution can be positively skewed, negatively skewed, or non-skewed based on whether Skew(X) is positive, negative, or 0. See Fig. 3.11. A commonly followed thumb rule states that skewness within ± 2.58 is acceptable for a normal curve.

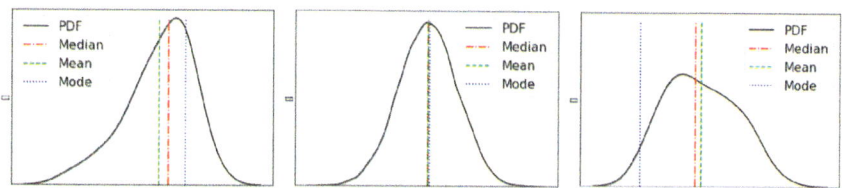

Fig. 3.11 Measures of Shape – Negative Skew, Symmetric, and Positive Skew

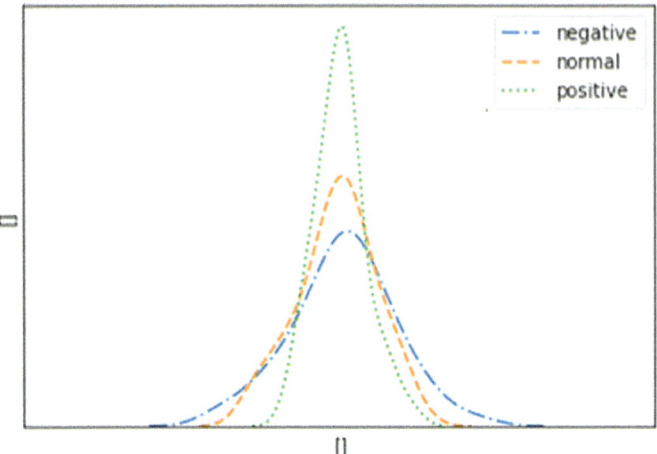

Fig. 3.12 Kurtosis

Kurtosis indicates the relative peaked-ness or flatness of the frequency distribution. Kurtosis is the fourth moment of the standardized score of X. (See Fig. 3.12)

$$\text{Kurtosis }(X) = E\left[\left(\frac{(X - \mu)}{\sigma}\right)^4\right]$$

The above computation for kurtosis always gives a positive value. The kurtosis of a standard normal distribution is 3. We use the equation below to compare other curves with the standard normal.

$$\text{Kurtosis }(X) - \text{Kurtosis (standard normal)} = E\left[\left(\frac{(X - \mu)}{\sigma}\right)^4\right] - 3$$

This equation returns a negative, zero, or positive value as seen in Fig. 3.12.

Positive kurtosis implies that the curve is more peaked than a normal distribution. A negative kurtosis implies that the curve is flatter. See Fig. 3.12. A commonly followed thumb rule states that kurtosis within ± 1.96 is acceptable for a standard normal curve. Kurtosis indicates the outliers' presence (and influence) in a distribution.

3.5.2 T-Distribution

The 't-distribution' or Student's t-distribution was developed by William Sealy Gosset (1876–1937). As a brewer at Guinness in Dublin, Ireland, he developed a method for quality control based on small samples. Since Guinness prohibited

Fig. 3.13 t-Distribution
Versus Standard Normal
Distribution

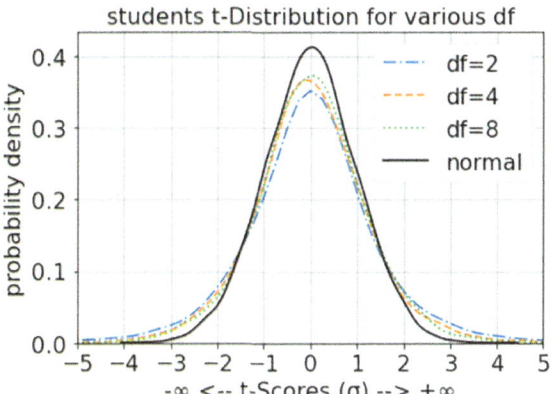

its employees from publishing their research findings, Gosset published his work
under the pseudonym 'Student'.

The t-distribution curve is a bell-shaped symmetric curve. As the sample size,
n increases, the t-distribution approaches the standard normal distribution. See
Fig. 3.13. The t-distribution has only one parameter—the number of degrees of
freedom (df).

df = n − 1, for a t-distribution, where n is the sample size.

The mean of the t-distribution is 0, and its standard deviation is $\sqrt{[df / (df - 2)]}$.

3.5.3 Chi-Squared Distribution

A Chi-squared ($\chi 2$) distribution is a right-skewed continuous probability distribution. Let 'n' be a positive integer. The probability density function of a random
variable x, having a Chi-squared distribution, with 'df' degrees of freedom, can be
expressed as.

$$f(x, df) = \frac{x^{\left(\frac{df}{2}-1\right)} e^{-\frac{x}{2}} 2^{\frac{df}{2}} \tau\left(\frac{df}{2}\right), \text{if } x > 0}{\text{if } x \ <= \ 0; \ f(x, \ df) = 0.}$$

(In the equation above, τ is the gamma function).

As the number of degrees of freedom increases, the Chi-square distribution
becomes approximately equal to a normal distribution, as shown in Fig. 3.14. (We
will discuss the degree of freedom of Chi-square distribution in Chapter 4). The
curve starts with a zero on the left side and extends to infinity on the right. The
total area under a Chi-square curve equals 1.

Fig. 3.14 Chi-Square
Distribution

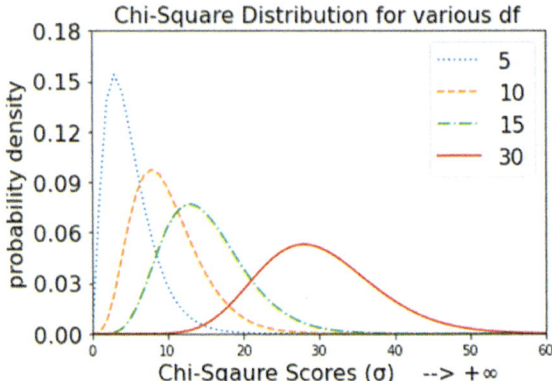

Tutorial 3.5.1 t Distribution

```
import numpy as np
import seaborn as sb
import matplotlib.pyplot as plt

n = 10000
xTicks  =np.arange(-5,6,1)
yTicks  =list(np.arange(0,0.5,0.1))
df = [3,5,7]
df_leg = ['df=2','df=4','df=8','normal']
ls=['dashdot','dashed','dotted','solid']
l=0
for i in df:
    x = np.random.standard_t(i, size=n)
    sb.kdeplot(x,linestyle=ls[l])
    l+=1
x = np.random.normal(0, 1, size=n)
sb.kdeplot(x, color='black')
plt.legend(df_leg, fontsize=16)
plt.xlim(-5,5)
plt.xticks(xTicks, fontsize=16)
plt.yticks(yTicks, fontsize=16)
plt.xlabel('-∞ <-- t-Scores (σ) --> +∞', fontsize=16)
plt.ylabel('probability density', fontsize=16)
plt.title ('students t-Distribution for various df', fontsize=16)
plt.grid()
```
See Figure 3-13: t-Distribution Versus Standard Normal Distribution

Tutorial 3.5.2 Chi-Square Distribution

```python
from scipy import stats
import numpy as np
import matplotlib.pyplot as plt

linestyles = [':', '--', '-.', '-']
yTicks  =list(np.arange(0,0.19,0.03))
x = np.linspace(0, 99, 100)
deg_of_freedom = [5, 10, 15, 30]
i = 0
for df in deg_of_freedom:
    plt.plot(x, stats.chi2.pdf(x, df), linestyles[i])
    i = i+1
plt.xlim(0, 60)
plt.ylim(0, 0.18)
plt.xlabel('x')
plt.title('Chi-Square Distribution for various df', fontsize=16)
plt.legend(deg_of_freedom)
plt.xlabel('  Chi-Sqaure Scores (σ)    --> +∞', fontsize=16)
plt.yticks(yTicks, fontsize=16)
plt.ylabel('probability density', fontsize=16)
plt.show()
See Figure 3-14 Chi-Square Distribution
```

3.6   Sampling Distributions and Central Limit Theorem

Understanding a phenomenon can be achieved by collecting information by examining every individual within the entire group (population) related to the issue. This approach is commonly known as complete enumeration or a census. However, in numerous instances, assessing or measuring every item within the population is neither feasible nor practical, especially for cost-value analysis. Consequently, information is derived from a subset of the population, termed a sample. This method is termed 'sampling.'

3.6.1   Sampling Methods

Methods used for sampling are categorized as (a) random sampling or probabilistic sampling or (b) non-random sampling or judgment sampling. See Fig. 3.15.

Let us explore various random sampling methods. Simple random sampling ensures that each potential sample has an equal likelihood of being chosen, and

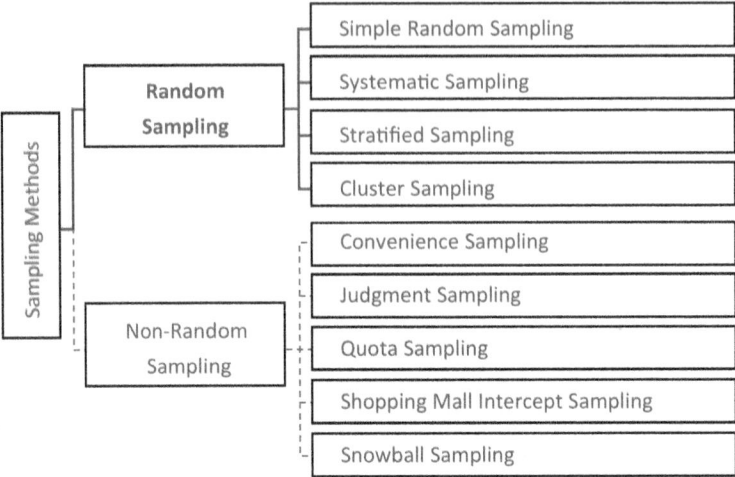

Fig. 3.15 Sampling methods

every item in the population has an equal opportunity to be part of the sample. In systematic sampling, starting from a random point, elements are selected at regular intervals, measured in time, order, or space. In stratified sampling, the population is categorized into relatively homogeneous groups (strata), and then elements are randomly selected from each stratum. Cluster sampling involves dividing the population into groups or clusters and selecting a random sample from these clusters. These individual clusters serve as representative subsets of the entire population.

3.6.2 The Central Limit Theorem

The probability distribution of all possible values of the sample statistic is known as the sampling distribution. For a normal population, the distribution of sample means follows a normal distribution with mean (μ) and standard error ($\sigma/\sqrt{n}$).

Central Limit Theorem (CLT) states that, as the sample size increases, the sampling distributions closely approximate the normal distribution and become clustered around the population mean, for all distributions of independent, identically distributed variables that have a finite variance (Frost, n.d.; Frost 2023; Levin, 2011; Zaki & Meira, 2014).

It may be also noted that, for a large sample, the sampling distribution of sample proportion will follow a normal distribution with mean 'p' and standard error '$\sqrt{[pq/n]}$'.

Figure 3.16 shows the result of an experiment in measuring chicken weight (in grams) from a large poultry. We take 10 samples of chicken and calculate

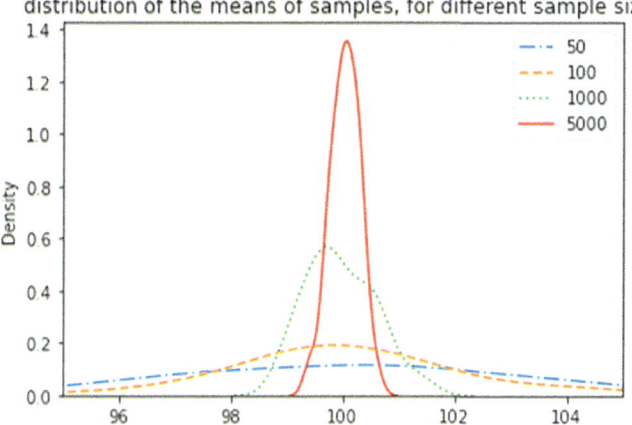

Fig. 3.16 Distribution of sample means, for different sample sizes

the means (10 means). We repeat the above experiment for sample sizes of 10, 50, 500,1000, and 5000 chickens per sample. We observe that with larger sample sizes, the distribution of sample means becomes more normal and more tightly clustered.

The standard deviation of the distribution of a sample statistic is known as the standard error of the statistic. Standard error decreases with an increase in sample size. Determination of appropriate sample size depends upon two criteria:

Degree of precision or extent of the permissible error (e).
Degree of confidence placed with the sample results $(1 - \alpha)$.

Tutorial 3.6.1 Central Limit Theorem Exercise

```
import numpy as np
import matplotlib.pyplot as plt

mu, sigma, nPoultry = 100, 20, 50
samples = [50,100,1000,5000]

import seaborn as sns
mu, sigma, nPoultry = 100, 20, 50
ls=['dashdot','dashed','dotted','solid']
l=0
for nChicks in samples:
    S = 100
    for i in range(1, nPoultry):
        s = np.random.normal(mu, sigma, nChicks)
        S = np.append(S,s.mean())
```

```
    plt.xlim(95,105)
    sns.kdeplot(S,linestyle=ls[l])
    l+=1
plt.legend(samples,loc='best',prop={"size":10}) #font size 10
t = 'distribution of the means of samples, for different sample sizes'
plt.title(t, fontweight=10)
plt.show()
See Figure 3-16: Distribution of Sample Means
```

3.7   Point and Interval Estimates

A statistic (e.g., mean) computed from a data sample gives a point estimate of the population from which the sample is drawn. However, the computed mean will vary from sample to sample. So the exact value of the population mean remains elusive. However, the population means lie within a certain interval, with a certain confidence. The researcher chooses the confidence level a priori based on his requirements. The confidence level can be expressed as

$$\text{Confidence level} = 100(1 - \alpha)\%$$

where α, the *level of significance*, is used in our computations.

The commonly used confidence levels are 90%, 95%, and 99%. The corresponding to α are 0.10, 0.05, and 0.01.

Confidence Interval (CI) = Point estimate $\pm$ Margin of Error.

The margin of error comprises two entities—critical value and measure of the variability of the sampling distribution. The critical value is a number that corresponds to α, the *level of significance*, set a priori. It is denoted as CV.

See Fig. 3.17. The solid line is the point estimate. The critical values are marked as dash-dot lines. In a two-sided test, the interval estimate of the population parameter will lie between the critical values. The dotted lines show the probable error in the estimate, for the given significance level.

Fig. 3.17 Point estimate, and level of significance (α)

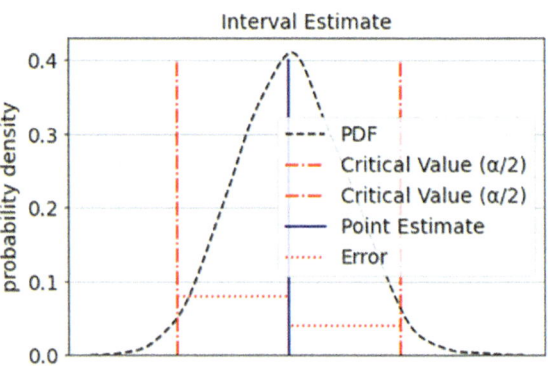

Table 3.4 Sample statistics and population parameters

Measurement	Symbol for sample statistic	Symbol for population parameter statistic
Mean	$\bar{x}$	μ
Proportion	p	π
Variance	s^2	σ^2
Standard deviation	s	σ
Size	n	N

Note that, for a normal population, the distribution of sample means follows normal probability distribution with mean (μ) and standard error ($\sigma/\sqrt{n}$). Therefore, the standard error ($\sigma/\sqrt{n}$) measures the variability of a normal population. The following equation can express the margin of error:

$$\text{The margin of error} = z_{\alpha/2} * \sigma/\sqrt{n}$$

The sample statistic can be used to infer an interval estimate of the population. For example, if $\hat{x}$ is the sample mean, we may infer the population mean μ with a margin of error, based on the required significance level. More specifically, the population mean μ is expected to fall in the interval bounded by ($\hat{x}$ - margin of error) and ($\hat{x}$ + margin of error), for the given significance level. This is the basis of inferential statistics.

Table 3.4 shows a list of statistics and their symbols in the sample space and the population. In subsequent chapters, we will explore how to infer the population parameters from the sample statistics.

Summary

The chapter covers the broad spectrum of statistical methods, dividing them into descriptive and inferential statistics. Descriptive statistics involves representing data through various means such as tables, graphs, and charts, employing techniques like box plots, histograms, and summary statistics. Measures like arithmetic mean, geometric mean, median, mode, and standard deviation are discussed with their characteristics and applications.

Inferential statistics focuses on estimating population parameters from sample data. Probability concepts include events, sample spaces, and probabilities expressed as fractions or decimals between zero and one. The distinction between discrete and continuous random variables is made, and probability distribution functions, binomial distribution, Poisson distribution, and normal distribution are explored. Skewness and kurtosis are explained as measures of distribution shape.

The t-distribution, chi-square distribution, and their parameters and applications are outlined. Sampling methods, both random (probabilistic) and non-random (judgment) sampling, are introduced. The concept of the sampling distribution, influenced by the Central Limit Theorem (CLT), is discussed. The CLT states that sampling distributions approximate the normal distribution as the sample size increases. The chapter also emphasizes point estimates, confidence intervals, and the variation of computed means across different samples for statistical inference.

Questions

Comprehensions

1. Write a short note on the main categories of summary statistics falling under descriptive statistics.
2. Describe skewness and kurtosis.
3. Describe the following, with proper association—A random experiment, Outcome, Sample space, and Event.
4. Define probability.
5. State the axioms of probability.
6. With respect to independent events, define the following: -

    a. Joint probability
    b. Conditional probability

7. With respect to mutually dependent events, define the following: -

    a. Conditional probability for statistically dependent events.
    b. Jsoint probability for statistically dependent events.

8. State and explain Bayes' theorem.
9. Define and explain a random variable with examples.
10. Define the expected value of a discrete random variable.
11. Define binomial distribution.
12. Define Poisson distribution.
13. Define normal distribution.
14. Write a note on standard normal distribution.
15. Write a note on random sampling methods.
16. State and explain the Central Limit Theorem.
17. Define an interval estimate for the mean, assuming a normal distribution.

Analysis and Application

18. What are the critical differences between descriptive and inferential statistics, and when is each type of statistic proper?
19. Compare and contrast the characteristics and use cases of the mean, median, and mode as measures of central tendency in statistics.
20. Explain the difference between a discrete random variable and a continuous random variable. Provide examples of each.

21. Explain how the shape of a binomial distribution changes as n increases while keeping p constant. Given a binomial distribution with n = 30 and p = 0.2, cssalculate the mean and standard deviation of the distribution. Interpret the results in the context of a real-world scenario.

22. Discuss the relationship between a Poisson distribution's mean (λ) and the standard deviation (σ). Using the Poisson distribution, calculate the probability of observing three or fewer events in an interval with an average rate (λ) of 5. Interpret the result.

23. Given two independent events A and B with P(A) = 0.4 and P(B) = 0.3, calculate the joint probability P(A and B) and the conditional probabilities P(A|B) and P(B|A).

24. Explain the concept of statistical dependence and how it impacts the calculation of conditional probabilities. Provide an example where one event's outcomes affect another's outcomes.

25. Compare and contrast simple random sampling and stratified sampling. In what research scenarios would you choose one over the other, and why?

26. Explain the advantages and disadvantages of cluster sampling. Provide examples of situations where cluster sampling is appropriate.

27. Compare the skewness and kurtosis values for a normally distributed dataset and a non-normally distributed dataset. Discuss the implications for statistical analysis.

28. Describe the Central Limit Theorem and its role in making statistical inferences about populations. How does the sample size affect the applicability of the Central Limit Theorem?

29. Analyze the differences between a t-distribution and a standard normal distribution. When would you use one over the other in statistical analysis?

30. You have a dataset of student exam scores. How can you apply descriptive statistics to summarize the performance of the students?

31. In a call center, the average number of calls received in an hour is 15. Calculate the probability of receiving exactly 10 calls in the next hour using the Poisson distribution.

32. Discuss situations where the Poisson distribution is suitable for modeling events, such as accident rates or website traffic.

33. Compare and contrast the characteristics of the binomial and Poisson distributions. Under what conditions would you choose one over the other for modeling a specific scenario?

34. You are conducting a survey to estimate the average income of a population. How would you use a point estimate and a confidence interval to report your findings?

35. Explain how knowledge of the normal distribution can be applied to quality control in a production environment.

36. In a manufacturing process, product weights are normally distributed with a mean of 500 g and a standard deviation of 10 g. Calculate the probability that a randomly selected product weighs more than 515 g.

37. A small sample of 12 students is selected to assess their performance on using t-test. Calculate the 95% confidence interval for the population mean test score, given a sample mean of 85 and a sample standard deviation of 5.

38. In a medical study, a researcher tests the hypothesis that two treatments have different success rates. Explain how the chi-square distribution can be used to analyze the data and draw conclusions about treatment effectiveness.

39. Describe a situation in which the chi-square test for independence would be applied and its significance.

Exercises

The questions in this section are based on two datasets shown below. Refer to Chapter 1 for the data description.

- Penguins (seaborn dataset)
- ChickWeight (statsmodels dataset - https://www.statsmodels.org/stable/index.html)

Exercise 3.1 Box Plot
Use the dataset penguins. Remove rows with missing values. Generate boxplots:-

(a) mass by species and sex
(b) mass by species and island

Exercise 3.2 Histogram
Use the dataset penguins. Remove rows with missing values. Generate histograms:-

(a) mass by species
(b) mass by island

Exercise 3.3 Scatter Plot
Use the dataset penguins. Remove rows with missing values. Generate scatter plot:

(a) bill length by flipper length by species

Exercise 3.4 Strip Plot

Use the dataset penguins. Remove rows with missing values.

(a) Body mass by island by sex

Exercise 3.5 Random Number Functions

(a) Generate a 3*2 matrix of random integers.
(b) Generate a 3*2 matrix of random floats.

Exercise 3.6 Probability Distribution of Admitted COVID Patients
Plot the probability distribution of COVID patients to be admitted, given the rate of COVID positive with co-morbidity is 0.01 over all those who are identified as COVID positive. Assume 300 patients arrive a day and 100 are found COVID positive. Use binomial and Poisson. What distribution would you prefer and why?

References

Fernandez-Granda C (2017) Probability and statistics for data science. In *Probability and Statistics for Data Science*

Frost J (n.d.) *Statistics by Jim*. https://Statisticsbyjim.Com/Basics/Central-Limit-Theorem/

Frost, J. (2023). *Statistics by Jim*. https://statisticsbyjim.com/basics/central-limit-theorem/

Hutten EH (1958) Probability and induction. *The British Journal for the Philosophy of Science*, 9(33). https://www.journals.uchicago.edu/doi/https://doi.org/10.1093/bjps/IX.33.43

Levin R (2011) Statistics for management. In *The Statistician* (Issue 1). https://doi.org/10.2307/2348398

Malhotra NK (2020) Marketing research an applied prientation seventh edition. *Pearson Education*

Zaki MJ, Meira W (2014) Data mining and analysis: fundamental concepts and algorithms. Cambridge University Press

Chapter 4
Hypothesis Testing

Learning Objectives

- Understand the fundamentals of hypothesis testing.
- Demonstrate t-test for comparison of means.
- Demonstrate the ANOVA test.
- Demonstrate the chi-square test.

Overview

In this chapter, we will discuss hypothesis testing fundamentals and explore t-test, ANOVA, and chi-square tests.

Definitions

ANCOVA: If the set of independent variables consists of categorical and metric variables, the technique is called the analysis of covariance (ANCOVA).

ANOVA (one-way analysis of variance): The one-way analysis of variance aims to compare the means of two or more independent groups to determine whether the associated population means are significantly different.

Chi-square test: The chi-square test of independence, also known as the chi-square test of association, is a nonparametric statistical test. Its purpose is to assess whether there is a significant association between categorical variables.

Supplementary Information   The online version contains supplementary material available at https://doi.org/10.1007/978-981-99-0353-5_4.

S. Sundararajan, *Multivariate Analysis and Machine Learning Techniques*,
Transactions on Computer Systems and Networks,
https://doi.org/10.1007/978-981-99-0353-5_4

Hypothesis testing: In hypothesis testing, we compare the statistics from a sample data distribution with the parameters of a model (t-distribution, chi-square distribution, etc.). If the sample is consistent with the model, the null hypothesis is not rejected; otherwise, the null hypothesis is rejected in favor of the alternative hypothesis.

Independent samples t-test: The independent samples t-test compares the means of two independent groups in a sample to determine whether the associated population means are significantly different.

Nonparametric tests: Nonparametric tests are used when the variables are categorical (nominal or ordinal), e.g., chi-square test.

One-sample t-test: The one-sample t-test examines whether the mean of a population is statistically different from a hypothesized value.

Paired samples t-test: Paired samples t-test compares the means of two measurements taken from the same subject. For example, the awareness of a product and purchase intention.

Parametric tests: Parametric tests are used for numeric variables (ratio scaled or interval scaled), e.g., t-test.

Test statistic: A test statistic is a number calculated from the data sample.

T-test: For comparison of means, when the sample size is small or the population standard deviation is not known, we will prefer the t-test over the z-test.

Type I error: Two types of errors can occur in hypothesis testing. A type I error occurs when the null hypothesis is rejected when it is true.

Type II error: Type II error occurs if the null hypothesis is rejected when we should not. Type I error is considered more severe than type II.

4.1   The Fundamentals of Hypothesis Testing

Jerzy Neyman and Egon Pearson formulated the theory of hypothesis testing. They hypothesized that regardless of the results of an experiment on data samples, a researcher could never be certain whether it holds good for the population. However, based on two parameters—α (significance level) and β ([$1 - \beta$] is called the power of the test), a researcher could make certain conclusions on the probability of success of the experiment. Probability, or p-value as it is commonly called, is attributed to Ronald Fisher, considered the father of modern statistics and experimental design.

We saw in the previous chapter that a statistic (e.g., mean) calculated from a random sample of data does not give the exact estimate of the population parameter, as the statistic will vary from sample to sample. However, we can hypothesize that the population parameter will lie within a certain interval, with a certain confidence. How do we infer that our hypothesis is true or false? One of the ways to achieve this is hypothesis testing.

It may be noted that there are two major categories of tests. Parametric tests are used for numeric variables (ratio scaled or interval scaled). Nonparametric tests are used when the variables are categorical (nominal or ordinal). T-test is a parametric test; chi-square test is a nonparametric test.

4.1.1 Hypothesis Testing Procedure

In hypothesis testing (Duchesnay 2021; Levin 1984; Navidi n.d.; Rice 2007), we compare the statistics from a sample data distribution with the parameters of a model (t-distribution, chi-square distribution, etc.). If the sample is consistent with the model, the null hypothesis is not rejected; otherwise, the null hypothesis is rejected in favor of the alternative hypothesis. A null hypothesis is a statement of the status quo, one of no difference or no effect. The alternative hypothesis is the opposite of the null hypothesis—it hypothesizes some difference or effect. The general procedure for hypothesis testing is given below (Malhotra 2020).

1. Define the problem.
2. Formulate the null (H_0) and the alternative (H_a) hypotheses.
3. Select the appropriate test.
4. Select the significance level (α) for testing H_0. Typically, α is set to 0.05.
5. Execute the statistical test based on the data sample.
   Typically, this step is executed by a statistical software package. It computes the test statistic using the data sample and finds out the p-value associated with the test statistic.
6. Analyze the test results. We can use any one of the following methods for this

   - If the value of the computed statistic falls within the critical region, reject H_0
     Or
   - If the p-value $< = \alpha$, reject H_0. Otherwise, do not reject H_0.

7. Inference: Interpret the results and state our conclusion.

4.1.2 Hypothesis Formulation

A hypothesis test starts with formulating the null hypothesis and the alternative hypothesis. Some examples concerning hypothesis testing of the population mean (μ) are given below.

One-Sample t-test

The null hypothesis for a hypothesis concerning a population mean (μ) is expressed as

$H_0: \mu = \mu_0$, where μ_0 is some number.

The formulation of the alternative hypothesis depends on the problem at hand. There are three ways to formulate an alternative hypothesis.

$H_a: \mu \neq \mu_0(1)$ Two-sided test.
$H_a: \mu < \mu_0(2)$ Left-tailed test.
$H_a: \mu > \mu_0(3)$ Right-tailed test.

Multiple Samples Test

$H_0: \mu_1 = \mu_2 = \ldots = \mu_k$,
Where

$\mu_1, \mu_2 \ldots \mu_k$ are the mean of different populations,
H_a: at least one mean is μ_i is different from other means.

4.1.3 Type I Error, Type II Error, and the Level of Significance

As mentioned in the introductory section, regardless of the results of an experiment on data samples, a researcher could never be certain whether it holds good for the population. Two types of errors can occur: type I error and type II error (Biau et al. 2010). A type I error occurs when the null hypothesis is rejected when it is true. Type II error occurs if we reject the null hypothesis when we should not. Type I error is considered more severe than type II. Drawing a courtroom analogy, punishing an innocent person is a type I error, and acquitting the guilty is a type II error. See Table 4.1.

The probability of type I error is α. It is also called the significance level. A researcher frames the hypothesis in a manner to have the null hypothesis rejected (when it is false). The probability of type II error is β. The power of the test is indicated by $(1-\beta)$. The significance level alpha (α) and power of the test $(1-\beta)$ are inversely related. If the power is close to 1, the hypothesis test is good at detecting a false null hypothesis. Ideally, α and β should both be small. The researcher can set the value of α, but β cannot be set.

See Fig. 4.1. The vertical line shows the critical value (CV). In this figure, CV is set to z-score $= 1.96$, or $\alpha = 0.05$ (equivalent to a confidence level of 95% for a right-sided test). Here, α is the region to the right side of the critical value in the

Table 4.1 Type I and type II error

	Concept			Example	
	H_0 true	H_0 is false		H_0: accused is innocent	H_0: accused is guilty
Do not Reject H_0	Yes	Type II error	*Acquit*	Yes	Type II error
Reject H_0	Type I error	Correct	*Punish*	Type I error	Correct

Fig. 4.1 H_0, H_a, critical value, level of significance, and power

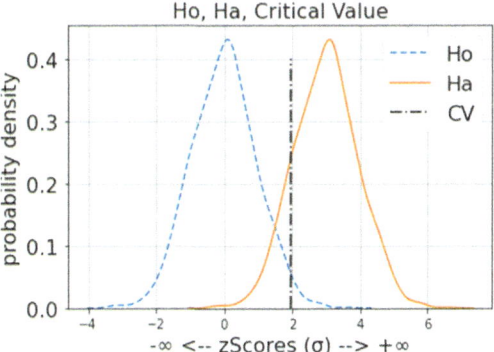

null hypothesis (H_0) curve, and β is the region that falls to the left of the critical value in the alternate hypothesis (H_a) curve. It may be observed that, if the CV line moves to the right, the significance level (α) increases, and power ($1-\beta$) decreases.

4.2   Comparison of Means Overview

In the subsequent sections, we will do tutorials on hypothesis testing regarding the comparison of means. When the sample size is small or the population standard deviation is unknown, we prefer t-test (assuming t-distribution) over z-test (assuming normal distribution).

We may be concerned with a single population parameter or wish to compare multiple populations. In the case of multiple populations, they may be independent or paired. Independent samples are drawn randomly from different populations—e.g., different groups of respondents such as males and females. Samples are said to be paired if different features are observed off the same sample—e.g., height and weight of the respondents.

Our tutorials will be based on the t-test and chi-square test. We will target a level of significance (α) of 0.05. The tutorials covered are shown in Figs. 4.2.

4.3   Comparison of Means—Independent Samples t-Test

The independent samples t-test compares the means of two independent groups in a sample to determine whether the associated population means are significantly different. Two variables are used in this:

- The dependent variable or test variable (body mass of penguins) is expected to be a numeric variable (integer or float).

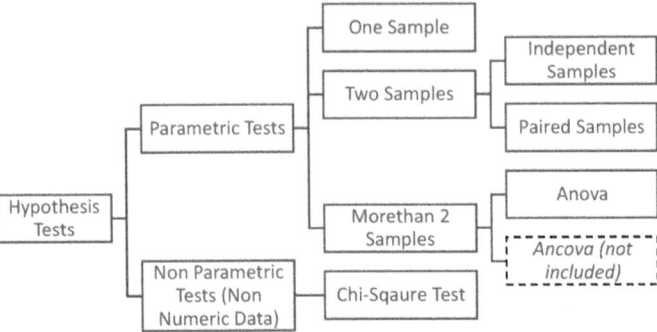

Fig. 4.2 Hypothesis testing—comparison of means—tutorial coverage

- The independent variable or grouping variable (penguin species—'Adelie', 'Chinstrap', and 'Gentoo') is expected to be a categorical variable (nominal or ordinal).

Variances in the two groups are assumed to be approximately equal. When the assumption of homogeneity of variances does not hold, a variation called the Welch t-test can be used. Outliers affect the estimates (see Sect. 2.8.2).

The Hypothesis on Two Independence Samples t-test

The hypothesis can be formulated in two ways. The first formulation is easy to understand. However, it is the second formulation that is generally tested

(1) H0: $\mu1 = \mu2$ (the two population means are equal).
      Ha: $\mu1 \neq \mu2$ (the two population means are not equal).
(2) H_0: $\mu_1 - \mu_2 = 0$ (the difference between the two population means $= 0$).
      H_a: $\mu_1 - \mu_2 \neq 0$ (the difference between the two population means $\neq 0$).

The Hypothesis on the Homogeneity of Variances

The independent samples t-test relies on the assumption of homogeneity of variance—i.e., both groups have the same variance. There are many tests for the homogeneity of variance (Bartlett's test, Levine's test, etc.). The hypotheses for homogeneity of variance are given below:

(a) H0: $\sigma12 - \sigma22 = 0$ (the population variances of groups 1 and 2 are equal).
(b) Ha: $\sigma12 - \sigma22 \neq 0$ (the population variances of groups 1 and 2 are not equal).

The t-Test Formula, Equal Variances Are Assumed

When two independent samples are assumed to be drawn from populations with identical variances $(\sigma 1^2 = \sigma 2^2)$, the t-statistic can be expressed as shown below (Malhotra 2020). The degree of freedom is $(n1 + n2 - 2)$ when equal variance is assumed.

$$t = \frac{\bar{x}_1 - \bar{x}_2}{s_p\sqrt{1/n_1 + 1/n_2}}$$

where

$\bar{x}_1, \bar{x}_2$ Are the means of sample-1 and sample-2,
n_1, n_2 are the sizes of sample-1 and sample-2,
s_1, s_2 are the standard deviation of sample-1 and sample-2,
s_p is the pooled standard deviation, expressed as follows:

$$S_p = \sqrt{\frac{[n_1 - 1]s_1^2 + (n_2 - 1)s_2^2}{n_1 + n_2 - 2}}$$

The t-Test Formula, Equal Variances not Assumed

We can pool the sample variances (sp) assuming equal population variances. Otherwise, we cannot pool the sample variances. When the two independent samples are assumed to be drawn from populations with unequal variances $(\sigma 1^2 \neq \sigma 2^2)$. The t-statistic is computed as shown below.

$$t = \frac{\bar{x}_1 - \bar{x}_2}{\sqrt{s_1^2/n_1 + s_2^2/n_2}}$$

When equal variance is not assumed, the degree of freedom is the weighted harmonic mean, the weights being the sample size n_1, n_2. It is computed as follows:

$$df = \frac{\left(\frac{s_1^2}{n_1} + \frac{s_2^2}{n_2}\right)^2}{\frac{1}{n_1-1}\left(\frac{s_1^2}{n_1}\right)^2 + \frac{1}{n_2-1}\left(\frac{s_2^2}{n_2}\right)^2}$$

Two Independent Samples t-test Example

```
Penguins Data Set Description

Data about 342 penguins from three islands, three species
RangeIndex: 344 entries, 0 to 343

 #    Column               Non-Null Count  Dtype
 0    species              344 non-null    ['Adelie', 'Chinstrap', 'Gentoo']
 1    island               344 non-null    ['Biscoe','Dream','Torgersen']
 2    bill_length_mm       342 non-null    float64
 3    bill_depth_mm        342 non-null    float64
 4    flipper_length_mm    342 non-null    float64
 5    body_mass_g          342 non-null    float64
 6    sex                  333 non-null    ['Female', 'Male']
```

Refer to the penguin dataset. We would like to know whether the body mass differs across penguin species.

Answer:

We will follow the schematic diagram for the t-test in Fig. 4.3. When the sample size is small or the population standard deviation (σ) is unknown, we prefer the t-test (z-test, otherwise). This tutorial will use a t-test since the population parameter (σ) is unknown. We will set the significance level α to 0.05.

We first need to get descriptive data statistics—See Tutorial Sects. 4.1, Fig. 4.4, and Penguin Dataset Description. We will follow the detailed steps for hypothesis testing shown below. Steps 5, 6, and 7 are explained as comments along with the program code.

1. Define the problem.
   We have data regarding penguins. There are three Penguin Species. We would like to know whether the body mass differs across penguin species.
2. Formulate the null (H_0) and the alternative (H_a) hypotheses.
   H_{01}: $\mu_1 = \mu_2$; H_{a1}: $\mu_1 \neq \mu_2$ (Comparing Adelie, Chinstrap).
   H_{02}: $\mu_1 = \mu_3$; H_{a2}: $\mu_1 \neq \mu_3$ (Comparing Adelie, Gentoo).

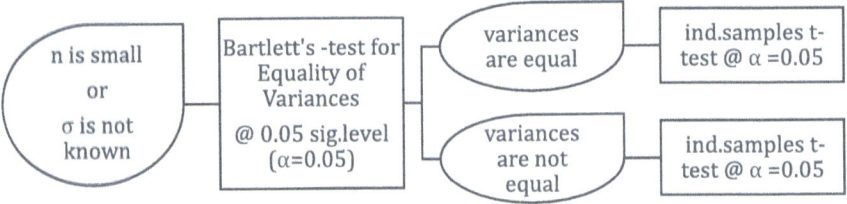

Fig. 4.3 Schematic diagram for two independent samples t-test

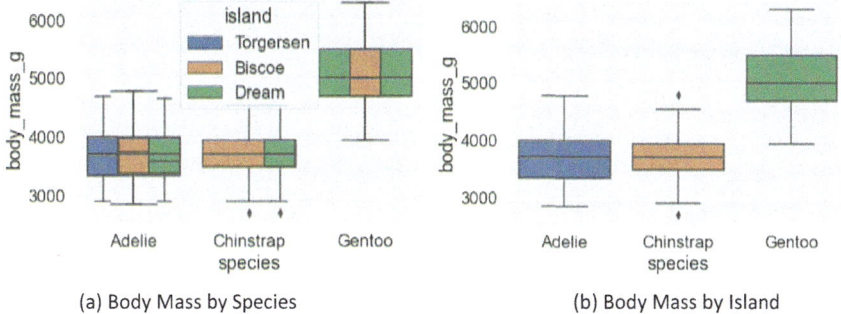

(a) Body Mass by Species   (b) Body Mass by Island

Fig. 4.4 Multiple box plots—body mass by species by Island

$H_{03}: \mu_2 = \mu_3 \, ; H_{a3}: \mu_2 \neq \mu_3$(Comparing Chinstrap, and Gentoo).

3. Select the appropriate test.

    We need to compare the means of body mass of penguin populations of three species (three independent samples).

    So, we select **independent samples t-test.**

4. Check for homogeneity of variances.

    The computational formulae for the t-test depend on whether the variance of the populations is the same or different. There are many tests for comparing variances. **Bartlett's test** is used when we are certain the underlying population has a normal distribution. **Levene's test** is used for non-normal distributions.

    The null hypothesis for the test is that the variances are equal for all samples. $H_0: \sigma_1^2 = \sigma_2^2 = \ldots = \sigma_k^2$.

    The alternate hypothesis is that the variances are not equal for one or more pairs.

    We will undertake **Bartlett's test** at a significance level (α) of 0.05.

    (If the p-value$< = \alpha$, reject H_0. Otherwise, do not reject H_0).

5. Select the significance level (α) for the t-test.

    $\alpha = 0.05$.

6. Do the statistical test, setting variances are equal$=$True or False.

7. Analyze the test results.

    If the p-value $< = \alpha$, reject H_0. Otherwise, do not reject H_0.

8. Inference: Interpret the results and state our conclusion.

Tutorial 4.3   Two Independence Samples t-test

```
import pandas as pd
import seaborn as sb
from scipy import stats
```

Tutorial 4.3.1 Data Setup; Descriptive statistics

```
d=sb.load_dataset('penguins')
d.info()
d.columns
['species', 'island',
   'bill_length_mm', 'bill_depth_mm', 'flipper_length_mm',
   'body_mass_g', 'sex']
    species = ['Adelie', 'Chinstrap', 'Gentoo']
    island  = ['Biscoe','Dream','Torgersen']
d.dropna(inplace=True)  # dropping rows having any cells with null value
d.shape
sb.set(font_scale=1.5)
sb.boxplot(x='species',y='body_mass_g',data=d)
sb.boxplot(x='species',y='body_mass_g',hue='island',data=d)
```
See Figure 4-4: Multiple Box Plots

Tutorial 4.3.2 Bartlet's test - checking the equality of variances

```
s_Ad=d[d['species']=='Adelie']      # species_Adelie
s_Ch=d[d['species']=='Chinstrap']   # species_Chinstrap
s_Ge=d[d['species']=='Gentoo']      # species_Gentoo
```

Bartlett-test: to check the equality of variances, when we are not sure whether the distributions are normal
H0: variances of the body_mass of samples are equal
Ha: variances of the body_mass of samples are NOT equal

```
bts, p = stats.bartlett(s_Ad.body_mass_g,s_Ch.body_mass_g)
p
```
$p = 0.097$; $p > 0.05$; failed to reject H0
variances of the body_mass of samples _Ad, _Ch are equal

```
bts, p = stats.bartlett(s_Ad.body_mass_g,s_Ge.body_mass_g)
p
```
$p = 0.307$; $p > 0.05$; failed to reject H0
variances of the body_mass of samples _Ad, _Ge are equal

```
bts, p = stats.bartlett(s_Ch.body_mass_g,s_Ge.body_mass_g)
p
```
$p = 0.017$; $p < 0.05$; H0 is rejected
variances of the body_mass of samples _Ch, _Ge are NOT equal

Tutorial 4.3.3 Independent Samples t-test -2 tailed - variance equal

Null hypothesis H0: Mean of the body_mass of samples s1, s2 are equal
Alternate hyp. Ha: Mean of the body_mass of samples s1, s2 are NOT equal

```
ti_Ad_Ch, p = stats.ttest_ind(s_Ad.body_mass_g,s_Ch.body_mass_g,
equal_var = True)
p
```

p = 0.674; p > 0.05; Failed to reject the null hypothesis
Mean of the body_mass of samples _Ad, and _Ch are equal
The body mass does not differ across the species Adelie and Chinstrap

```
ti_Ad_Ge, p = stats.ttest_ind(s_Ad.body_mass_g,s_Ge.body_mass_g,
equal_var = True)
p
```
p = 0.000; p < 0.05; The null hypothesis is rejected
Mean of the body_mass of samples _Ad, and _Ch are equal
The body mass of the species Adelie and Gentoo differ

Tutorial 4.3.4 Independent samples t-test - 2 tailed - variance not equal

Null hypothesis H0: Mean of the body_mass of samples s1, s2 are equal
Alternate hyp. Ha: Mean of the body_mass of samples s1, s2 are NOT equal

```
ti_Ch_Ge, p = stats.ttest_ind(s_Ch.body_mass_g,s_Ge.body_mass_g,
equal_var = False)
round(p,3)
p
```
p ~ 0.000. That is p < 0.05; The null hypothesis is rejected.
Mean of the body_mass of samples s_Ch, s_Ge are NOT equal.
Therefore, the body mass of the species Gentoo and Chinstrap differ

Inference: -
1) The body mass of the species Adelie and Chinstrap do NOT differ
2) The body mass of the species Adelie and Gentoo differ
3) The body mass of the species Gentoo and Chinstrap differ

4.4 One Sample T-Test

The one-sample t-test examines whether the mean of a population is statistically different from a hypothesized value. The test variable (e.g., body mass of penguins) is expected to be a numeric variable (integer or float). Homogeneity of variances is assumed (i.e., variances of the sample data and population are the same). Outliers affect the estimates.

The test statistic for a one-sample t-test is computed as follows:

$$t = \frac{\overline{x} - \mu_0}{S_{\overline{x}}}$$

where
$\overline{x}$ is the sample mean,
μ_0 is the test value.
$s_{\overline{x}}$ is the estimated standard error of the mean, computed as

$$s_{\overline{x}} = \frac{s}{\sqrt{n}}$$

s = sample standard deviation,
n is the sample size.

One-sample test for the population mean (μ) takes one of the three forms listed below.

(1) Hypothesis: Population Mean is Equal to the Test Value ($\mu = \mu_0$)

H_{01}: $\mu = \mu_0$; (H_{a1}: the population mean is equal to the given number μ_0).
H_{a1}: $\mu \neq \mu_0$(H_{a1}: the population mean is not equal to the given number μ_0).
The null hypothesis expects the population mean to fall between $\mu - \alpha/2$ and $\mu + \alpha/2$. If the computed statistic falls within t-statistic (t) for the given degree of freedom t—$\alpha/2$ and t+$\alpha/2$, the null hypothesis cannot be rejected.

This is diagrammatically shown in Fig. 4.5. Here α is the significance level, usually set to 0.05; μ is the population mean; μ_0 is the test value. The area bounded by the red lines indicates the non-rejection region. The area to the left of the red line at ($\mu - \alpha/2$) and the area to the right of the red line at ($\mu + \alpha/2$) are the rejection regions. The null hypothesis is rejected if the value of the t-statistic computed from the dataset is beyond the critical value for the given degree of freedom. Note that, in the case of t-tests, the degree of freedom is sample size—1, (i.e., n − 1).

(2) Hypothesis: Population Mean is Greater Than the Test Value ($\mu > \mu_0$)

H_{02}: $\mu_= \mu_0$; (H_{02}: the population mean is equal to the given number μ_0).
H_{a2}: $\mu_{\ >}\mu_0$(H_{a2}: the population mean is greater than the given number μ_0).
This is diagrammatically shown in Fig. 4.6a. Here α is the significance level, usually set to 0.05; μ is the population mean; μ_0 is some number. The null hypothesis is rejected if the t-statistic computed from the dataset is less than the critical value for the given degree of freedom.

Fig. 4.5 Hypothesis testing: two-sided t-test

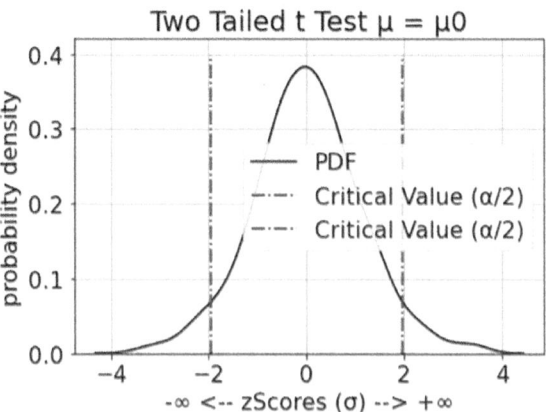

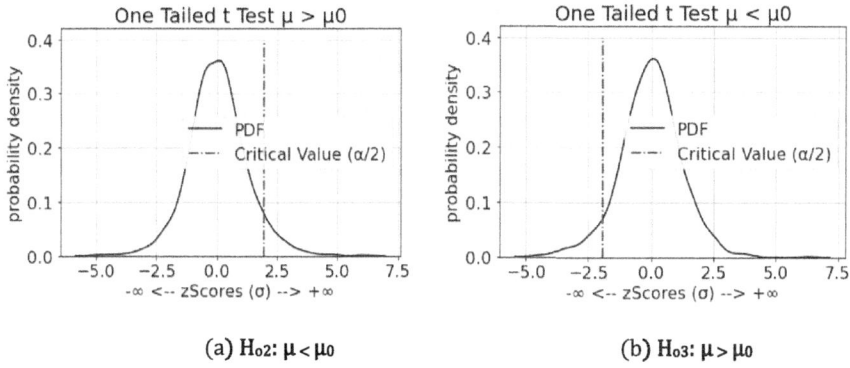

Fig. 4.6   Hypothesis testing: one-sided t-test

(3) Hypothesis: Population Mean is Greater Than the Test Value ($\mu < \mu_0$)

H_{03}: $\mu = \mu_0$; (H_{02}: the population mean is equal to the given number μ_0).
H_{a3}: $\mu > \mu_0$ (H_{a2}: the population mean is less than the given number μ_0).
This is diagrammatically shown in Fig. 4.6. Here α is the significance level, usually set to 0.05; μ is the population mean; μ_0 is some number. If the value of the t-statistic computed from the dataset is greater than the critical value for the given degree of freedom, the null hypothesis is rejected.

One-Sample t-test Example

Refer to the penguin dataset. Do the following tests:

- Explore whether the body mass of the Adelie species of penguins is greater than 3.6 kg.
- Explore whether the body mass of the Adelie species of penguins is less than 3.6 kg.
- Give an estimate of the body mass of the Adelie species of penguins.

Answer:

We will follow the method illustrated in 'Sect. 4.3' to compare means. Look at the schematic diagram for the t-test in Figs. 4.3. Since the population standard deviation (σ) is unknown, we prefer the t-test, for a significance level of $\alpha = 0.05$.

Tutorial 4.4   One-Sample t-test

Tutorial 4.4.1 Data Setup

```
import seaborn as sb                    # for loading the iris dataset
import matplotlib.pyplot as plt         # for graphics
from scipy import stats
import numpy as np

d  = sb.load_dataset('penguins')
d.dropna(inplace=True)
s_Ad=d[d['species']=='Adelie']      # species_Adelie
s_Ch=d[d['species']=='Chinstrap']   # species_Chinstrap
s_Ge=d[d['species']=='Gentoo']      # species_Gentoo
```
See Figure 4-4, for the box plots on body mass by species

Tutorial 4.4.2 One Sample t-test: Is bodymass mean > 3600

Null hypothesis Ho: mean of penguin bodymass = 3600
Alt. hypothesis Ha: mean of penguin bodymass > 3600
Note the parameter alternative (Ha) ='greater'
```
   tscore, pvalue = stats.ttest_1samp(s_Ad.body_mass_g,
          popmean = 3600, alternative='greater')
```
pvalue = 0029; p < 0.05; The null hypothesis is rejected
Therefore, Mean is > 3600

Tutorial 4.4.3 One Sample t-test: Is bodymass mean < 3800

Null hypothesis Ho: mean of penguin bodymass = 3800
Alt. hypothesis Ha: mean of penguin bodymass < 3800
Note the parameter alternative (Ha) ='less'
```
   tscore, pvalue = stats.ttest_1samp(s_Ad.body_mass_g,
          popmean = 3800, alternative='less')
```
pvalue = 0.007; p < 0.05; The null hypothesis is rejected
Therefore, Mean is < 3800

Tutorial 4.4.4 Cross Verification of the one-tailed t-tests

According to central limit theorem, the population mean is expected to lie
between sample mean +/- std. deviation / sqrt(n), where n is the sample size

```
   import math
   mu = np.round(s_Ad.body_mass_g.mean(),2)  # 3706.16
   s = s_Ad.body_mass_g.std()    # 458.62
   n = len(s_Ad.body_mass_g)     # 146
   sigma = s/math.sqrt(n)
   sigma = np.round(sigma,2)
   print('estimate of population mean: ', mu,  '+/-',sigma)
```

```
estimate of population mean: 3706.16 +/- 37.96

Inferences: -
1. Mean is > 3600
2. Mean is < 3800
3. The Interval Estimate of the Mean is 3706.16 +/- 37.96
```

4.5 Comparison of Means—Paired Samples

This test compares the means of two measurements taken from the same subject. Some applications of paired measurements are:

- The awareness of a product and purchase intention.
- Loyalty to a brand and customer satisfaction.
- The yield before and after treatment (at two different points in time).
- The health parameters before and after medical treatment (at two different points in time).

The paired samples t-test aims to ascertain if statistical evidence indicates that the mean value of the differences between paired observations is significantly distinct from zero. The data consists of two continuous numeric variables (integer or float). We may use the Wilcoxon Signed-Ranks Test to compare paired means of non-normal distributions. Like other t-tests, outliers affect the estimates of paired t-test as well.

The test statistic for a one-sample t-test is computed as follows:

$$t = \frac{x_{diff}}{S_{\bar{x}}}$$

where
x_{diff} is the mean of the difference between the paired observations,

$t = \frac{x_{diff}}{s_{\bar{x}}}$

$s_{\bar{x}}$ is the estimated standard error of the mean given by

$$s_{\bar{x}} = \frac{s_{diff}}{\sqrt{n}}$$

S_{diff} is the standard deviation of the difference between the paired observations,
n is the sample size.

Hypothesis Formulation

The hypothesis can be formulated in two ways. The first formulation is easy to understand. However, it is the second formulation that is actually tested. This can be visualized as in Fig. 4.6.

(1) H0: $\mu1 = \mu2$ (the two population means are equal).
    Ha: $\mu1 \neq \mu2$ (the two population means are not equal).
(2) H_0: $\mu_1 - \mu_2 = 0$ (the difference between the two population means $= 0$).
    H_a: $\mu_1 - \mu_2 \neq 0$ (the difference between the two population means $\neq 0$).

Paired Samples t-test Example

A retail store manager surveyed 30 customers in preparation for upcoming seasonal sales. Four characteristics—purchase intention, product awareness, brand
awareness, and brand loyalty—were measured on an interval scale of $\{0...7\}$.
Assume a significance level of 0.05. Explore the characteristics that influence a
purchase decision.

Tutorial 4.5 Paired Samples t-test

```
from scipy import stats
import numpy  as np

purchase_intent=[5,6,5,5,2,4,4,4,5,6,5,4,2,5,6,3,6,3,5,5,2,5,5,5,5,5,5,5,5,5]
product_aware=[5,6,5,5,5,3,5,5,5,2,6,4,4,5,6,3,6,6,5,5,5,4,5,5,5,5,4,5,5,6]
brand_aware=[5,6,5,5,5,4,5,5,5,6,6,6,5,6,6,6,6,5,5,5,4,5,5,4,5,5,5,5,3,6]
brand_loyal=[5,6,5,5,5,4,6,5,6,5,6,6,5,5,6,6,6,5,5,5,5,5,5,4,5,5,5,6,5,5]
```

H$_0$: means of brand loyalty and purchase intention are equal
H$_a$: means of brand loyalty and purchase intention are NOT equal
```
  tps, p = stats.ttest_rel(brand_loyal,purchase_intent)
  p # 0.004
```
p-value is less than 0.05; H$_0$ is rejected
Therefore, brand loyalty and purchase intention are NOT equal

H$_0$: means of brand awareness and purchase intention are equal
H$_a$: means of brand awareness and purchase intention are NOT equal
```
  tps, p = stats.ttest_rel(brand_aware, purchase_intent)
  p # 0.012
```
p-value is less than 0.05; H$_0$ is rejected
Therefore, brand awareness and purchase intention are NOT equal

H0: means of product awareness and purchase intention are equal
Ha: means of product awareness and purchase intention are NOT equal
```
  tps, p = stats.ttest_rel(product_aware, purchase_intent)
  p # 0.284
```
p-value is greater than 0.05; We fail to reject H$_0$
product awareness and purchase intention are equal

```
Inferences: -
1. Brand loyalty does not influence purchase intention.
2. Brand awareness does not influence purchase intention.
3. Product awareness influences purchase intention
Recommendation: -
Undertake promotional activity to convert brand loyalty and / or brand aware-
ness to purchase intention.
```

4.6   One-Way Anova

The objective of the one-way analysis of variance is to compare the means of two or more independent groups to determine whether the associated population means are significantly different. The dependent variable to be analyzed is expected to be numeric (e.g., body mass of penguins). The independent variable is expected to be categorical (e.g., penguin species). The independent variable divides the sample into two or more mutually exclusive factor levels or groups. The variances exhibited by the different groups are assumed to be approximately equal. Outliers affect the estimate.

We may use the Kruskal–Wallis test if the normality, homogeneity of variances, or the assumption regarding outliers are not met. One-way analysis of variance involves only one categorical variable or a single factor. If two or more factors are involved, the analysis is termed 'n-way analysis of variance'. If the set of independent variables consists of both categorical and metric variables, the technique is called the analysis of covariance (ANCOVA). In this case, the categorical independent variables are still considered factors, whereas the metric-independent variables are referred to as covariates.

The hypotheses concerning one-way ANOVA can be expressed as follows:

$H_0: \mu_1 = \mu_2 = \mu_3 = ... = \mu_k$ ('all the population means are equal').

H_a: at least one of the k population means (μ_i) is not equal to the others.

The one-way ANOVA uses an F-statistic that evaluates whether the group means differ significantly. F-statistic can be expressed as (Malhotra 2020):

$$F = \frac{SSR/(k-1)}{SSE/(n-k-1)}$$

where

SSR = the regression sum of squares, with k-1 degrees of freedom,
SE = the error sum of squares, with n-k-1 degrees of freedom,
k is the number of groups (factor levels of the categorical variable),
n = sample size.

Fig. 4.7 Chicken weight
by diet

One Way Anova Example

The 'ChickWeight' dataset contains data on the body mass of chicken from day
1 to day 21. They were divided into four groups and given four different 'diets'.
Explore the influence of 'diet' on 'chicken weight' See Fig. 4.7.

Tutorial 4.6   Anova

Tutorial 4.6.1 Data Setup

```
import pandas as pd
import seaborn as sb
import numpy as np
import scipy.stats as ss
import matplotlib.pyplot as plt

import statsmodels.api as sm
d =sm.datasets.get_rdataset("ChickWeight").data
d.columns
d.info()
```

```
RangeIndex: 578 entries, 0 to 577
Data columns (total 4 columns):
    #   Column  Non-Null Count  Dtype
---  ------  --------------  -----
 0   weight  578 non-null    int64
 1   Time    578 non-null    int64
 2   Chick   578 non-null    int64
 3   Diet    578 non-null    int64
```

```
sb.boxplot(y='weight', x='Diet', data=d)
plt.xlabel('Diet Category', fontsize=16)
plt.ylabel('Weight', fontsize=16)
plt.title('Chicken Weight on Four Different Diet', fontsize=16)
```

See Figure 4-7: Chicken Weight by Diet

Tutorial 4.6.2 One Way Anova / F test

```
c1=d[d.Diet== 1].weight
c2=d[d.Diet== 2].weight
c3=d[d.Diet== 3].weight
c4=d[d.Diet== 4].weight
```

f-test to check whether the mean of all the distributions are equal
Ho: c1mean = c2mean = c3mean = c4 mean
Ha: the mean of at least one of the groups is different from others
```
    f, p = ss.stats.f_oneway(c1, c2, c3, c4)
    round(p,3) #0.000
```
p value < 0.05. null hypothesis is rejected
The mean of at least one of the group is different from the others

Tutorial 4.6.3 Bartlett's test

bartlett's test: to check the equality of variances, when no assumptions are
made about the normality of the distributions
```
    bts, p = ss.bartlett(c1, c2, c3, c4)
    round(p,3) #p = 0.000; p < 0.05
```
The variance of at least one of the group is different from the others

Tutorial 4.6.4 Checking mean pairwise, assuming normal distribution

```
    ss.stats.f_oneway(c1,c2)    #p <=0.05;  means are equal
    ss.stats.f_oneway(c1,c3)    #p <=0.05;  means are equal
    ss.stats.f_oneway(c1,c4)    #p <=0.05;  means are equal
    ss.stats.f_oneway(c2,c3)    #p <=0.05;  means are equal
    ss.stats.f_oneway(c2,c4)    #p = 0.166; means are NOT equal
    ss.stats.f_oneway(c3,c4)    #p = 0.449; means are NOT equal
```

Tutorial 4.6.5 Pairwise t-test

Based bartlett's test done above, equal variance is set to True or False
Ho: mi = mj
Ha: mi != mj

```
    ss.stats.ttest_ind(c1,c2,equal_var = True)  #(1) #p < 0.05
    ss.stats.ttest_ind(c1,c3,equal_var = True)  #(2) #p < 0.05
    ss.stats.ttest_ind(c1,c4,equal_var = True)  #(3) #p < 0.05
    ss.stats.ttest_ind(c2,c3,equal_var = True)  #(4) #p < 0.05
    ss.stats.ttest_ind(c2,c4,equal_var = False) #(5) #p = 0.166
    ss.stats.ttest_ind(c3,c4,equal_var = False) #(6) #p = 0.448
```

```
Interpretation: -
```
The difference of means was statistically significant in four paired comparisons, as the p-value < 0.05
However, there is insufficient evidence to conclude a significant difference
between the means of the following two groups, as the p-value > 0.05.
```
C2-weight not = C4-weight
C3-weight not = C4-weight
```

For Sanity Check, let's print the mean weights of each group.
```
   print(c1.mean(),c2.mean(),c3.mean(),c4.mean())
102.64    122.61    142.95    135.26
```

From statistical tests, and the mean value of weights, we observe that Diet
4 is better than Diet 1. However, Diet 3 is better than Diet 1, Diet 2. So
Diet 3 is recommended.

4.7 Chi-Square Test of Independence

The chi-square test of independence, or the chi-square test of association, is a
nonparametric test. The objective of the test is to determine whether there is an
association between categorical variables. This test uses a contingency table (a
cross-tabulation or two-way table). The rows represent the categories of one variable, while the columns represent the categories of the other variable. Each variable
must have two or more categories. Each cell in the table indicates the total count
of cases for a particular combination of categories.

The hypothesis of the chi-square test can be expressed in two ways. They are
both useful in different circumstances.

(1) H_0: 'Variable 1 is independent of Variable 2'.
    H_a: 'Variable 1 is dependent on Variable 2'.
(2) H_0: 'Variable 1 is not associated with Variable 2'.
    H_a: 'Variable 1 is associated with Variable 2'.

The test statistic for the chi-square test of independence is denoted χ^2 and is
computed as follows (Malhotra 2020). Please look at Table 4.2 and Tutorial 4.5 for
a better understanding.

$$\chi^2 = \Sigma \frac{(f_o - f_e)^2}{f_e}$$

where

f_o is the observed value,
fe is the expected value $= n_r n_c/n$,
n_r is the sum of all the cells in the row,
n_c is the sum of all the cells in the column,
n is the sample size.

Table 4.2 Cross Tab—restaurant footfall by day-wise/gender-wise

f_o (observed value) cross tab: day/gender				f_e (expected value) $= n_r n_c / n$				$\frac{(f_o - f_e)^2}{f_e}$		
	Male	Female	Total		Male	Female	Total		Male	Female	Total
Thu	30	32	62	Thu	40	22	62	Thu	2.50	4.55	
Fri	10	9	19	Fri	12	7	19	Fri	0.33	0.57	
Sat	59	28	87	Sat	56	31	87	Sat	0.16	0.29	
Sun	58	18	76	Sun	49	27	76	Sun	1.65	3.00	
	157	87	244		157	87	244	$\chi^2 = \sum \frac{(f_o - f_e)^2}{f_e} =$			13.05

Fig. 4.8 Critical value for a given df in chi-square test

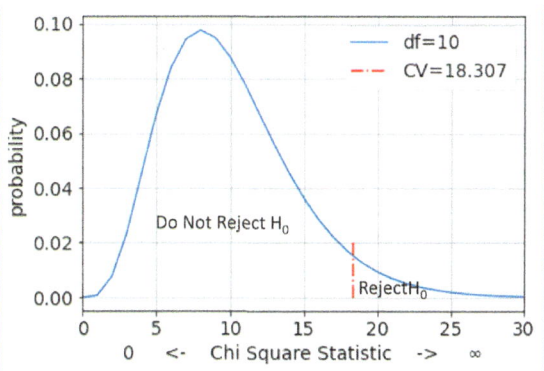

The computed χ^2 value is compared to the critical value from the χ^2 distribution table with degrees of freedom df $= (R - 1)(C - 1)$ and the confidence level chosen. We reject the null hypothesis if the computed χ^2 value > the critical χ^2 value (available from χ^2 tables). The χ^2 probability distribution of degree of freedom 10 (18.307) is shown in Fig. 4.8. If the χ^2 computed from the experimental data is greater than 18.307, we reject the null hypothesis with 95% confidence level.

Alternatively, we reject the null hypothesis if the p-value associated with the computed χ^2 is < 0.05. The rejection implies a significant association, or dependence, between the two categorical variables.

Example Refer to Table 4.2. A restaurant manager wants to know whether the footfall across the week differs by gender. Table 4.2 shows a cross-tabulation of restaurant footfall day-wise and gender-wise. For example, the first data cell in the first table has a value of 30, which shows that 30 males were present in the restaurant on Thursday. Let us apply the critical value method of hypothesis testing.

Interpretation and inference:

- df $= (r-1) * (c-1) = 3 * 1 = 3$
- Chi-Square Critical Value for (df $= 3$) is 7.8.
- Chi-Square Statistic Computed $= 13$.

- Since the computed Chi-Square Statistic (13)>Critical Value (7.81), the null hypothesis of independence is rejected.
- The distribution of Male footfall and Female footfall across the days is different.

Chi-Square Test Example

The 'tips' dataset contains data collected from a restaurant, over two and a half months. The variables include the day of service—Thursday to Sunday, the footfalls by gender, bill amount, tips, gender of the person paying the bill, etc. Investigate whether the footfall across the week differs by gender. Apply two methods for hypothesis testing (a) the p-value method and (b) the critical value method.

Tutorial 4.7   Chi-square test - restaurant tips - footfall

Tutorial 4.7.1 Data Setup

```
from scipy import stats
import matplotlib.pyplot as plt
import seaborn as sb
import pandas as pd

d=sb.load_dataset('tips')
d.info( )
```

```
RangeIndex: 244 entries, 0 to 243
Data columns (total 7 columns):
 #    Column       Non-Null Count  Dtype
---   ------       --------------  -----
 0    total_bill   244 non-null    float64
 1    tip          244 non-null    float64
 2    sex          244 non-null    category
 3    smoker       244 non-null    category
 4    day          244 non-null    category
 5    time         244 non-null    category
 6    size         244 non-null    int64
dtypes: category(4), float64(2), int64(1)
```

```
d.columns #'total_bill','tip','sex','smoker','day','time','size']
d.day.unique()   # ['Sun', 'Sat', 'Thur', 'Fri']
d.time.unique()  # ['Dinner', 'Lunch']
d.sex.unique()   # ['Female', 'Male']
```

Tutorial 4.7.2 Cross Tab of plot footfalls vs gender

```
ct = pd.crosstab(d.day, d.sex)
print(ct)
```

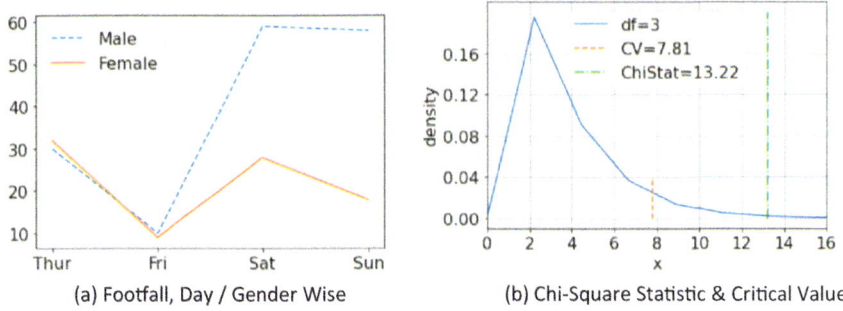

Fig. 4.9 Footfalls—associated chi-square statistic and critical value

```
sex    Male   Female
day
Thur    30      32
Fri     10       9
Sat     59      28
Sun     58      18
The table has 4 rows and 2 columns.
The degree of freedom (df) = (r-1)x(c-1) = 3x1 = 3

   plt.rcParams.update({'font.size': 16})
   plt.plot(ct.Male,linestyle='--')
   plt.plot(ct.Female,linestyle='-')
   plt.legend(['Male','Female'])
See Figure 4-9 (a): Footfalls - Associated Chi-Square Statistic
```

Tutorial 4.7.3 Chi-square test / p-Value method

```
Ho: footfall is not associated with gender
   chi2, p, df, exp =  stats.chi2_contingency(ct)
   print (chi2, df)
chi square statistic = 13.22 at degree of freedom = 3
   print (p)          # 0.004

Inference based on p-value:
Since p < 0.05, the null hypothesis of independence is rejected.
The footfall is associated with gender.
The footfalls of males and females differ across the days.
```

Tutorial 4.7.4 Chi-Square test / Critical Value Method

```
alpha = 0.05-> Confidence_Interval_Coefficient = 0.95
The degree of freedom df is computed above (df=3)
Compute Critical Value
   CV = stats.chi2.ppf(Confidence_Interval_Coefficient, df) # 7.814
```

```
Inference based on Critical Value: -
See Figure 4-9 (b): Footfalls - Associated Chi-Square Statistic
df = (r-1)*(c-1) = 3*1 =3
Chi-Square Critical Value for (df=3) is 7.81
Chi-Square Statistic Computed = 13.22
```
Since the computed Chi-Square Statistic (13.22) > Critical Value (7.81), the null hypothesis of independence is rejected. Therefore, we infer that the distribution of Male and Female footfall across the days differs.

The footfalls of males and females differ across the days. Male footfalls are higher than female footprints – Crosstab is shown below: -

```
    Male   Female
Thur    30       32
Fri     10        9
Sat     59       28
Sun     58       18
```

Data Analytics in Action

Statistical Studies in Brewery and Agricultural Fields

Karl Pearson, William Sealy Gosset, and Ronald A Fisher are considered pioneers of modern statistics. Karl Pearson (1857–1936) founded the world's first statistics department at University College, London in 1911. His contributions include the **correlation coefficient**, method of moments, **chi-square** distribution (in the year 1900), and **principal component analysis**. He contributed significantly to the field of biometrics and meteorology.

William Sealy Gosset (1876–1937) was a brewer for Guinness Brewery in Dublin. He observed that the existing statistical techniques using large sample sizes were not useful for the small sample sizes he had to work with. This led to the development of **t-test** and t-distribution (in 1908). From 1906 to 1907, Gosset was employed at Pearson's laboratory in London, where he focused on topics such as the Poisson limit to the binomial distribution, the sampling distribution of the mean, standard deviation, and correlation coefficient.

Ronald A Fisher (1890–1962) developed the 'analysis of variance (**ANOVA**) method in 1918, which generalized the t-test for comparing more than two means. He developed the randomized experimental design for his work in the agricultural fields at Roth Amsted Research—the UK, from 1919 to 1933. He introduced the well-known Iris flower dataset to exemplify discriminant analysis in 1936. Other major contributions from Fisher include works on the method of maximum likelihood, fiducial inference, the derivation of various sampling distributions, etc. Together with J. B. S. Haldane and Sewall Wright, Fisher is one of the three principal founders of population genetics.

Summary

Jerzy Neyman and Egon Pearson formulated the theory of hypothesis testing. They hypothesized that regardless of the results of an experiment on data samples, a researcher could never be certain whether it holds good for the population. However, based on two parameters, significance level and the power of the test, a researcher could make specific conclusions on the probability of the experiment's success. The probability value, commonly referred to as the p-value, is often associated with Ronald Fisher, acknowledged as the pioneer of modern statistics and experimental design.

In hypothesis testing, we compare the statistics from a sample data distribution with the parameters of a model (t-distribution, chi-square distribution, etc.). If the sample is consistent with the model, the null hypothesis is not rejected; otherwise, the null hypothesis is rejected in favor of the alternative hypothesis. The criterion for rejecting the null hypothesis involves a test statistic. A test statistic is a number calculated from the data sample.

There are two major categories of tests. Parametric tests are used for numeric variables (ratio scaled or interval scaled). Nonparametric tests are used when the variables are categorical (nominal or ordinal). T-test is a parametric test, chi-square test is a nonparametric test.

Two types of errors can occur in hypothesis testing. A type I error occurs when the null hypothesis is rejected when it is true. Type II error occurs by rejecting our null hypothesis when we should not. Type I error is considered more severe than type II. The probability of type I error is a. It is also called the significance level. The probability of type II error is ß. The power of the test is indicated by $(1 - ß)$.

In comparing means, the t-test is preferred over the z-test when dealing with small sample sizes or unknown population standard deviations. The independent samples t-test assesses if the means of two independent groups in a sample are significantly different, assuming equal variances. When variances are not assumed to be equal, the Welch t-test is an alternative. Outliers can impact estimates. The one-sample t-test examines if the mean of a population differs from a hypothesized value. The paired samples t-test compares the means of two measurements from the same subject, such as product awareness and purchase intention.

The one-way analysis of variance aims to compare the means of two or more independent groups to determine if associated population means differ significantly. If assumptions like normality or homogeneity of variances are not met, the Kruskal–Wallis test may be used. One-way analysis of variance involves a single categorical variable, while multiple factors are considered in 'n-way analysis of variance'. Analysis of covariance (ANCOVA) incorporates categorical and metric variables in the independent variables.

The chi-square test of independence, a nonparametric test, determines if there is an association between categorical variables using a contingency table. Rows represent categories for one variable, and columns represent categories for the other variable.

Questions

Comprehension:

1. Compare and contrast test statistics to population parameters.
2. Write a short note on the need for hypothesis testing.
3. Describe the key steps involved in the hypothesis testing procedure. How does the significance level (α) relate to hypothesis testing?
4. In the context of hypothesis formulation, what are the null and alternative hypotheses? Provide examples of null and alternative hypotheses for a one-sample t-test and a two-sample t-test.
5. What is the critical value, and how does it impact the outcome of a hypothesis test? How does it relate to the level of significance (α)?
6. When conducting a two-sample t-test, what are the assumptions about variances? How can you determine whether you should assume equal variances or not?
7. In what context is the t-test preferred over z-test?
8. Compare and contrast independent samples t-test with paired samples t-test.
9. In which context does the assumption of homogeneity of variance assume importance in the t-test?
10. State the hypothesis and explain the assumptions in the ANOVA test.
11. State the hypothesis and explain the assumptions in the chi-square test.
12. State the hypothesis and explain the assumptions in various categories of t-test.

Analysis:

13. What is the purpose of hypothesis testing, and how does it help researchers conclude population parameters based on sample data?
14. Explain the difference between parametric tests and nonparametric tests. When and why would you choose one over the other?
15. What are type I and type II errors in hypothesis testing? Why is it important to understand and control these errors in statistical analysis?
16. How does the significance level (α) relate to the power of a statistical test? Explain the trade-off between these two parameters.

Application:

17. Imagine you have collected data on the heights of two groups of people: Group A and Group B. Formulate null and alternative hypotheses for a two-sample t-test to determine if there is a significant difference in the heights of these two groups.
18. You are conducting a study to compare the test scores of students who received tutoring (Group 1) and those who did not (Group 2). What type of t-test would you use, and why? Formulate the null and alternative hypotheses for this scenario.

19. Given a dataset with body mass measurements of penguins from different species (Adelie, Chinstrap, Gentoo) explain how you would perform a two-sample t-test to determine if there are significant differences in body mass between these species.
20. Discuss a real-world scenario where controlling type I error is crucial. Provide an example of a study or experiment where such control is needed.
21. You are conducting a survey to test the effectiveness of a new advertising campaign on consumer purchase intentions. Which statistical test would you use, and how would you set the significance level? Formulate the null and alternative hypotheses for this test.
22. Consider a situation where you have data on the ages of customers who purchased online versus those who made an in-store purchase. What statistical test would you use to compare the means of these two groups, and why?
23. When conducting a two-sample t-test, you find that the p-value is less than the chosen significance level (α). What conclusion can you draw from this result, and what implications does it have for your study or experiment?
24. Explain the concept of power in hypothesis testing and provide a real-world example where high statistical power is essential for drawing meaningful conclusions.

Exercises

The questions in this section are based on the datasets listed below:

1. Penguins (seaborn dataset—https://seaborn.pydata.org/).
2. Tips (seaborn dataset—https://seaborn.pydata.org/).
3. ChickWeight (https://www.statsmodels.org/stable/index.html).
4. Retail Store—Brand Loyalty (Data provided in this chapter).

Exercise 4.1 Independent Samples t-Test

This has reference to the penguin's dataset. Is the body mass of penguins different across islands—find out using independent samples t-test.

Exercise 4.2 One-Sample t-test and Interval Estimate

This has reference to the ChickWeight dataset.

1. Using independent samples t-test, determine whether the body mass of chicken having Diet 4 is expected to be greater than 100gm. Use the step-by-step procedure for hypothesis testing.
2. Get an interval estimate of the body mass of chicken feeding Diet 4.

Exercise 4.3 Paired Samples t-Test
Consider the retail store case discussed in this section.

```
#
import pandas as pd
import seaborn as sb
from scipy import stats
#
# Tutorial 4.1.1 Descriptive statistics#
```

A retail store manager surveyed 30 customers in preparation for an upcoming seasonal sales. Four characteristics were measured on an interval scale of 0..7. Assume a significance level of 0.05. Identify the characteristics that influence brand loyalty using paired samples t-test.

Exercise 4.4 One-Way ANOVA

This has reference to the penguin's dataset. Investigate the following using one-way ANOVA.

1. Is the body mass of penguins different across species?
2. Is the body mass of penguins different across islands?
3. Is the body mass of penguins different across genders?

Exercise 4.5 Chi-Square Test—Footfalls by Time

This has reference to the tips dataset. The dataset contains data collected from a restaurant over two and a half months. Refer to Tutorial 4.5 for details. Using the chi-square test, identify whether a significant difference exists in the footfalls during lunch/dinner over the days in the week. Demonstrate both the p-value method and the critical value method.

Exercise 4.6 Chi-Square Test—Footfalls by Gender

This has reference to the tips dataset. The dataset contains data collected from a restaurant over two and a half months. Refer to Tutorial 4.5 for details. Using the chi-square test, identify whether a significant difference exists in the footfalls of male/female customers over the days in the week. Demonstrate both the p-value method and the critical value method.

References

Biau DJ, Jolles BM, Porcher R (2010) P value and the theory of hypothesis testing: an explanation for new researchers. Clin Orthop Relat Res 468(3). https://doi.org/10.1007/s11999-009-1164-4

Duchesnay ETLFY (2021) Statistics and machine learning in Python: Release 0.5

Levin RI (1984) Statistics for management. Prentice Hall

Malhotra NK (2020) Marketing research an applied orientation, 7th edn. Pearson Education

Navidi W (n.d.) Statistics for engineers and scientists

Rice JA, RJA (2007) Mathematical statistics and data analysis. Thomson Brooks Cole

Chapter 5
Regression Analysis

Learning Objectives

- Explain the concepts of correlation and regression with practical examples.
- Discuss the procedure for regression analysis.
- Understand the regression model.
- Illustrate data validation associated with regression analysis.
- Apply regression analysis to solve a business case.
- Demonstrate stepwise regression method.
- Demonstrate feature ranking and selection.
- Examine polynomial regression.

Overview

The chapter starts with a theoretical exposition of correlation and regression. A detailed discussion of the regression analysis procedure follows which involves 10 steps, including the examination of the regression model, data validation associated with regression analysis, visual inspections, stepwise regression, as well as feature ranking and selection.

The above steps are illustrated using a case study. Finally, we also discuss polynomial regression and demonstrate it through a Python application. It may be noted that regression is discussed further in Chap. 11—Machine Learning.

Supplementary Information The online version contains supplementary material available at https://doi.org/10.1007/978-981-99-0353-5_5.

S. Sundararajan, *Multivariate Analysis and Machine Learning Techniques*,
Transactions on Computer Systems and Networks,
https://doi.org/10.1007/978-981-99-0353-5_5

Definitions

Akaike information criterion (AIC): AIC indicates model fit. It is used for comparing models. The lower the AIC, the better the model.

Autocorrelation: Autocorrelation is the similarity over a sequence of observations or observations in a time series. This may result in underestimates of the standard error, leading to a misjudgment of the significance of the predictors.

Coefficient of determination: The coefficient of determination R^2 is the proportion of the explained variation to the total variation. R^2 must be adjusted for the number of independent variables (k) and the sample size (n).

Correlation: Correlation measures the nature and strength of association between two variables.

Durbin-Watson test statistic: The Durbin-Watson test statistic measures autocorrelation. A value around '2' is acceptable.

Feature Ranking: We may need to rank and select fewer variables based on criteria such as the partial regression coefficient or the p-value associated with the t statistic. The features can be handpicked if there are only a few of them. However, we may need to automate the process if there are numerous features.

Homoscedasticity: Homoscedasticity assumes that the variance of the error term is constant. Homoscedasticity is an important assumption in many multivariate techniques.

Multicollinearity: Multicollinearity arises when intercorrelations among the predictors are very high.

Ordinary Least Squares (OLS): Among linear unbiased estimators, the ordinary least squares (OLS) estimator demonstrates the least sampling variance when the errors in the linear regression model are uncorrelated, possess equal variances, and have an expectation value of zero.

Part correlation: Part correlation measures the strength of the association of a single independent variable with the dependent variable when the effect of the other independent variables in the regression model is removed.

Partial correlation coefficient: The partial correlation coefficient measures the strength of the association of a single independent variable with the dependent variable when the effects of the other independent variables in the model are held constant. It measures the incremental predictive effect of an independent variable.

Polynomial regression: If the relationship between the predictors and dependent variable is not linear, we may add additional 'interaction' terms to build an effective regression model. Though the underlying relationship is polynomial, it may be thus modeled as a linear model.

Regression analysis: Regression analysis examines associative relationships between a metric dependent variable (y) and one or more independent variables (X).

Stepwise regression: Stepwise regression is a method for discarding independent variables that are not significant. In the backward regression method, we start with all the independent variables in the model and remove the variables from the regression equation, one at a time. In the forward method, we add independent variables to the regression equation, one at a time. Multicollinearity is a major challenge in variable selection.

Variance Inflation Factors: Variance inflation factor (VIF) measures multicollinearity. VIF is directly related to the tolerance value (VIF = 1/Tolerance). A value around <5 is acceptable in social science research.

5.1 Correlation

Correlation measures the nature and strength of association between two variables. Pearson correlation coefficient (or product-moment correlation), r, indicates the linear association between two metrics (interval or ratio scaled) variables (Levin 2011; Zaki and Meira 2014).

Assume a sample of n observations of variables x and y. The correlation coefficient, r, is calculated as shown below.

$$r = \sigma(xy)/\sigma(x)\sigma(y),$$

where

$$\sigma(xy) = \sum_{i=1}^{n} (x_i - x_{mean})(y_i - y_{mean})/(n-1) \text{ is the covariance of x and y}$$

$$\sigma(x) = \sqrt{\sum_{i=1}^{n} (x_i - x_{mean})^2/(n-1)} \text{ is the standard deviation of x}$$

$$\sigma(y) = \sqrt{\sum_{i=1}^{n} (y_i - y_{mean})^2/(n-1)} \text{ is the standard deviation of y}$$

The value of 'r' varies from -1 to $+1$. In social sciences, 0.3 $|r| < 0.8$ is considered moderate, and $|r| > 0.8$ good (Kendall rank correlation and Spearman rank correlation are nonparametric tests, for estimating the correlation of two variables that are ordinal or metric). In Fig. 5.1 the blue line shows a positive correlation between two variables—planning versus outcome; the yellow line shows a negative correlation between two variables—experience versus monitoring.

Correlation between project planning and project outcome.

An example of the calculation of the Pearson correlation coefficient is shown in Table 5.1. The data consists of measurements of two attributes—project planning (P) and project outcome (O), from 15 software projects. A positive correlation of 0.72 is observed between them. This implies that project planning positively influences the project outcome.

For calculating sample statistics such as the mean, the observed measurements are summed up and the sum is divided by 'n', the sample size. We use the divider $n - 1$, rather than n to estimate the population mean. This is for adjusting

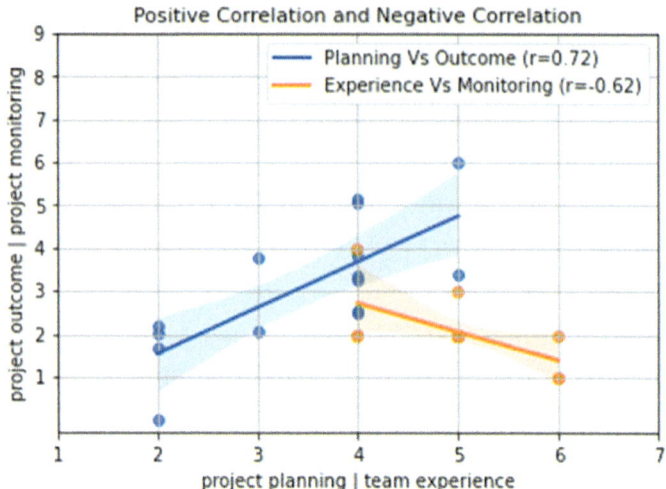

Fig. 5.1 Positive and negative correlations

Table 5.1 Correlation between project planning and project outcome

P	O	$r_{(P,O)} = r_{(O,P)} = \sigma_{(P,O)}/(\sigma_P * \sigma_O)$				0.72
project_ plan	project_ out- come	$P_i - P_{mean}$	$(P_i - P_{mean})^2$	$O_i - O_{mean}$	$(O_i - O_{mean})^2$	$(P_i - P_{mean})$ * $(O_i - O_{mean})$
5	3	1.53	2.35	−0.20	0.04	−0.31
2	2	−1.47	2.15	−1.20	1.44	1.76
4	5	0.53	0.28	1.80	3.24	0.96
4	3	0.53	0.28	−0.20	0.04	−0.11
3	2	−0.47	0.22	−1.20	1.44	0.56
5	6	1.53	2.35	2.80	7.84	4.29
2	1	−1.47	2.15	−2.20	4.84	3.23
2	2	−1.47	2.15	−1.20	1.44	1.76
4	5	0.53	0.28	1.80	3.24	0.96
2	2	−1.47	2.15	−1.20	1.44	1.76
4	3	0.53	0.28	−0.20	0.04	−0.11
4	4	0.53	0.28	0.80	0.64	0.43
4	3	0.53	0.28	−0.20	0.04	−0.11
4	3	0.53	0.28	−0.20	0.04	−0.11
3	4	−0.47	0.22	0.80	0.64	−0.37
$P_{mean} =$ 3.47	$O_{mean} =$ 3.20	$Var_P =$ $\sigma_P =$	1.12 1.06	$Var_O =$ $\sigma_O =$	1.89 1.37	$\sigma_{(P,O)} =$ 1.04

the degree of freedom, which can be explained like this. Assume that there are 15 observations, we are free to choose 14 out of 15 at random; but for selecting the 15th and last element, we would be left with no choice. In our example, note that the sample statistics—covariance, sigma, and r, are divided by $n - 1$ rather than n. The statistical significance of the correlation coefficient needs to be ascertained based on the associated p-value.

Let us run this using the Python code shown below. The numpy library is used in the first method, which provides the correlation coefficient. In the second method, we use scipy.stats library is used, which provides the p-value or level of significance, in addition to the correlation coefficient. The results show that $r = 0.716$ @ 0.003 level of significance. Since the p-value is ≤ 0.05, the correlation is significant.

Tutorial 5.1   Correlation

Tutorial 5.1.1 Correlation Using NumPy Function

```
import numpy as np

a = np.zeros([15,2])
a[:,0] = np.array([5,2,4,4,3,5,2,2,4,2,4,4,4,4,3])
a[:,1] = np.array([3,2,5,3,2,6,1,2,5,2,3,4,3,3,4])
r=np.corrcoef(a[:,0], a[:,1])
print( r) # 0.71637516 # positive correlation
See the blue line in Figure 5-1: Positive and Negative Correlations
```

Tutorial 5.1.2 Correlation Using SciPy Function

```
import scipy.stats
import numpy as np

a = np.zeros([15,2])
a[:,0] = np.array([5,2,4,4,3,5,2,2,4,2,4,4,4,4,3])
a[:,1] = np.array([3,2,5,3,2,6,1,2,5,2,3,4,3,3,4])
Refer:
docs.scipy.org/doc/scipy/reference/generated/scipy.stats.pearsonr.html
r, p = scipy.stats.pearsonr(a[:,0], a[:,1])
print('r=',round(r,3),'@',round(p,3),'level of significance')
r= 0.716 @ 0.003 level of significance -> positive correlation
```

Correlation between team experience with monitoring needs.

Another example of the Pearson correlation coefficient calculation is shown in Table 5.2. The data consists of measurements of two attributes—project monitoring (M) and experience (E), from 15 software projects. A negative correlation of -0.62 was observed between them. This suggests that monitoring becomes less necessary when the team is more experienced. The associated p-value was 0.014, which is less than 0.05. Therefore, the correlation is significant—a statistically

Table 5.2 Correlation between team experience and monitoring needs

M	E	$r_{(M,E)} = r_{(E,M)} = \sigma_{(M,E)}/(\sigma_M * \sigma_E)$				-0.62
project_ monitoring	team_ experience	$M_i - M_{mean}$	$(M_i - M_{mean})^2$	$E_i - E_{mean}$	$(E_i - E_{mean})^2$	$(M_i - M_{mean}) * (E_i - E_{mean})$
2	5	-0.07	0	0	0	0
2	4	-0.07	0	-1	1	0.07
2	5	-0.07	0	0	0	0
2	5	-0.07	0	0	0	0
1	6	-1.07	1.14	1	1	-1.07
3	5	0.93	0.87	0	0	0
2	5	-0.07	0	0	0	0
2	5	-0.07	0	0	0	0
2	5	-0.07	0	0	0	0
1	6	-1.07	1.14	1	1	-1.07
2	5	-0.07	0	0	0	0
2	4	-0.07	0	-1	1	0.07
2	5	-0.07	0	0	0	0
2	6	-0.07	0	1	1	-0.07
4	4	1.93	3.74	-1	1	-1.93
$M_{mean} =$ 2.07	$E_{mean} =$ 5	$\mathrm{Var}_M =$ $\sigma_M =$	0.49 0.7	$\mathrm{Var}_E =$ $\sigma_E =$	0.43 0.65	$\sigma_{(M,E)} =$ -0.29

significant negative correlation exists between project monitoring (*M*) and experience (*E*).

5.2   Regression

Regression analysis examines associative relationships between a metric dependent variable and one or more independent variables.

5.2.1   *Mathematical Insights on Linear Regression Model*

Those who are not particular about mathematical insights may skip the rest of this sub-section and proceed to the 'Linear Regression Model'. Consider the data points (O_i, P_i) shown in Table 5.3. The project outcome O_i can be predicted from project plan P_i, using the following set of equations, $O_i = P_i b_1 + b_0$.

We need two equations to solve an equation of two variables, b_1 and b_0. However, we have 15 different equations. There could be a unique solution, in the rare case when all the points are aligned!. However, in the general case, we should

Table 5.3 A set of 15 linear equations to solve for $O_i = P_i b_1 + b_0$	project_ outcome (O_i)	project_ plan (P_i)	$O_i = P_i b_1 + b_0$
	3	5	5 $b1 + b0$
	2	2	2 $b1 + b0$
	5	4	4 $b1 + b0$
	3	4	4 $b1 + b0$
	2	3	3 $b1 + b0$
	6	5	5 $b1 + b0$
	1	2	2 $b1 + b0$
	2	2	2 $b1 + b0$
	5	4	4 $b1 + b0$
	2	2	2 $b1 + b0$
	3	4	4 $b1 + b0$
	4	4	4 $b1 + b0$
	3	4	4 $b1 + b0$
	3	4	4 $b1 + b0$
	4	3	3 $b1 + b0$

presume that no single line, $P_i b_1 + b_0 = O_i$, can cut through the scatter plot of P, and O. We instead look for a solution that minimizes the error (residual). We will examine the solution further in subsequent sections.

5.2.2 Multiple Linear Regression Model

Multiple linear regression analysis examines associative relationships between a target variable (y) and feature variables $X \{X_1 \ldots X_k\}$, where k is the number of features. The following equation shows one instance of y, $y^{(i)}$

$$y^{(i)} = \beta_0 + \beta_1 X_1^{(i)} + \beta_2 X_2^{(i)} + \cdots + \beta_k X_k^{(i)} + \varepsilon^{(i)}$$

where
$y^{(i)}$ is the observed value of an instance of the target variable,
X is the feature vector, a set of 'k' independent variables,
β's are the '$k+1$' weights, or coefficients,
$\varepsilon^{(i)}$ is the error or the difference between the observed and predicted values of an instance of y, $y^{(i)}$,
$i = 1 \ldots n$, 'n' being the number of data points in the sample.
In practice, we use sample data to estimate y. The following equation estimates y, based on the best estimates of coefficients $b_0 \ldots b_k$ computed from the sample data.

$$y^{(i)} = b_0 + b_1 X_1^{(i)} + b_2 X_2^{(i)} + \cdots + b_k X_k^{(i)} + e^{(i)} \qquad (5.1)$$

$$\widehat{y}^{(i)} = b_0 + b_1 X_1^{(i)} + b_2 X_2^{(i)} + \cdots + b_k X_k^{(i)} \tag{5.2}$$

where $\widehat{y}^{(i)}$ is the estimated value of y, by the regression model and error in the estimate, $e^{(i)} = y^{(i)} - \widehat{y}^{(i)}$.

If only one independent variable exists, it is known as simple linear regression. If the number of independent variables is more than one, we call it multiple linear regression. To solve an equation of k variables, we need k equations. In practice, we have a large number of observations $n \gg k$ and the solution is usually not unique. There is no single hyperplane $y = f(X)$, that can cut through the hyperspace space of dimension $k + 1$ dimension formed by y and vector X. We instead look for a solution that minimizes the error (residual) between the observed value of y and the value estimated from the equation $\widehat{y} = f(X)$. Matrix method, ordinary least squares (OLS) method, and gradient descent are three common approaches used to solve linear regression.

The matrix method involves expressing the linear regression problem as a matrix equation and solving it using linear algebra. It provides an exact mathematical solution, which can be computationally efficient for small to moderate-sized datasets. It requires the inverse of the matrix $(X^T X)$, which can be computationally intensive. In the ordinary least squares (OLS) method, we minimize the sum of squared differences between the predicted and actual values. It is more straightforward to implement, as inverse matrix computation is not involved. However, it is unsuitable for situations with nonlinear relationships between variables. Moreover, OLS is sensitive to outliers. Machine learning usually employs another method—gradient descent (or its variations). Gradient descent is an iterative optimization technique to find the coefficients that minimize the cost function. A typical cost function considered in regression problems is a mean squared error. Gradient descent is suitable for large datasets and high dimensional feature spaces. However, it may require careful tuning of hyperparameters, such as the learning rate for convergence.

We will discuss the matrix method and OLS method in the following sections. Discussion of gradient descent is deferred to Chap. 11.

The matrix method solution

The equation-1 above can be represented in matrix form as

$$Y = XB + E$$

where
   B is column vector representing $b_0 \dots b_k$,
   X is $n \times (k + 1)$ matrix,
   Y is $n \times 1$ column vector.
   To minimize the error $|E|$,
   Set, $XB = Y$
   Multiplying both sides by X^T

$$X^T XB = X^T Y$$

$$B = \left(X^T X\right)^{-1} X^T$$

Multiplying both sides by X^T

$$X^T X B = X^T Y$$

$$B = \left(X^T X\right)^{-1} X^T Y$$

OLS method solution

The most popular method is to minimize the sum of squared errors $\Sigma(y - \hat{y})^2$ which, according to Gauss, is the best solution for a normal population. The Gauss-Markov theorem asserts that the ordinary least squares (OLS) estimator exhibits the smallest sampling variance among linear unbiased estimators when the errors in the linear regression model are uncorrelated, have equal variances, and have an expectation value of zero (see Eq. 5.2).

$$\hat{y}^{(i)} = b_0 + b_1 X_1^{(i)} + b_2 X_2^{(i)} + \cdots + b_k X_k^{(i)}$$

$$e^{(i)} = y^{(i)} - \hat{y}^{(i)}$$

Let 'SSE' be the sum of squared errors.

$$SSE = \sum_{i=1}^{n} \left(y^{(i)} - \hat{y}^{(i)}\right)^2$$

$$SSE = \sum_{i=1}^{n} \left(y^{(i)} - f(x)\right),^2$$

where

$$f(x) = \hat{y}^{(i)} = b_0 + b_1 X_1^{(i)} + b_2 X_2^{(i)} + \cdots + b_k X_k^{(i)} + e^{(i)}$$

From the above equation, we can derive the values of the coefficients $b_0 \ldots b_k$ and build the linear regression model shown in Eq. 5.1. For example, consider the OLS method for one independent variable (simple linear equation).

$$\hat{y}^{(i)} = b_0 + b_1 x^{(i)} + e^{(i)}$$

$$e^{(i)} = y^{(i)} - \hat{y}^{(i)}$$

$$SSE = \sum_{i=1}^{n} [y^{(i)} - \left(b_0 + b_1 x^{(i)}\right)]^2$$

From the above equation, we can derive the following:

$$b_1 = \frac{n(\Sigma xy) - \Sigma x \Sigma y}{n(\Sigma x^2) - (\Sigma x)^2}$$

$$b_0 = \frac{(\Sigma y)(\Sigma x)^2 - (\Sigma x)(\Sigma xy)}{n(\Sigma x^2) - (\Sigma x)^2}$$

5.2.3   The Coefficient of Determination

Consider the association between project_planning and project_outcome, dis-cussed in Tutorial 5.1. Here, the 'project_outcome' (y) depends on 'project_plan-ning' (x). Figure 5.2 shows the scatter plot of y over x. The yellow line indicates the mean value of y. The blue line shows the linear equation that predicts y_i for a given x_i

$$y_i = b_0 + b_1 x_i$$

Take the case of the data instance (4, 5). Here the observed y-value is 5. The corresponding y-value predicted by the regression equation (blue line) is 3.7, rep-resented by (4, 3.7). The mean value of y is 3.2, which is represented by the corre-sponding point (4, 3.2). In other words, the observed y-value is 5, which is higher than the predicted y-value of 3.7, which is higher than the mean value of y (3.2).

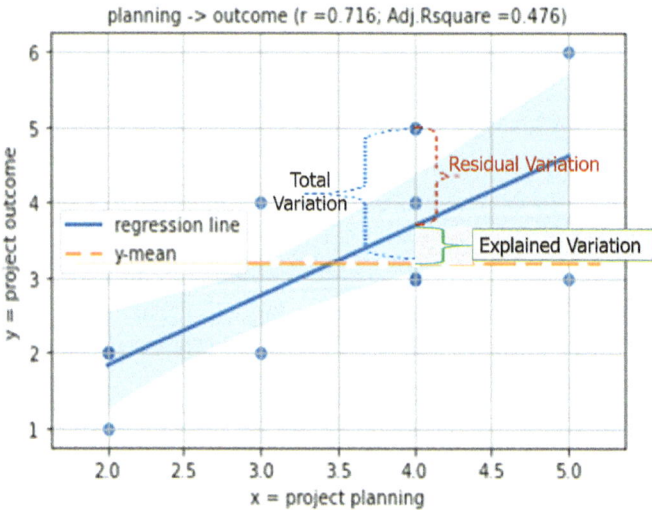

Fig. 5.2 Explained and unexplained variation in regression model

The coefficient of determination R^2 indicates the strength of the association of y and x. It is determined as follows:

Coefficient of determination R^2

$= \text{Explained Variation}/\text{Total Variation}$

$= \text{Sum of Squares Regression}/\left(\begin{array}{l}\text{Sum of Squares Regression}\\ + \text{Sum of Squares Regression Residual}\end{array}\right)$

$= SS_{\text{regression}}/SS_{\text{total}}$

R^2 must be adjusted for the number of independent variables (k) and the sample size (*n*). Check Sect. 5.1 for an explanation of the value adjustment (Malhotra 2020).

$$\text{Adjusted } R^2 = R^2 - k\left(1 - R^2\right)/(n - k - 1)$$

5.2.4 Manual Calculation of R^2

Consider Tutorial 5.1. Here, the data consists of two measurements—project planning (P) and project outcome (O), from 15 software projects. Here, we compute the Adjusted R^2 for project outcome (*O*). Detailed calculations are in Table 5.4. The first column shows the project outcome (O_i). The mean value of the project outcome is $U_{\text{mean}} = 3.2$. The second column shows the project outcome predicted (O_{pred}) based on the rating for project planning. The python code for that is shown below. The third column shows the predicted outcome's deviation from the project mean, and the fourth column shows the corresponding variance. The fifth column is the deviation of the observed value of the project outcome from the predicted value, and the sixth column shows the corresponding variance. The bottom row is the summation of the values in the column.

Explained Variation $=$ Sum of Squares Regression $= SS_{\text{reg}} = 13.48$

Un $-$ Explained Variation $=$ Sum of Squares Residuals $= SS_{\text{res}} = 12.85$

Total Variation $=$ Sum of Squares Regression $+$ Sum of Squares Residuals

SS Total $= SS_{\text{reg}} + SS_{\text{res}} = 13.48 + 12.85 = 26.33$

$R^2 = SS_{\text{reg}}/SS_{\text{total}} = 13.48/26.33 = 0.51$

Given the number of independent variables $k = 1$, sample size $n = 15$, and $R^2 = 0.51$.

Adjusted $R^2 = R^2\text{-}k(1\text{-}R^2)/(n\text{-}k\text{-}1)$

$= 0.51\text{-}1(1\text{-}0.51)/(15\text{-}1\text{-}1) = 0.47$

Adj.R^2 is 0.47. This implies that project_planning can explain 0.47 or 47% of the variance in project_outcome. That means, if we know the rating for project_planning, we can predict the project_outcome in 47% of the cases. The section below shows the Python code for computing Adj.R^2 and predicting project_outcome based on the project_planning score.

Table 5.4 Computation of adjusted R^2

project_ outcome (O_i)	project_ outcome_ predicted (O_{pred})	$O_{pred} - O_{mean}$	Explained = $(O_{pred} - O_{mean})^2$	$O_i - O_{pred}$	Unexplained = $(O_i - O_{pred})^2$
3	4.62	1.42	2.02	−1.62	2.62
2	1.84	−1.36	1.85	0.16	0.03
5	3.69	0.49	0.24	1.31	1.72
3	3.69	0.49	0.24	−0.69	0.48
2	2.77	−0.43	0.18	−0.77	0.59
6	4.62	1.42	2.02	1.38	1.90
1	1.84	−1.36	1.85	−0.84	0.71
2	1.84	−1.36	1.85	0.16	0.03
5	3.69	0.49	0.24	1.31	1.72
2	1.84	−1.36	1.85	0.16	0.03
3	3.69	0.49	0.24	−0.69	0.48
4	3.69	0.49	0.24	0.31	0.10
3	3.69	0.49	0.24	−0.69	0.48
3	3.69	0.49	0.24	−0.69	0.48
4	2.77	−0.43	0.18	1.23	1.51
$O_{mean} =$ 3.20		$SS_{reg} =$ 13.48			$SS_{res} =$ 12.85

Tutorial 5.2 Linear regression - Using statsmodels / ols

```
import pandas as pd
```

refer: https://www.statsmodels.org/stable/regression.html
```
from statsmodels.formula.api import ols
# creating the dataframe
d  = pd.DataFrame()
# explicitly assign column names, load data
d['project_plan'] = [5,2,4,4,3,5,2,2,4,2,4,4,4,4,3]
d['project_outcome'] = [3,2,5,3,2,6,1,2,5,2,3,4,3,3,4]
```

ols linear regression (y, x)
```
m = ols("project_outcome ~ project_plan", data=d).fit()
```

Predict Project Outcome score based on the rating for Project Planning
```
m.predict(d.project_plan)
```

```
m.rsquared_adj
```
Adjusted R^2 = 0.47; Since the value is less than 0.5, the model is not significant

```
m.fvalue      # 13.704 : F statistic
m.f_pvalue    # 0.0027 : Probability of F statistic
m.tvalues     # The significance of independent variable, project planning
```

5.2.5   Significance of the Overall Linear Regression Model

F-test is used to test the null hypothesis that the coefficient of multiple determination of the population:

H_o: R^2population $= 0$.

The test statistic has an F distribution with k and $(n - k - 1)$ degrees of freedom.

$$F = (SS_{regression}/k)/(SS_{residual}/(n - k - 1))$$

$$\text{Alternatively, } F = \frac{SS_{regression}}{SS_{residual}} * \frac{(n - k - 1)}{k}$$

If the F-statistic is significant (e.g., p-value ≤ 0.05), the model is considered statistically significant. In the above example, the F-statistic is 13.7, for 1, 13 degrees of freedom. The associated p-value is 0.0027, which is less than equal to the sig. level assumed '0.05'. This implies that the null hypothesis is rejected, and R^2 population > 0. The coefficient of regression is significant. In summary, the regression model that predicts project outcomes based on project planning is significant.

5.3   The Regression Analysis Procedure

The steps to be followed in regression analysis are listed below (Hair et al. 2010). Some of the underlying terms and concepts are discussed in the subsequent sections.

5.3.1   Primary Inspection of the Results of Regression Analysis

The basic checks in regression analysis include the statistical significance of the overall model, the statistical significance of the variables involved, and the percentage of variation explained by the model. If they are satisfactory, the regression model (or regression equation) can be constructed. We need to do the following steps [S1]–[S10] in order.

[S1]   Do the independent variables together explain the variation in the dependent variable in a statistically significant manner? If we aim for a level of significance (α) of 0.05, check whether the p-value of the F-statistic ≤ 0.05.

[S2]   Do each of the independent variables have a statistically significant association with the dependent variable? We may use one of the following methods.

(a) Check the significance level of the 't' statistic of the coefficients 'bi'. Identify the variables that are not significant (p-value >0.05). Discard them one at a time (based on how high their p-values are) and re-estimate the model iteratively.

(b) Stepwise regression: This is a more rigorous method. Check the significance of the coefficient bi of each variable through an incremental F-test, retain the most contributing ones—refer to the section on stepwise regression. This method is discussed in a separate section. It may be noted that this method is also not fail-proof, as multicollinearity can affect variable selection.

[S3] Check the coefficient of determination. This is also called adjusted R^2 or the explanatory power of the model. This is the amount of variation in the dependent variable that is explained by the independent variables.

[S4] Construct the regression model (also called the regression equation).

5.3.2   Some Concepts Associated with Regression Analysis

The researcher needs to consider various aspects of the data and the model to develop an effective regression model. The common aspects that are inspected include autocorrelation of the observations, multicollinearity of the variables, homoscedasticity of residual error, part and partial correlation of each variable with the outcome, normality of the distribution, model fit, etc. They are discussed here.

Standardized Regression Equation

Data standardization helps us compare variables measured on different scales, e.g., pounds and kilograms. There are many methods for data standardization—min–max, z-score, decimal scaling, etc. z-score standardization is popularly used in statistical data mining. A variable is standardized by subtracting its sample mean from it and dividing it by its standard deviation. An observation x_i can be transformed to z-score as follows:

$$z_i = (x_i - x_{\text{mean}})/, s$$

where 's' is the standard deviation of the sample.

Multicollinearity and Variance Inflation Factor (VIF)

Multicollinearity arises when intercorrelations among the predictors are very high. Variance Inflation Factors (VIF) is a measure of multicollinearity (Hair et al. 2010). VIF is directly related to the tolerance value (VIF $= 1/$Tolerance). VIF value of 3–5 is normally acceptable for an independent variable. Multicollinearity can result in several problems:

- The partial regression coefficients may not be estimated precisely.
- The standard errors are likely to be high.
- It becomes difficult to assess the relative importance of the independent variables in explaining the variation in the dependent variable.
- Predictor variables may be incorrectly included or removed in stepwise regression.

How to deal with challenges from multicollinearity?

- Remove some of the highly correlated independent variables or use only one of the variables in a highly correlated set of variables.
- Combine the highly correlated independent variables into one.
- Use dimension reduction techniques designed for highly correlated variables, such as principal components analysis or partial least squares regression to transform a set of independent variables into fewer components.
- Use ridge regression or latent root regression, which are advanced forms of regression analysis that can handle multicollinearity.

Autocorrelation and the **Durbin-Watson** test **statistic**

Autocorrelation is the similarity over a sequence of observations or the observations in a time series. This may result in underestimates of the standard error, leading a researcher to think that the predictors (independent variables) are significant when they are not (Malhotra 2020). The Durbin-Watson test statistic is a numerical measure employed in statistical analysis to assess the presence of autocorrelation in the residuals of a regression model. The range of values for this statistic and a thumb rule for their interpretation is shown below:

- A value of '2' indicates no autocorrelation.
- A value between 0 and 2 indicates positive autocorrelation (common in time series).
- A value>2, indicates negative autocorrelation.

Homoscedasticity

Homoscedasticity assumes that the variance of the error term is constant. Otherwise, the data are said to be heteroscedastic. Homoscedasticity is an important assumption in many multivariate techniques.

Partial correlation coefficient

The partial correlation coefficient measures the strength of the association of a single independent variable with the dependent variable when the effects of the other independent variables in the model are held constant. It measures the incremental predictive effect of an independent variable (Hair et al. 2010). It can be expressed as follows:

$$r_{xy.z} = \frac{\left[r_{xy} - \left(r_{xz}r_{yz}\right)\right]}{\sqrt{(1 - r_{xz}^2)(1 - r_{yz}^2)}}$$

The partial correlation coefficient plays a crucial role in variable selection during the construction of regression models, including methods like forward addition or backward elimination. It helps identify the independent variable that adds the most predictive power beyond what is already accounted for by the independent variables already incorporated into the regression model.

Part correlation

Part correlation measures the strength of association of a single independent variable with the dependent variable when the effect of the other independent variables in the regression model is removed (Malhotra 2020). It can be expressed as follows:

$$r_{y(x.z)} = \frac{\left[r_{xy} - \left(r_{xz}r_{yz}\right)\right]}{\sqrt{(1 - r_{xz}^2)}}$$

Skewness/Kurtosis

The Omnibus and Jarque-Bera (JB) tests are used to check whether the skewness and kurtosis are within the limits expected of a normal distribution. The Jarque-Bera test works only works for large data samples (>2000). The p-value of the test statistic must be within acceptable limits (e.g., ≤ 0.05).

Akaike information criterion (AIC)

AIC is an indicator of model fit. It is used for comparing models. According to this criterion, the best fit model is the model that explains the highest amount of variation using the minimum possible independent variables. The lower the AIC, the better the model.

5.3.3   Advanced Inspection of the Results of Regression Analysis

Check for autocorrelation of the observations, Skewness/Kurtosis of the dependent variables, multicollinearity of the variables, randomness of residual errors, and the relative importance of the independent variables.

[S5]   Is the autocorrelation of the observations within acceptable limits? A value
       of 2 for the Durbin-Watson statistic indicates that there is no autocorrela-
       tion among the observations.
[S6]   Are the Skewness/Kurtosis of the dependent variable within acceptable
       limits? p-values of Omni Bus/JB Test must be ≤ 0.05.
[S7]   Inspect the multicollinearity of the variables (variable to variable interac-
       tion). Is each variable's variance inflation factor (VIF) within acceptable
       limits? A thumb rule, for social science problems, states that VIF must be
       ≤ 5.
[S8]   Is the residual plot homoscedastic.
[S9]   Are the partial regression plots of each independent variable with the
       dependent variable homoscedastic?
[S10] Determine the relative importance of the independent variables.

5.4 Case Study on Regression Analysis

A study was conducted to explore the variables that influence software project
performance ('project_perf'). Eleven variables were hypothesized to influence
project performance. 100 responses were collected from IT consultants working
on various projects (Sundararajan et al. 2019). The distributions of the 11 independent variables are shown in Fig. 5.3. Please refer (Sundararajan 2023) for data
and data description.

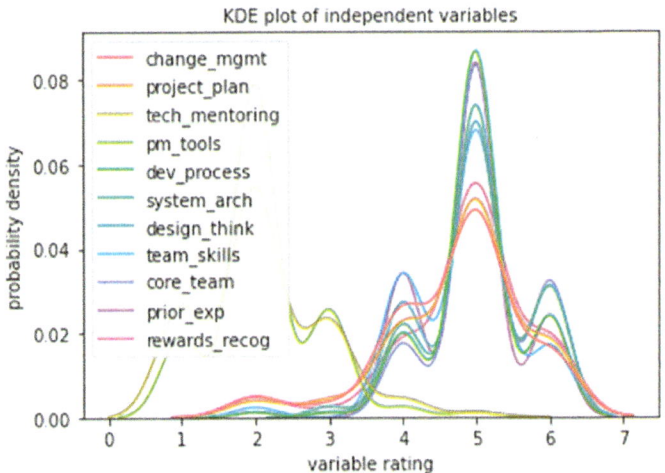

Fig. 5.3 Distribution of the variables influencing software project performance

5.4.1   Case Study—Part I—Number Checks

Perform regression analysis. Inspect the p-value of the F-statistic for significance, the level of significance of the 't' statistic of the coefficients 'bi', the coefficient of determination, the regression model, the autocorrelation of the observations for acceptable limits, skewness/kurtosis of the dependent variable for acceptable limits the variance inflation factor (VIF) of each variable for acceptable limits, the residual plot for homoscedasticity, the partial regression plots of the independent variables with the dependent variable for homoscedasticity, and [S10] the relative importance of the independent variables.

Tutorial 5.4   Detailed Regression Analysis - Using statsmodels / OLS

Tutorial 5.4.1 Data Setup

```
import pandas as pd
```

For data and data description please refer {(Sundararajan, 2023)}. If the data has missing values, they must be removed / imputed

```
d=pd.read_csv('itprojects.csv')
d.columns
```

```
'Case_No',
'change_mgmt', 'project_plan', 'tech_mentoring', 'pm_tools',
'dev_process', 'system_arch', 'design_think', 'team_skills',
'core_team', 'prior_exp', 'rewards_recog',
'project_type', 'project_perf'
```

```
X = d[['change_mgmt', 'project_plan', 'tech_mentoring', 'pm_tools',
       'dev_process', 'system_arch', 'design_think', 'team_skills',
       'core_team', 'prior_exp', 'rewards_recog']]
```

Tutorial 5.4.2 The Distribution of the Feature Variables (X)

A kernel density estimate plot shows the frequency distribution of vars, like a histogram. But it uses a continuous probability density function

```
import matplotlib.pyplot as plt
import seaborn as sb

sb.kdeplot(data=X)
plt.xlabel('variable rating')
plt.ylabel('probability density')
plt.title ('KDE plot of independent variables', size=10)
plt.show()
```

See Figure 5-3 Distribution of the variables influencing software project performance

Tutorial 5.4.3 Linear Regression (using OLS)

We will use 'OLS' (Refer statsmodels.api). Note that the module name is in capital letters, it is not 'ols', which we used earlier
```
import statsmodels.api as sm  # for regression
```

Do Data Setup as in 5.4.1. Read in the DataFrame d, Setup features in X

Add a column of constants, as a place holder for intercept
```
X['intercept'] = 1
```

The dependent variable is project performance
```
y = d.project_perf
```

```
model = sm.OLS(y,X,missing='drop') #discard missing values if any
result=model.fit()
```

```
round(result.f_pvalue,3) # 0.0 p-value of F test
round(result.pvalues,3)  # p-value of t-Tests, last one is intercept
change_mgmt        0.359
project_plan       0.000
tech_mentoring     0.008
pm_tools           0.614
dev_process        0.734
system_arch        0.899
design_think       0.771
team_skills        0.005
core_team          0.033
prior_exp          0.149
rewards_recog      0.243
intercept          0.000
```

```
result.rsquared_adj    #0.536
```

Discussion: -

result.summary() gives Complete Result

result.params gives the coefficients of independent variables followed by the intercept

[S1] Overall Significance of the model - p(F statistic)

The Prob of F-statistic < 0.05; This implies that regression model is significant at 0.05 levels

[S2] Significance of individual predictors p(t statistic) must be < 0.05

Only 4 variables have p-value less than 0.05. They are:- project_plan, team_skills, tech_mentoring, core_team. The intercept is not a variable. so do not check its p-value.

[S3] The strength of association (Adj.R^2 = 0.536 = 53.6%) is Above 50%. Therefore, the regression model is significant

Tutorial 5.4.4 Discarding the Variables that are Not Significant

Note:- This is a simple approach. If any of the variables are interdependent, the method will not be effective

```
import statsmodels.api as sm  # for regression
```

Do Data Setup as in 5.4.1. Read in the DataFrame d; Setup features in X. We choose only four independent variables that were found to be significant

```
X = d[['project_plan', 'tech_mentoring','team_skills','core_team' ]]
```

Add a column of constants, as a place holder for intercept

```
X['intercept'] = 1
```

The dependent variable is project performance

```
y = d.project_perf
```

```
model  = sm.OLS(y,X,missing='drop') # discard missing values if any
result = model.fit()
print(result.summary())  # Note the () after summary. It is a method
```

```
                    OLS Regression Results
----------------------------------------
R-squared:                        0.562
Adj. R-squared:                   0.544    (with all var, it was 0.536)
Method:                   Least Squares
Prob (F-statistic):            2.47e-16
No. Observations:                   100
AIC:                              211.1    (with all var, it was 219.3)
Df Residuals:                        95
Df Model:                             4
----------------------------------------------------------
                    coef     std err        t       P>|t|
----------------------------------------------------------
project_plan      0.7151       0.077     9.291       0.000
tech_mentoring    0.2430       0.090     2.689       0.008
team_skills       0.3196       0.107     2.996       0.003
core_team         0.2589       0.111     2.334       0.022
intercept        -6.7584       0.797    -8.484       0.000
----------------------------------------------------------
Durbin-Watson:                    1.760
Prob(Omnibus):                    0.001
Prob(JB):                         0.119
Skew:                            -0.111
Kurtosis:                         2.014
----------------------------------------
```

The Model:-

[S1] Overall Significance of the model - p(F statistic)
The Prob of F-statistic < 0.05. This implies that the regression model is significant at 0.05 levels
[S2] Significance of individual predictors - p-value of t statistics. All 4 independent variables have p-values less than 0.05. (Intercept is not a variable. so do not check its p-value)
[S3] The strength of association (Adjusted R^2)

(Adjusted R^2) = 0.544 or 54.4%. This is above 50%. So, the explanatory power of the model is acceptable

[S4] The Regression Model

```
      project_perf =
        -6.7584 + 0.7151 * project_plan + 0.2430 * tech_mentoring +
        0.3196 * team_skills + 0.2589 * core_team
```

Other Validity Checks:-

[S5] Durbin-Watson tests autocorrelation. A value of 2 is ideal. In this case, the value is 1.760, which is acceptable.

durbin watson statistic can be also computed as follows, separately:-

```
   from statsmodels.stats.stattools import durbin_watson
   durbin_watson(x)
```

[S6] Omnibus and JB indicate whether the skewness and kurtosis are within the limits expected of a normal distribution. The Jarque-Bera test works for large data samples (>2000) only. Omnibus test p-value = 0.001, which is < 0.05. Therefore, skewness and kurtosis are within expected limits

Tutorial 5.4.5 Variance Inflation Factor (VIF)

[S7] Variance Inflation Factor (VIF)

Refer: https://patsy.readthedocs.io/en/latest/quickstart.html

```
   from    patsy import dmatrices
   from    statsmodels.stats.outliers_influence import
   variance_inflation_factor
```

Do Data Setup as in 5.4.1. Read in the DataFrame d.

Using patsy.dmatrices split data into y,X

```
   y, X = dmatrices('project_perf ~ project_plan + tech_mentoring +
   team_skills + core_team',data=d, return_type='dataframe')
```

Create an empty DatFrame

```
   vifmatrix = pd.DataFrame()
```

Save column names in varlist, except the column intercept

```
   varlist    = X.columns.drop('Intercept')
   nc         = X.shape[1]  # number of columns = 5
```

Create columns and name the columns in the DataFrame vifmatrix

```
   vifmatrix['variable'] = varlist
```

Compute vif for the feature variables
We do not need vif of the first column of X (Intercept). So, use range(1,nc)

```
   vifmatrix['vif'] = [variance_inflation_factor(X.values, i) for i in
   range(1,nc)]
   print(vifmatrix)
```

```
          variable           vif
0      project_plan    1.109133
1    tech_mentoring    1.229132
2       team_skills    1.383400
3         core_team    1.211738
```

[S7] Variance Inflation Factor (VIF) is almost equal to 1, for all the independent variables. This is acceptable.

5.4.2   Case Study—Part II—Visual Inspections

In this section we will do visual inspections of the residual plot and partial
regression plot of the dependent variable with each independent variable.

The Residual Plot and Homoscedasticity [S8]

The difference between an observed value of y_i and the value predicted by the
regression equation is called residual (error). To ensure that the residuals are random (and are not due to hidden variables or the sequencing of observations), the
following assumptions must be validated (Malhotra 2020):

- The error term is normally distributed.
- The mean value of the error term is zero.
- The variance of the error term is constant.
- The error variance does not change over time or sequence of the observations
  (Homoscedasticity).

Figure 5.4a shows the results from Tutorial 5.10. It shows a residual plot that is
somewhat homoscedastic, as it does not show any specific trend across the observations on the x-axis. From Fig. 5.4b, we observe that the mean is zero and the
variance is somewhat constant (the variance does not show any specific trend
across the observations).

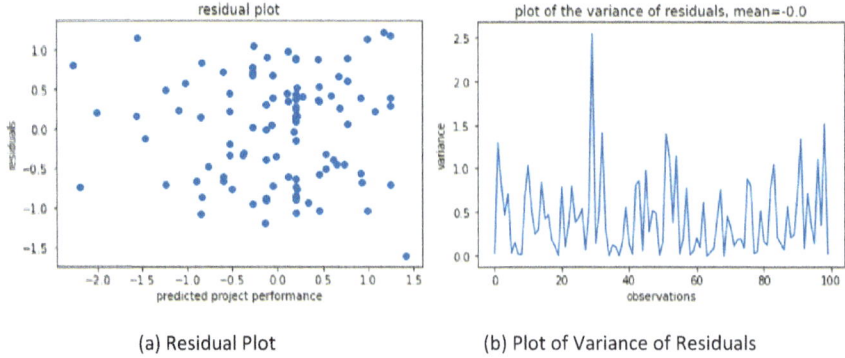

(a) Residual Plot   (b) Plot of Variance of Residuals

Fig. 5.4   Residuals plot

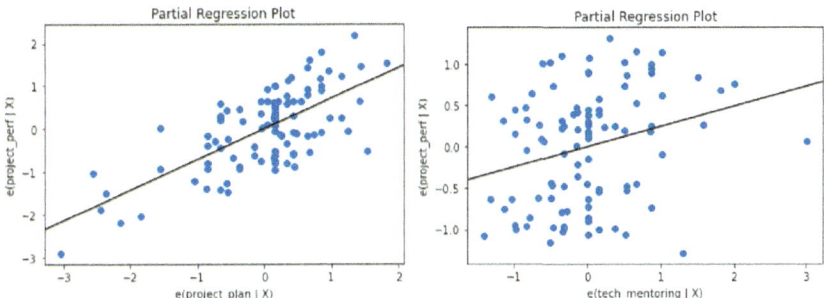

Fig. 5.5   Partial regression plots

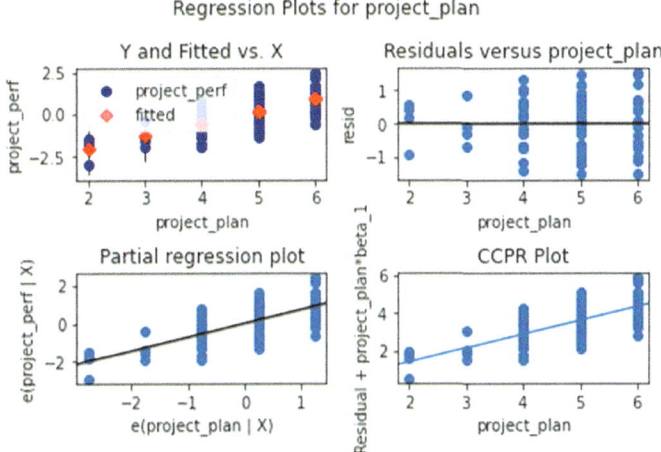

Fig. 5.6   Partial regression and residuals plots for 'project plan'

The Homoscedasticity of the Partial Regression Plots [S9]

The partial regression plot of the dependent variable with each independent variable, must be homoscedastic. Partial regression plots of the outcome variable for four predictor variables are shown in Fig. 5.5. The partial regression plots are somewhat homoscedastic, as there is no prominent trend across the x-axis.

Visual Inspections

The following visual inspections are carried out here.

[S8] Is the residual plot homoscedastic?

[S9] Are the partial regression plots of each independent variables with the dependent variable homoscedastic?

Tutorial 5.4.6 The Residual Plot

[S8] The residual plot must be homoscedastic.
The residual error distribution of input variables, over all the values, is expected to have a mean value of zero and constant variance. Figure 5-4 (a) shows that the mean value is not zero. Figure 5-4 (b) shows that the variance is somewhat normal.
Homoscedasticity implies that the variance of the residual error, in a regression model is constant. From a visual examination of Figure 5-4 (a), and Figure 5-4 (b), the residual error plot appears somewhat homoscedastic.

```
import pandas as pd
from statsmodels.formula.api import ols
import matplotlib.pyplot as plt

d=pd.read_csv('itprojects.csv')
X = d[['project_plan', 'tech_mentoring','team_skills','core_team' ]]
```

Add a column of constants, as a place holder for intercept
```
X['intercept'] = 1
```

The dependent variable is project performance
```
y = d.project_perf

model = ols('y ~ project_plan + tech_mentoring + team_skills +
core_team', data=d).fit()

yp = model.predict(X)
residual = y - yp
plt.scatter(x=yp, y=residual)
plt.title('residual plot')
plt.xlabel('predicted project performance')
plt.ylabel('residuals')
plt.show()

mean ='mean=' + str(round(residual.mean(),4))
var = residual ** 2
plt.plot(var)
plt.title ('plot of the variance of residuals, '+mean)
plt.xlabel('observations')
plt.ylabel('variance')
plt.show()
```
See Figure 5-4 Residuals Plot

Tutorial 5.4.7 Partial Regression Plots - Brief

[S9] Inspecting the homoscedasticity of partial regression plots
The partial regression plot shows the relationship between the response (target) and an explanatory variable (feature), after removing the effect of all other explanatory variables (features)
Interpretation of the Results: -
Check Figure 5-5, partial regression plots of the outcome variable for each predictor variable. The variance of project_perf over different values of project_plan is somewhat constant. Therefore, this partial regression plot is somewhat homoscedastic. The variance of project_perf over different values of tech_mentoring is not constant; the variance being high in the middle. Therefore, this partial regression plot cannot be considered as homoscedastic.

```
import statsmodels.api as sm
import pandas as pd
import matplotlib.pyplot as plt

d=pd.read_csv('itprojects.csv')

sm.graphics.plot_partregress(
    endog='project_perf', exog_i='project_plan',
    exog_others=['tech_mentoring', 'team_skills', 'core_team'],
    data = d, obs_labels=False)
plt.show()
```
See Figure 5-5: Partial Regression Plots

```
sm.graphics.plot_partregress(
    endog='project_perf', exog_i='tech_mentoring',
    exog_others=['project_plan', 'team_skills', 'core_team'],
    data = d, obs_labels=False)
plt.show()
```
See Figure 5-5: Partial Regression Plots

```
sm.graphics.plot_partregress(
    endog='project_perf', exog_i='team_skills',
    exog_others=['project_plan', 'tech_mentoring', 'core_team'],
    data = d, obs_labels=False)
plt.show()

sm.graphics.plot_partregress(
    endog='project_perf', exog_i='core_team',
    exog_others=['project_plan', 'team_skills', 'tech_mentoring'],
    data = d, obs_labels=False)
plt.show()
```

Tutorial 5.4.8 Partial Regression Plot - Detailed

Plot regression results against one regressor in a 2 by 2 figure - endog (target) versus exog (feature); residuals vs exog (feature); fitted (target) vs exog (feature); fitted plus residual exog (feature). See Figure 5-6 Partial Regression and Residuals Plots for 'project plan'

```
import statsmodels.api as sm
import statsmodels.formula.api as sf
import pandas as pd
import matplotlib.pyplot as plt

d=pd.read_csv('itprojects.csv')
variables = 'project_perf ~ project_plan'
m = sf.ols(variables, data=d).fit()

plot = sm.graphics.plot_regress_exog(m, 'project_plan')
plot.tight_layout(pad=1.0)
plt.show()

variables = 'project_perf ~ team_skills'
m = sf.ols(variables, data=d).fit()
plot = sm.graphics.plot_regress_exog(m, 'team_skills')
plot.tight_layout(pad=1.0)
plt.show()

variables = 'project_perf ~ tech_mentoring'
m = sf.ols(variables, data=d).fit()
plot = sm.graphics.plot_regress_exog(m, 'tech_mentoring')
plot.tight_layout(pad=1.0)
plt.show()
```

5.5   Case Study—Stepwise Regression

Assume that there are many independent variables in our model. We want to discard trivial variables and select the significant ones. Stepwise regression is a method for achieving that. In stepwise regression, the independent variables are entered or removed from the regression equation one at a time, based on the effect on R^2 (Hair et al. 2010). The partial correlation coefficient is a good measure for variable selection.

Table 5.5 Partial correlation of project outcome vs variables taken one at a time

Forward add	Variable	Partial correlation	Incremental R^2	Backward remove
1	project_plan	0.622	0.387280	
2	team_skills	0.296	0.087467	
3	tech_mentoring	0.278	0.077043	
4	core_team	0.225	0.050565	
	prior_exp	−0.153	0.023482	7
	rewards_recog	0.124	0.015476	6
	change_mgmt	0.098	0.009561	5
	pm_tools	0.054	0.002904	4
	dev_process	0.036	0.001314	3
	design_think	0.031	0.000966	2
	system_arch	0.014	0.000183	1

The Forward Regression Method

We add the variables, one at a time based on the partial correlation coefficient, as per the order shown under the column 'Forward add'. The Python code for computing the partial correlation is shown below.

The Backward Regression Method

Initially, regression analysis is done with the entire set of variables (all the independent variables, versus the dependent variable). If the model is not significant, we remove the independent variables one at a time, based on its partial correlation coefficient. We repeat this process till the entire model is significant. The order in which the variables are removed is shown under the column 'Backward remove' in Table 5.5. Initially, we have 11 variables. We remove the variables, system_arch, design_think, pm_tools, dev_process, change_mgmt, rewards_recog, and prior_exp in order. After 7 iterations, 4 independent variables remain, all of which are significant. The selected variables are project_plan, tech_mentoring, team_skills, and core_team.

In a particular iteration, if there is a conflict in removing a variable, we may look at the significance of the t-test of the b_i's (Note that the b_i's are the coefficients of the independent variables in the regression equation). Also, inspect the contribution to Adjusted R^2. We will remove the variable with the lowest contribution to Adjusted R^2 or the highest p-value associated with the t-test.

Tutorial 5.5.1 Forward Regression Method

In stepwise regression, the independent variables are entered or removed from the regression equation one at a time, based on the effect on R^2. The partial correlation coefficient is a good measure for variable selection.

```
import pandas as pd
import numpy  as np
```

Install pingouin package. You may refer:
https://pingouin-stats.org/generated/pingouin.partial_corr.html

```
!pip install pingouin
import pingouin as pin
```

Download the file from GitHub (See Chapter-1.3)

```
d = pd.read_csv('itprojects.csv')

X = d[['change_mgmt', 'project_plan', 'tech_mentoring',
       'pm_tools', 'dev_process', 'system_arch', 'design_think',
       'team_skills', 'core_team', 'prior_exp', 'rewards_recog']]

PC = pd.DataFrame()
PC['variable'] = X.columns
PC['partial_corr'] = np.ones(len(PC.variable))
PC['incremental_R2'] = np.ones(len(PC.variable))

j = 0
for i in PC.variable:
    Xcov = X.drop(i, axis=1)
    pc   = pin.partial_corr(data=d, y='project_perf', x=i,
covar=list(Xcov.columns))
    PC['partial_corr'][j]   = round(pc['r'][0],3)
    PC['incremental_R2'][j] = pc['r'][0] ** 2
    j = j+1

PC = PC.sort_values(by='incremental_R2',ascending=True)
print(PC)
```
The result is shown in Table 5-5 Partial Correlation

Tutorial 5.5.2 The Backward Regression Method

```
import statsmodels.api as sm  # for regression

X = d[['change_mgmt','project_plan','tech_mentoring','pm_tools','dev_process'
,'system_arch','design_think','team_skills','core_team','prior_exp','rewards_
recog']]
X['intercept']=1

y= d.project_perf
```

```
    for i in PC.variable:
        X.drop(i,inplace=True,axis=1)
        print(X.columns)
        model = sm.OLS(y,X)
        result=model.fit()
        modelfit = False
        for j in range(len(result.pvalues)-1):

If p-value > 0.05, we found a variable that is NOT significant
            if result.pvalues[j] > 0.05:
                break

The last one is intercept. It can be discarded
            if j == (len(result.pvalues)-2):

If the other variables are significant, the entire model is GOOD!
                modelfit = True
                break
        if modelfit == True:
We found that the independent variables are all significant!
Let us quit the 'for' loops! Job Over
            break

    print(X.columns)
'project_plan','tech_mentoring','team_skills','core_team','intercept'

Use the features selected in stepwise regression to build a model
```

5.6   Case Study—Feature Ranking

If you have too many variables, you may need to rank and select fewer variables based on criteria such as the partial regression coefficient or p-value associated with the t statistic, etc. The researcher may maintain some variables even
though they are less important if he/she posits that they are important to the model
from a theoretical or practical perspective. The features can be handpicked if there
are only a few of them. However, if there are numerous features, we may automate
the process.

Feature Ranking—Manual Determination

[S10]   Determine the relative importance of the independent variables. Some of
the methods for this are mentioned below.

1. Observe the effect on R^2, with the entry or exit of a variable. See the Section on Stepwise Regression for more details.
2. Inspect the partial correlation coefficient.
3. Inspect the absolute values of partial regression coefficients $|'b_i'|$, in the standardized regression equation.
4. Inspect the absolute values of the bivariate $(X_i \sim y)$ correlation coefficient.
5. Inspect the part correlation coefficient.

Automatic Feature Selection

In this tutorial, we use the 'scikitlearn' machine learning library to rank and select features (refer: https://scikit-learn.org/stable/modules/linear_model.html). The estimator is 'trained' on the initial set of features. Based on the feature ranking that emerges, some of the least prominent ones are removed. The above procedure is recursively repeated on the pruned set until the desired number of features is achieved. Chapter 11 Machine learning will throw more light on the underlying concepts.

Tutorial 5.6   Feature Ranking - Manual Determination

Tutorial 5.6.1: Inspect the partial regression coefficients $|'b_i'|$

Standardize y, and X and build regression model analysis.
Compare the significance of the independent variables, by checking the p-values associated with them thrown by the t-Test

```
import pandas as pd
import statsmodels.api as sm   # for regression
from scipy import stats        # for z-score standardisation
```

Do Data Setup as in 5.4.1. Read in the DataFrame d

Setup Features in X
```
Xz = stats.zscore(X)
y = d.project_perf

model  = sm.OLS(y,Xz,missing='drop') # discard missing values if any
result = model.fit()
print(result.summary())
```

	bi	p-Value
project_plan	0.6312	0
team_skills	0.2489	0.0044
tech_mentoring	0.2218	0.0077
core_team	0.197	0.0321
rewards_recog	0.099	0.24
change_mgmt	0.0777	0.3565
pm_tools	0.0396	0.6119
dev_process	0.0315	0.733

```
design_think      0.0286            0.7699
system_arch       0.0133            0.8986
prior_exp        -0.1451            0.147#
```

Interpretation: -
The variables listed above are sorted in the descending order of importance

Tutorial 5.6.2 Bivariate (X_i ~ y) Correlation

Inspect the absolute values of the bivariate (X_i ~ y) corr. coefficient
Do Data Setup as in 5.4.1. Read in the DataFrame d; Setup Features in X

```python
import numpy as np
varlist   = X.columns
y = d.project_perf
corrmatrix                 = pd.DataFrame()
corrmatrix['variable'] = varlist
nvar                       = len(varlist)
corrmatrix['corr'] = [np.corrcoef(y,
        d[varlist[i]])[0,1] for i in range(nvar)]
corrmatrix.sort_values(by='corr',ascending=False)
```

```
         variable     corr
1      project_plan  0.675742
0       change_mgmt  0.447929
7       team_skills  0.392090
10    rewards_recog  0.273900
8         core_team  0.217314
5       system_arch  0.217010
9         prior_exp  0.193188
6      design_think  0.167782
4       dev_process  0.116861
2    tech_mentoring -0.088914
3          pm_tools -0.130020
```

Interpretation: -
Each method will give a different result, due to multicollinearity (or the
interdependence of the feature variables)

Tutorial 5.6.3 Compute Partial Correlations

Each feature versus target
Save the absolute value of the partial correlations. Sort it

```python
import pandas as pd
import pingouin as pin
import numpy as np

d=pd.read_csv('itprojects.csv')
X = d[['change_mgmt', 'project_plan', 'tech_mentoring', 'pm_tools',
        'dev_process', 'system_arch', 'design_think', 'team_skills',
        'core_team', 'prior_exp', 'rewards_recog']]
```

```
PC = pd.DataFrame()
PC['variable'] = X.columns
PC['partial_corr'] = np.ones(len(PC.variable))

j = 0
for i in PC.variable:
    Xcov = X.drop(i, axis=1)
    pc   = pin.partial_corr(data=d, y='project_perf',
x=i,covar=list(Xcov.columns))
    PC['partial_corr'][j]   = abs(pc['r'][0])
    j = j+1

PC.sort_values(by='partial_corr',ascending=True, inplace=True)
print(PC)
            variable   partial_corr
5         system_arch       0.013543
6         design_think      0.031087
4          dev_process      0.036254
3             pm_tools      0.053885
0          change_mgmt      0.097780
10       rewards_recog      0.124403
9            prior_exp       0.153238
8            core_team       0.224867
2        tech_mentoring      0.277566
7           team_skills      0.295748
1          project_plan      0.622318
```

Analysis: -

Note that the rank of the independent variables obtained from the bi-variate correlation from that of the partial correlation. Partial correlation assesses the association between two variables, while controlling (or accounting for) the influence of other variables. Therefore, it is more reliable.

Tutorial 5.6.4 Recursive Feature Elimination

Specify the number of desired features as an input

```
import pandas as pd
import numpy as np
from scipy import stats
from sklearn.linear_model import Ridge
from sklearn.linear_model import ElasticNet
from sklearn.feature_selection import RFE
from sklearn.feature_selection import RFECV

d=pd.read_csv('itprojects.csv')

X = d[['change_mgmt', 'project_plan', 'tech_mentoring', 'pm_tools',
        'dev_process', 'system_arch', 'design_think', 'team_skills',
        'core_team', 'prior_exp', 'rewards_recog']]
```

```
z-score standardization
   X = stats.zscore(X) # standarize X

   y = d.project_perf
```

Tutorial 5.6.5 RFE using elastic net

Elastic net is a regularization method in machine learning, that helps in feature elimination/feature selection. It is a combination of ridge and lasso regression. More details will be covered in Chapter 11

```
   m=RFE(estimator=ElasticNet(), n_features_to_select=4, step=1,
         importance_getter='coef_')
   m.fit(X,y)
   selected_var = pd.DataFrame()
   selected_var ['variable'] = X.columns
   selected_var ['ranking'] =  m.ranking_
   selected_var ['selected'] = m.support_
   selected_var
```

	variable	ranking	selected
0	change_mgmt	8	False
1	project_plan	1	True
2	tech_mentoring	7	False
3	pm_tools	6	False
4	dev_process	5	False
5	system_arch	4	False
6	design_think	3	False
7	team_skills	2	False
8	core_team	1	True
9	prior_exp	1	True
10	rewards_recog	1	True

```
   selected_var[selected_var.selected == True]
selected_var
```

1	project_plan	1	True
8	core_team	1	True
9	prior_exp	1	True
10	rewards_recog	1	True

Tutorial 5.6.6 RFE using Ridge Regression

Ridge regression is a method for regularization in machine learning. More details will be covered in Chapter 11.

```
   m=RFE(estimator=Ridge(), n_features_to_select=4, step=1,
         importance_getter='coef_')
   m.fit(X,y)
   selected_var = pd.DataFrame()
   selected_var ['variable'] = X.columns
   selected_var ['ranking'] =  m.ranking_
   selected_var ['selected'] = m.support_
   selected_var
```

```
          variable  ranking  selected
0       change_mgmt        4     False
1      project_plan        1      True
2    tech_mentoring        1      True
3          pm_tools        6     False
4       dev_process        7     False
5       system_arch        8     False
6      design_think        5     False
7       team_skills        1      True
8         core_team        1      True
9          prior_exp        2     False
10    rewards_recog        3     False
```

```
selected_var[selected_var.selected == True]
          variable  ranking  selected
1     project_plan        1      True
2   tech_mentoring        1      True
7      team_skills        1      True
8        core_team        1      True
```

Tutorial 5.6.7 RFECV: Rank & Select optimum number of variables

RFECV (Recursive feature elimination with cross-validation) is used to select the best number of features

```
m=RFECV(estimator=Ridge(),step=1, importance_getter='coef_')
m.fit(X,y)

selected_var = pd.DataFrame()
selected_var ['variable'] = X.columns
selected_var ['ranking'] =  m.ranking_
selected_var ['selected'] = m.support_
selected_var[selected_var.selected == True]
```

```
          variable  ranking  selected
1     project_plan        1      True
2   tech_mentoring        1      True
7      team_skills        1      True
8        core_team        1      True
9         prior_exp        1      True
```

5.7   Polynomial Regression

If the relationship between the predictors and dependent variable is not linear, we may add additional 'interaction' terms to build a linear regression model. Though the underlying relationship is polynomial, it may be modeled using the above technique in some cases. Assume that we have two variables, x and y. If the relationship between them is not linear, we can introduce additional terms such as x^2, y^2, and xy as additional terms called interaction terms. We can model the relationship with a quadratic equation as shown below.

$$(x+y)^2 = x^2 + 2xy + y^2$$

The above equation can be considered as a linear relationship involving five variables, x, y, x^2, y^2, and xy. Similarly, if the relationship between x and y is a polynomial of higher degree, that can be modeled by including more interaction terms, such as x^3, y^3, x^2y, and xy^2.

Tutorial 5.7   Polynomial Regression

Tutorial 5.7.1 Diamonds - Data Pre-processing

```
import seaborn as sb
d  = sb.load_dataset('diamonds')
d  = d.dropna()

d['cuti']=d.cut.astype("category").cat.codes
d['colori']=d.color.astype("category").cat.codes
d['clarityi']=d.clarity.astype("category").cat.codes

X = d[['carat', 'cuti', 'colori', 'clarityi',
       'depth','table', 'x', 'y', 'z']]
y = d.price
```

Tutorial 5.7.2 Diamonds - OLS Linear Regression

```
from scipy import stats
```

Standardize features using zscore transformation
```
Xz = stats.zscore(X)
```

Add an intercept column Xz['intercept'] = 1, or standardise y
```
y = stats.zscore(y)
```

```
Perform OLS linear regression
import statsmodels.api as sm  # for regression
model = sm.OLS(y,Xz,missing='drop').fit()
model.rsquared  # 0.907
```

Tutorial 5.7.3 Polynomial Regression

```
from sklearn.linear_model import LinearRegression
from sklearn.preprocessing import StandardScaler
from sklearn.preprocessing import PolynomialFeatures
from scipy import stats
```

Do Datasetup as shown in 5.14.1. Standardize the features
```
  Xz = stats.zscore(X)

  y = stats.zscore(y)
```

Create polynomial features X^2
```
  polynomial = PolynomialFeatures(degree=2, include_bias=False)
  XzPoly = polynomial.fit_transform(Xz)
```

Fit the linear regression
```
  model = LinearRegression().fit(XzPoly, y)
```

Perform Polynomial Regression
```
  model.score(XzPoly, y)  # 0.963 (R²)
```

Explanatory power of polynomial regression (R² = 0.963) is observed to be higher compared to linear regression (R² = 0.907)

Summary

Correlation is a measure of the nature and strength of association between two variables. Pearson correlation coefficient (or product-moment correlation), r, indicates the linear association between two metrics (interval or ratio scaled) variables. r varies between -1 and $+1$.

Regression analysis examines associative relationships between a metric dependent variable (y) and one or more independent variables (X). We look for a solution that minimizes the error between the observed value of y and the value estimated from the equation $y = f(X)$. The most popular method is to minimize the sum of squared errors. The Gauss-Markov theorem asserts that the ordinary least squares (OLS) estimator exhibits the smallest sampling variance among linear unbiased estimators when the errors in the linear regression model are uncorrelated, have equal variances, and have an expectation value of zero.

The coefficient of determination R^2 is an indication of the strength of the association of y and x. It is the ratio of the explained variation to the total variation (sum of squares regression/sum of squares regression + sum of squares residual). R^2 needs to be adjusted for the number of independent variables (k) and the sample size (n).

In the regression analysis procedure, we inspect (Hair et al. 2010) the p-value of the F-statistic for significance, (Levin 2011) the level of significance of the 't' statistic of the coefficients 'bi', (Sundararajan 2023) the coefficient of determination, (Sundararajan et al. 2019) the regression model, (Sundararajan et al. 2019) the autocorrelation of the observations for acceptable limits, (Zaki and Meira 2014) skewness/kurtosis of the dependent variable for acceptable limits, [7] the variance inflation factor (VIF) of each variable for acceptable limits, [8] the residual plot for homoscedasticity, [9] the partial regression plots of the independent variables with the dependent variable for homoscedasticity, and [10] the relative importance of the independent variables.

There are many methods for data standardization: min–max, z-score, decimal scaling, etc. Z-score standardization is popularly used in statistical data mining.

Multicollinearity arises when intercorrelations among the predictors are very high. The Variance Inflation Factor (VIF) is a measure of multicollinearity. VIF is directly related to the tolerance value (VIF = 1/Tolerance). A value around <5 is acceptable in social science research. Multicollinearity gives rise to many challenges such as skewing of the estimates and skewing of the relative importance of the independent variables. One of the solutions is principal components analysis to transform highly correlated variables into fewer components.

Autocorrelation is the similarity over a sequence of observations or the observations in a time series. It can lead to underestimates of the standard error and result in the misjudgment of the significance of the predictors. The Durbin-Watson test statistic provides an assessment of autocorrelation. A value around '2' is acceptable.

Homoscedasticity assumes that the variance of the error term is constant. It is an important assumption in many multivariate techniques.

The partial correlation coefficient is a measure of the strength of the association of a single independent variable with the dependent variable when the effects of the other independent variables in the model are held constant. It measures the incremental predictive effect of an independent variable.

Part correlation is a measure of the strength of the association of a single independent variable with the dependent variable when the effect of the other independent variables in the regression model is removed.

Akaike information criterion (AIC) is an indicator of model fit. It is used for comparing models. According to this criterion, the model that explains the highest amount of variation using minimum possible independent variables is the best fit model. The lower the AIC, the better the model.

Stepwise regression is a method for discarding independent variables that are not significant. The independent variables are entered or removed from the regression equation one at a time, based on the effect on R^2. The partial correlation coefficient is a good measure for variable selection. In the backward regression method, we start with all the independent variables in the model and remove the variables from the regression equation, one at a time. In the forward method, we start adding independent variables to the regression equation, one at a time. Multicollinearity is a major challenge in variable selection.

We may need to rank and select fewer variables based on criteria such as the partial regression coefficient or p-value associated with the t statistic, etc. The features can be handpicked if there are only a few of them. However, if there are numerous features, we may automate the process.

If the relationship between the predictors and dependent variable is not linear, we may add additional 'interaction' terms to build an effective regression model. Though the underlying relationship is polynomial, it may be thus modeled as a linear model.

Questions

Comprehension

1. State how correlation and covariance are related?
2. Provide applications for multiple linear regression.
3. Explain polynomial regression with an example.
4. Define standardized regression equation and its application.
5. Define the coefficient of multiple regression.
6. Compare the partial correlation coefficient with the partial regression coefficient.
7. State and explain the hypothesis in testing the overall significance of the multiple regression model.
8. Explain the objective in examining residuals.
9. Write a note on multicollinearity. How does multicollinearity affect prediction accuracy?
10. Write a brief note on the approach to assess the relative importance of predictors in multiple regression. Mention the approaches one may take in the regression analysis of a model with several variables.
11. Explain a linear regression model in mathematical terms.
12. Mention the assumptions underlying linear regression.
13. Write a brief note on data validation in Regression Analysis.
14. Write a note briefly mentioning each of the 10 steps in the multiple regression analysis procedure.
15. Describe stepwise regression analysis.

Application

16. You are working on a real estate dataset to predict house prices. What independent variables (features) would you consider including in a multiple linear regression model, and why?
17. You are analyzing a dataset of stock prices over time. How would you address the issue of autocorrelation, and why is it important in this context?
18. Given a dataset with several highly correlated independent variables, describe the steps you would take to mitigate multicollinearity before building a multiple linear regression model.

Exercises

Exercise 5.1 Correlation Plot
Determine the strength and significance of the association between the following variables. Also, plot the joint distribution.

project_plan = [5, 2, 4, 4, 3, 5, 2, 2, 4, 2, 4, 4, 4, 4, 3]
project_outcome = [3, 2, 5, 3, 2, 6, 1, 2, 5, 2, 3, 4, 3, 3, 4]

Exercise 5.2 Regression Plot
Compute the coefficient of determination for the following variables, and the significance of their association. Also plot the joint distribution.

independent variable: project_plan = [5, 2, 4, 4, 3, 5, 2, 2, 4, 2, 4, 4, 4, 4, 3]
dependent variable: project_outcome = [3, 2, 5, 3, 2, 6, 1, 2, 5, 2, 3, 4, 3, 3, 4]

Exercise 5.3 Regression Analysis
Do a regression analysis on the 'mpg' dataset to determine the factors that determine mileage. The dataset can be accessed from seaborn library.

Exercise 5.4 Multicollinearity
Consider the regression analysis of the 'mpg' dataset done in Exercise 5.3. Comment on the multicollinearity of the independent variables involved in regression.

Exercise 5.5 Relative importance of the predictor variables
Consider the regression analysis of the 'mpg' dataset done in Exercise 5.3. Comment on the relative importance of the variables.

Exercise 5.6 Residual Plot
Consider the regression analysis of the 'mpg' dataset done in Exercise 5.3. Plot the 'residual plot' and interpret the plot.

Exercise 5.7 Partial correlation plots
Consider the regression analysis of the 'mpg' dataset done in Exercise 5.3. Plot the 'partial correlation plots' and interpret the plots.

Exercise 5.8 Rank and Select a specific number of parameters automatically
Consider the dataset 'mpg'. Show a method for selecting a specific number of independent variables that determine the mileage.

Exercise 5.9 Identify the optimum number of parameters automatically
Consider the dataset 'mpg'. Show a method for selecting a specific number of independent variables that determine the mileage.

Exercise 5.10 Perform regression analysis to predict diamond prices
See Sect. 1.6 for a brief description of the diamonds dataset. Build a regression model using the 'diamonds' dataset to predict diamond prices. Perform detailed regression analysis.

Exercise 5.11 Regression Analysis of MPG

In a study of factors influencing the speed ('mpg') of old model cabs, the following factors were identified as significant predictors: horsepower and weight. The results of the regression analysis are shown below. Interpret the results.

```
OLS Regression Results
==============================================================================
Dep. Variable:                    mpg   R-squared:                       0.706
Model:                            OLS   Adj. R-squared:                  0.705
Method:                 Least Squares   F-statistic:                     467.9
Date:                Fri, 01 Oct 2021   Prob (F-statistic):           3.06e-104
Time:                        17:44:24   Log-Likelihood:                 -1121.0
No. Observations:                 392   AIC:                             2248.
Df Residuals:                     389   BIC:                             2260.
Df Model:                           2
Covariance Type:            nonrobust
==============================================================================
                 coef    std err          t      P>|t|      [0.025      0.975]
------------------------------------------------------------------------------
horsepower    -0.0473      0.011     -4.267      0.000      -0.069      -0.026
weight        -0.0058      0.001    -11.535      0.000      -0.007      -0.005
intercept     45.6402      0.793     57.540      0.000      44.081      47.200
==============================================================================
Omnibus:                       35.336   Durbin-Watson:                   0.858
Prob(Omnibus):                  0.000   Jarque-Bera (JB):               45.973
Skew:                           0.683   Prob(JB):                     1.04e-10
Kurtosis:                       3.974   Cond. No.                     1.15e+04
==============================================================================
```

References

Hair JF, Black WC, Babin BJ, Anderson RE (2010) Multivariate data analysis. Vectors. https://doi.org/10.1016/j.ijpharm.2011.02.019

Levin R (2011) Statistics for management. The Statistician (1). https://doi.org/10.2307/2348398

Malhotra NK (2020) Marketing research an applied prientation, 7th edn. Pearson Education

Sundararajan S (2023) MVA-ML. https://github.com/sun-sri/MVA-ML

Sundararajan S, Marath B, K. Vijayaraghavan P (2019) Variation of risk profile across software life cycle in IS outsourcing. Softw Qual J 27(4). https://doi.org/10.1007/s11219-019-09451-8

Zaki MJ, Meira W (2014) Data mining and analysis: fundamental concepts and algorithms. Cambridge University Press

Chapter 6
Classification

Learning Objectives

- Understand the basic concepts and principles of classification.
- Demonstrate primary classification methods such as logistic regression, linear discriminant analysis (LDA), decision trees, and support vector machines, and demonstrate their usage through Python-based programs.
- Acquire familiarity with other classification methods—Bayes classification methods, Bayesian belief network, rule-based classification, K nearest neighbors, backpropagation, and genetic algorithms.

Overview

The chapter begins with a discussion of the principles and concepts of classification—an overview of popular classification methods, metrics, and model performance. We then explore various classification methods such as logistic regression, linear discriminant analysis (LDA), decision trees, and support vector machines. We will demonstrate these methods through Python-based tutorials. We end the chapter with a cursory glance at other classification methods—Bayes classification, Bayesian belief network, rule-based classification, K nearest neighbors, backpropagation, and genetic algorithms. Classification is further discussed in the Chapters 'Machine Learning' and 'Artificial Intelligence and Deep Neural Networks'. The learnings in this chapter give a solid foundation for those discussions.

Supplementary Information The online version contains supplementary material available at https://doi.org/10.1007/978-981-99-0353-5_6.

Definitions

Backpropagation: A neural network learns by adjusting the weights to optimize its ability to predict the class labels correctly.

Bayesian belief network: Multi-collinearity of features is a major challenge in multivariate analysis. Bayesian belief networks are probabilistic graphical models that represent dependencies among subsets of features and are modeled on joint conditional probability distributions.

Classification: Classification is an extensively used statistical technique in data mining. It is used for the prediction of categorical class labels.

Decision tree: A decision tree is a collection of nodes arranged as a binary tree. The interior nodes evaluate a logical condition for classification. The leaf nodes carry the actual data.

F1-score: The common performance measures for a classifier include—precision, recall, and F1-score. The F1-score or F score is the harmonic mean of precision and recall. F1 Score $= 2PR / (P+R)$.

Genetic algorithms: Genetic algorithms lend themselves well to parallelization. They find applications in classification and diverse optimization problems. In data mining, they can additionally be utilized to appraise the efficacy of alternative algorithms.

Gini index: In practice, we come across data with numerous feature variables, and there is a need to identify, rank, and select the vital ones. Information Gain and Gini index are popular measures for selecting the best features.

Kernel functions: Support vector machines (SVM) use nonlinear mapping to transform the original training data into a higher dimension to make classification feasible. The data transformation functions used by SVM are called Kernel functions, which include a variety of functions such as polynomial, radial basis function (RBF), and sigmoid.

KNN (K nearest neighbors): The nearest neighbor maintains a set of training data points described by n features or dimensions. Given an unknown tuple, a K nearest neighbor classifier searches for K nearest training data points. Nearness is measured using a distance metric, such as Euclidean distance.

Linear discriminant analysis: Linear discriminant analysis (LDA) is a technique for analyzing data when the target variable is categorical and the features are numeric. If there are c target classes, we need a maximum of c-1 discriminant functions using the feature variables, which will best discriminate between the target categories.

Logistic regression: Binary logistic regression uses a set of numeric features for a two-way classification. This can be extended for multi-class classification; if we have c classes, we need a maximum of c-1 classifiers.

Naive Bayes classification: Bayesian classifiers predict class membership probabilities. Bayesian classifiers assume that the effect of a feature on classification is independent of the other features, which helps to simplify the computations. This assumption is called class conditional independence, leading to the 'Naïve' Bayes classification method.

Overfitting: A decision tree may grow into too many branches. This may be attributable to noise or outliers, among other reasons. Such overfitting causes poor accuracy for unseen samples. Moreover, the process may be computationally prohibitive for big datasets. There are two approaches to avoid overfitting. In pre-pruning, we halt the brach construction early if the goodness of fit falls below a threshold. In post-pruning, we remove branches from a 'fully grown' tree.

Precision: The common performance measures for a classifier include—precision, recall, and F1-score. Precision represents the exactness—what percentage of the cases the classifier labeled as positive are positive? Precision $= TP / (TP + FP)$.

Recall: The common performance measures for a classifier include—precision, recall, and F1-score. The recall is a measure of completeness—what percentage of positive cases did the classifier label as positive? Recall $= TP / (TP + FN)$.

Rule-based classification: If the conditions in the intermediate nodes of a decision tree are written as {if … then … else} rules, it is called rule-based classification. One rule is created for each path from the root to a leaf. Rules must be mutually exclusive and collectively exhaustive.

Stepwise discriminant analysis: In stepwise discriminant analysis, the features are entered (or removed) one after the other based on their ability to discriminate between the groups.

Support vector machines: Support vector machines (SVM) employ a nonlinear mapping to convert the initial training data into a higher dimensional space. A hyperplane can always be identified by employing a suitable mapping to an adequately high dimensionality to effectively separate the data from two classes. SVM accomplishes the determination of this hyperplane by utilizing support vectors and margins. Support vectors, denoting the data points close to the decision boundary, play a crucial role in defining the margins.

Tree pruning: See overfitting.

6.1 Introduction

Classification is an extensively used statistical technique in data mining. It is used for the prediction of categorical class labels. A regression model uses continuous-valued functions for numeric prediction—e.g., sales. A classification model predicts categorical (discrete or unordered) class labels—e.g., credit/loan approval, fraud detection, and medical diagnosis. See Fig. 6.1.

There are various classification methods and researchers, and newer methods are emerging for classifying big data and network graphs. Some of the popular methods are listed below.

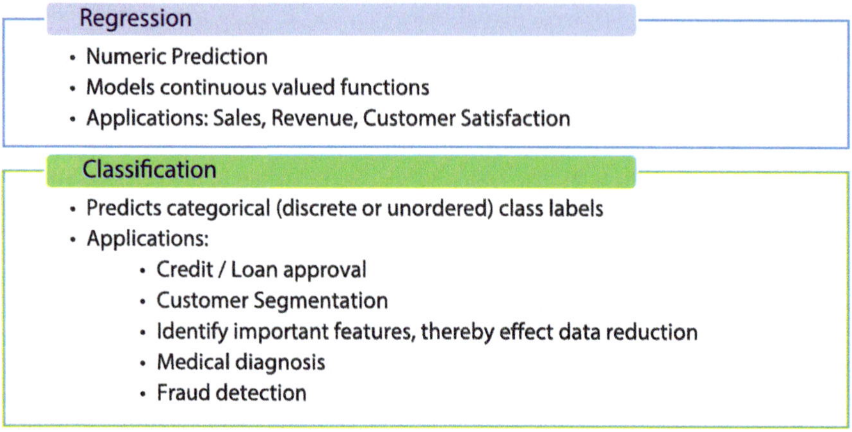

> **Regression**
> - Numeric Prediction
> - Models continuous valued functions
> - Applications: Sales, Revenue, Customer Satisfaction

> **Classification**
> - Predicts categorical (discrete or unordered) class labels
> - Applications:
>   - Credit / Loan approval
>   - Customer Segmentation
>   - Identify important features, thereby effect data reduction
>   - Medical diagnosis
>   - Fraud detection

Fig. 6.1 Numeric prediction and classification

- Binary Logistic Regression/Two-Way Classification.
- LDA—Linear Discriminant Analysis.
- Bayes Classification Methods.
- Decision Tree Induction.
- SVM—Support Vector Machine.
- Rule-Based Classification.
- Bayesian Belief Network.
- KNN (K Nearest Neighbors).
- Backpropagation (Neural networks).
- Genetic Algorithms.

Let us consider a simple classification example. The iris dataset consists of measurements on 150 iris flowers belonging to three categories—setosa, versi-color, and virginica. Four features are measured—length and width of sepals and petals. Based on the features, can we classify a flower into the correct category? We will explore that using various classification methods. A visual description of the data is given in Fig. 6.2, and the corresponding Python code is provided below.

Tutorial 6.1   The Iris Flowers Data Description

```
import pandas as pd
import seaborn as sb
pdf = sb.load_dataset('iris')
import matplotlib.pyplot as plt

pdf.info()
sb.set(font_scale = 1.5)

# probability density function of all variables, super-imposed
pdf.plot(kind='kde')
```

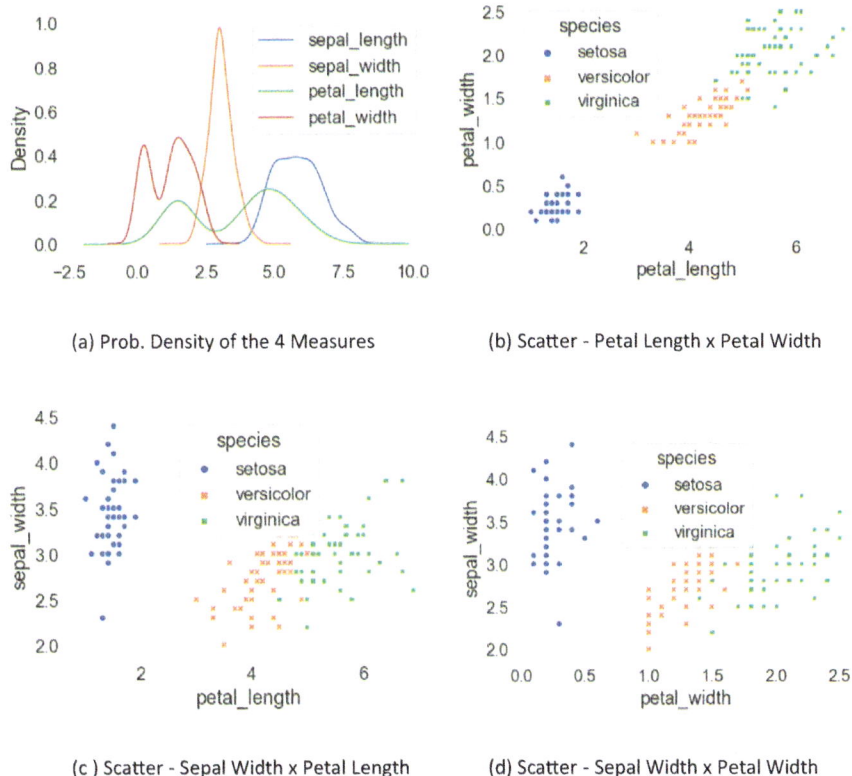

(a) Prob. Density of the 4 Measures   (b) Scatter - Petal Length x Petal Width

(c) Scatter - Sepal Width x Petal Length   (d) Scatter - Sepal Width x Petal Width

Fig. 6.2   The Iris flowers data description

```
sb.scatterplot(x='petal_length', y='petal_width', data=pdf, hue='species')
sb.scatterplot(x='petal_length', y='sepal_width', data=pdf, hue='species')
sb.scatterplot(x='petal_width',  y='sepal_width', data=pdf, hue='species')
```

6.2   Binary Logistic Regression

Binary logistic regression is a classification method that uses a set of numeric features for a two-way classification. This can be extended for use in multi-class classification.

Assume a binary classification problem with the target $y \in \{0, 1\}$. Assume k features (or variables or predictors) $X_1 \dots X_k$, and associated coefficients $b_1 \dots b_k$

$$z = b_0 + b_1 X_1 + b_2 X_2 + \cdots + b_k X_k$$

In the equation above, 'z' can take any real value. To use 'z' for binary classification, we may map the value of 'z' to $\{0,1\}$ by *inverse* logarithmic transformation.

$$P(y = 1) = \frac{1}{1 + e^{-z}}$$

where

$P(y=1)$ is the probability that the input belongs to class '1',

z is the score or logit.

More specifically, *inverse logit* or sigmoid function can be expressed as follows:

$$P(y = 1 | X_i, b_i) = \frac{1}{1 + e^{-z}}$$

We make use of the sigmoid transformation for binary logistic regression. In binary logistic regression, the target $y \in \{0, 1\}$. Assume that $\widehat{y}$ is the best estimate of y. We can express $\widehat{y}$ as follows:

$$\widehat{y} = 1, \text{if } P(y = 1 | Xi, \beta i) \geq 0.5; \widehat{y} = \text{otherwise}$$

The parameters (b_i) of the logit model are estimated using the method of maximum likelihood (Note that the b_i's are computed from the data sample and they are expected to be the best estimates of the population parameters β_i's). Logit, Sigmoid (or inverse logit), and Binary logistic regression method are diagrammatically depicted in Fig. 6.3.

Logistic regression is a binary classification algorithm that models the probability of an observation belonging to one of two classes. This can be extended for multi-class classification. Let us take a look at two popular approaches. In the One-vs-All (OvA) or One-vs-Rest (OvR) approach, we create a separate binary logistic regression model for each class in the dataset. For each model, one class is treated as the 'positive' class, and all other classes are grouped as the 'negative' class. During prediction, we apply all the models to an input. The class associated with the model that gives the highest probability is chosen as the predicted class.

In softmax regression (multinomial logistic regression), a single logistic regression model is used—but modified to handle multiple classes directly. Instead of modeling a binary outcome (0 or 1), the model assigns a probability to each class and then normalizes these probabilities using the softmax function. The softmax function ensures that the probabilities sum up to 1, and the class with the highest probability is selected as the prediction. This approach is suitable for problems with a larger number of classes. This method is extensively used in machine learning. Softmax function returns a vector of probabilities of an object belonging to each class as expressed by

$$P(y = i) = \frac{e^{-zi}}{\sum_{j=1}^{k} e^{zj}}$$

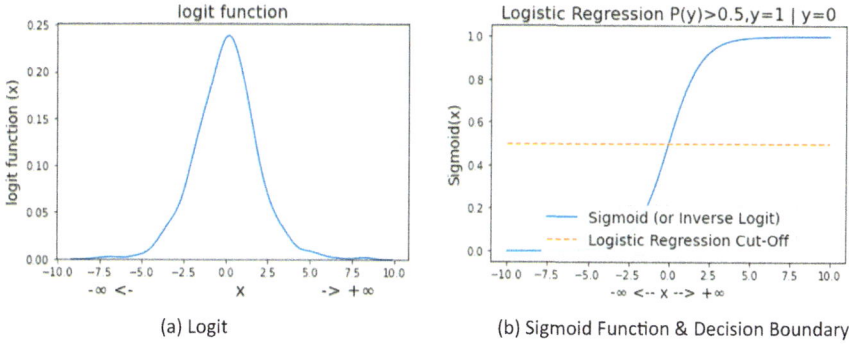

Fig. 6.3 Logit function and inverse logit (or sigmoid) function

where
P(y=i) is the probability that the input belongs to class i,
zi is the score or logit associated with class i,
k is the total number of classes.

6.3 Model Performance

A classifier must be able to predict the class label of unknown objects precisely. There are several measures to rate the performance of a classifier. Table 6.1 shows some commonly used measures. Tutorial 6.2 demonstrates an example.

- Precision: Precision represents the exactness—what percentage of the cases the classifier labeled as positive are actually positive? Precision = TP / (TP+FP).
- Recall: Recall is a measure of completeness—what percentage of positive cases did the classifier label as positive? Recall = TP / (TP+FN).
- F1-score: The F1-score or F score is the harmonic mean of precision and recall. F1 Score = 2PR / (P+R) (Table 6.1).

A trained classifier must be robust and accurate in predicting unknown objects. The following methods assess and improve a model's prediction accuracy—cross-validation and ensemble. Cross-validation aims to assess the performance and generalisability of a single machine learning model—to foresee how well a model may perform on unseen data and identify potential issues like overfitting or underfitting. It involves splitting the training dataset into multiple subsets or 'folds' to train and test the model iteratively. Ensemble methods combine the predictions of multiple base machine learning models to create a more robust and accurate predictive model. A detailed discussion on this topic is available in Sect. 11.6.5, Ensemble Methods (Table 6.2).

Table 6.1 Performance measurement in classification models

		Predicted Labels		
		1	0	
Observed Labels	1	True Positives (TP)	False Negatives (FN)	R (**Recall**)$=$ TP / (TP+FN) True Positive Rate
	0	False Positives (FP)	True Negatives (TN)	**F1 Score** $=$2PR / (P+R)
		P (**Precision**)$=$ TP / (TP+FP)	**Accuracy**$=$(TP+TN)/ (TP+TN+FP+FN)	Support$=$(TP+TN+FP+FN)

Table 6.2 Calculation of precision and recall

	Predicted class		Actual Class	Recall (R)$=$
	Versicolor	Not Versicolor		
Versicolor	47	3	50	47 / 50$=$0.94
Not Versicolor	0	50		(Row wise Ratio)
Total	47			
Precision (P)$=$	47 / 47$=$1	(Column-wise ratio)		

Tutorial 6.3.1 Binary Logistic Regression

In this tutorial, we are using the binary logistic regression method for classifying iris flowers into two categories - versicolor and virginica

```
import pandas as pd
import seaborn as sb
from sklearn.linear_model import LogisticRegression
from sklearn.metrics import classification_report, confusion_matrix

d = sb.load_dataset('iris')
d.info()

d.columns
# 'sepal_length', 'sepal_width', 'petal_length', 'petal_width', 'species'

d.species.unique() # 'setosa', 'versicolor', 'virginica'
```

Select data pertaining to two classes - versicolor, virginica
```
d2class = d[(d.species=='versicolor') | (d.species=='virginica')]
```

Load X with the four features
```
X = d2class[['sepal_length', 'sepal_width', 'petal_length', 'petal_width']]
```

Load y with the target labels - 'species'
```
y = d2class.species
```

Setup the classifier
```
model = LogisticRegression(solver='liblinear', random_state=0)
```

Train the model
```
model.fit(X, y)

model.classes_       # ['versicolor', 'virginica']

model.coef_.round(2) # [-1.71, -1.53,  2.47,  2.56]
```
[-1.71, -1.53, 2.47, 2.56] are the coefficients (b1, b2, b3, b4) corresponding to the four features x1,x2,x3,x4

```
model.intercept_.round(3) # [-1.216] # intercept or b0

confusion_matrix(y, model.predict(X))
```
array([[47,   3], # 47 versicolor flowers classified correctly
       [0, 50]   # 3 versicolor flowers classified wrongly

F Score = 2*P*R / (P + R) = 2*1* 0.94 / (1 + 0.94) = 0.97

```
print(classification_report(y, model.predict(X)))
```

	precision	recall	f1-score	support
versicolor	1.00	0.94	0.97	50
virginica	0.94	1.00	0.97	50
accuracy (overall)			0.97	100
macro avg (overall)	0.97	0.97	0.97	100
weighted avg	0.97	0.97	0.97	100

Weightage is given based on support (sample size). Here 50 instances each of versicolor and virginica are selected; so, the weights are equal. This is reflected in the 'weighted avg'.

Tutorial 6.3.2 Logistic Regression - Multi-class classification

```
import pandas as pd
import seaborn as sb
from sklearn.linear_model import LogisticRegression
from sklearn.metrics import classification_report, confusion_matrix

d = sb.load_dataset('iris')
d.info()
d.columns
d.species.unique() # 'setosa', 'versicolor', 'virginica'

X = d[['sepal_length', 'sepal_width', 'petal_length', 'petal_width']]
y = d.species

model = LogisticRegression(solver='liblinear', random_state=0)
model.fit(X, y)
```

```
model.classes_         # ('setosa', 'versicolor', 'virginica')
model.coef_.round(2)
[[ 0.41,  1.46, -2.26, -1.02],   Logistic discriminant function-1 coeff
 [ 0.43, -1.61,  0.58, -1.41],   Logistic discriminant function-2 coeff
 [-1.71, -1.53,  2.47,  2.56]])  Logistic discriminant function-3 coeff

model.intercept_.round(3) # [ 0.264,  1.094, -1.215]
The intercepts corresponding to the three logistic discriminant functions

confusion_matrix(y, model.predict(X))
array([[50,  0,  0],
       [ 0, 45,  5],
       [ 0,  1, 49]])

print(classification_report(y, model.predict(X)))
              precision    recall  f1-score   support
      setosa       1.00      1.00      1.00        50
  versicolor       0.98      0.90      0.94        50
   virginica       0.91      0.98      0.94        50

    accuracy                          0.96       150
```

6.4   Linear Discriminant Analysis (LDA)

Discriminant analysis is a technique for analyzing data when the target variable is categorical and the features are numeric. If there are c target classes, we need a maximum of c-1 discriminant functions to discriminate between the target categories (Malhotra 2020). Assume that we have k feature variables (X_i ... X_k). The discriminant analysis model can be expressed as follows:

$$D_i = b_0 + b_1X_1 + b_2X_2 + ... + b_kX_k$$

where

D_i is the discriminant score for ith discriminant function,

b_i's are the discriminant coefficients or weights in the ith discriminant function,

X_i's are the ith feature variable.

The coefficients, or weights (b), are estimated so that the groups differ as much as possible. This occurs when the ratio of the between-group sum of squares to the within-group sum of squares for the discriminant function is maximum.

We may include all the features in the discriminant function. Alternatively, we may include only vital features. This can be achieved through stepwise discriminant analysis, which is analogous to stepwise multiple regression. In this method, the features are entered (or removed) one after the other based on their ability to discriminate between the groups. We analyze variance for each feature, taken one at a time, considering the target variable as the category or grouping variable. The discriminant function includes the feature with the highest F ratio, given that it is statistically significant. Following this, the next feature is added based on the highest adjusted F ratio, considering all the features, including those previously selected. This iterative process persists until all feature variables are encompassed.

Tutorial 6.4.1 Linear Discriminant Analysis

In this tutorial, we are using the linear discriminant analysis method for classifying iris flowers

```
import pandas as pd
import seaborn as sb
from sklearn.discriminant_analysis import LinearDiscriminantAnalysis
from sklearn.metrics import classification_report, confusion_matrix

d = sb.load_dataset('iris')
d.info()

X = d[['sepal_length', 'sepal_width', 'petal_length', 'petal_width']]
y = d.species
```

Setup LDA classifier
```
model = LinearDiscriminantAnalysis(solver='svd')
```
Notes on Solver
'svd':   Singular value decomposition (default).
         SVD does not compute the covariance matrix. So, this solver is
         recommended for data with a large number of features.
'lsqr': Least squares solution.
         Can be combined with shrinkage or a custom covariance estimator
'eigen': Eigenvalue decomposition.
         It can be combined with shrinkage or a custom covariance estimator.

Train the LDA classifier
```
model.fit(X, y)

model.classes_   # ['setosa', 'versicolor', 'virginica']
```

Coefficients of the Discriminant Function 1
```
model.coef_.round(2)
array([[  6.31,  12.14, -16.95, -20.77],
       [ -1.53,  -4.38,   4.7 ,   3.06],   # Discriminant function1 Coeff
       [ -4.78,  -7.76,  12.25,  17.71]])  # Discriminant function1 Coeff
```

The intercepts for the three discriminant functions
```
model.intercept_.round(3)
# [-15.47, -2.02, -33.53]

confusion_matrix(y, model.predict(X))
array([[50,  0,  0],    all setosa flowers were correctly classified
       [ 0, 48,  2],    two versicolor flowers were wrongly classified
       [ 0,  1, 49]])   one virginica  flower is wrongly classified
```

```
print(classification_report(y, model.predict(X)))
              precision    recall  f1-score   support
      setosa       1.00      1.00      1.00        50
  versicolor       0.98      0.96      0.97        50
   virginica       0.96      0.98      0.97        50
    accuracy                           0.98       150
```

6.5   Decision Trees

The decision tree method uses a tree-based classification approach using categorical or metric attributes for node split. We will discuss a simple algorithm that generates a binary tree using metric attributes.

6.5.1   The Basic Decision Tree Algorithm

A binary decision tree is a collection of nodes arranged as a binary tree. The tree has a root node. From every node, two branches emerge. The terminal nodes are called leaves, all other nodes are called interior nodes. The interior nodes evaluate a logical condition for classification. If the condition is true, we traverse to the left branch; otherwise to the right branch. Tree traversal ends in a leaf node. The leaf nodes carry actual data. See Fig. 6.4.

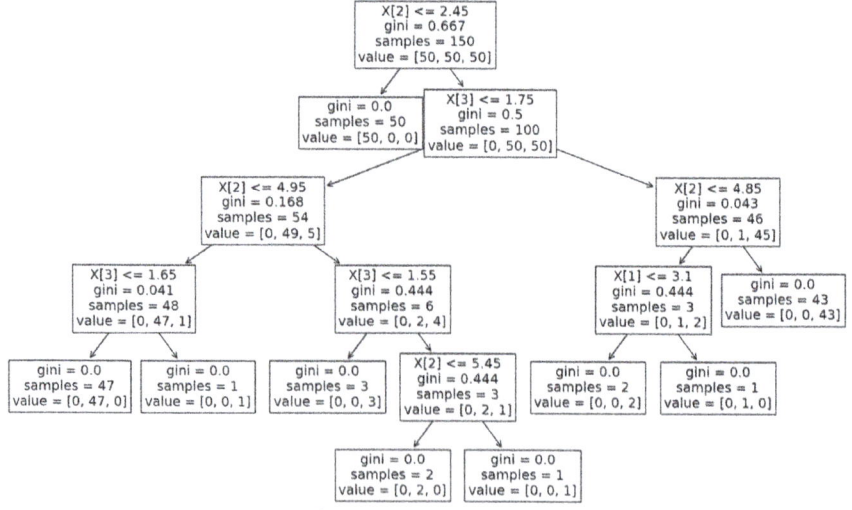

Fig. 6.4 Complete decision tree diagram

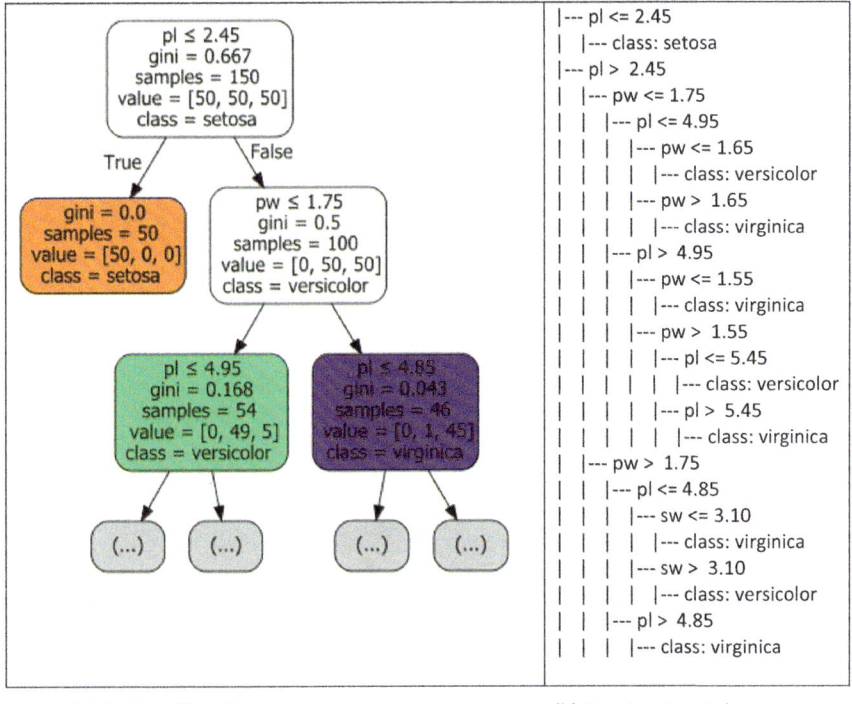

(a) Decision Tree Diagram   (b) Decision Tree Rules

Fig. 6.5 Decision tree diagram and corresponding rules

We aim to classify iris flowers into three labels—setosa, versicolor, and virginica, using four features—length and width of sepals and petals. The root node evaluates the condition—is the petal length $< = 2.45$ cm? If the condition is true, we traverse to the left branch; otherwise to the right branch. The leaf node carries a subset of data and the associated target label (setosa, versicolor, and virginica).

If all the data in the leaf belong to one category label, then we will call the leaf a pure node. The purity of leaf nodes increases with depth. If the classifier is 100% accurate, the full-blown tree's leaves will all be pure. Conditions for terminating the tree build are as follows:

- All samples in all leaf nodes are pure (belong to the same class).
- Split a node only if it results in a sufficient reduction in impurity.
- Set a minimum number of samples to create a leaf node.
- There are no samples left.

However, exhaustive tree construction may lead to overfitting. Moreover, the process may be computationally prohibitive for big datasets. Therefore, the tree needs to be pruned. This is discussed later (Fig. 6.5).

6.5.2 *Feature Selection*

The classifier accuracy is determined by the constituent features—how comprehensively the given set of features defines the phenomena at hand and whether the features are uncorrelated. In practice, we come across data with numerous feature variables, and there exists a need for identifying, ranking, and selecting the vital ones. Assume a dataset D, of class-labeled data objects. To build a decision tree, we must split D into subsets based on a split criterion. The ultimate objective is to get pure partitions (all the data in a subset belong to one category label). Information Gain and the Gini index are some of the popular measures to select the best feature for the next node split. The Gini index is costly when the number of classes is large. Those not particular about mathematical insights may skip the rest of this sub-section and proceed to 'Overfitting and Tree Pruning'.

Information Gain

Consider a classification problem where each object is characterized by a set of features {A ... K}. Assume that there are 'm' distinct classes. C_i (i = 1, ..., m). Let D be a dataset of labeled objects to be classified.

The probability 'pi' that an object in D belongs to class Ci.

= Number of objects of Class-i in D / Number of objects in D

$$p_i = \frac{|C_i, D|}{|D|}$$

The following equation (Jiawei Han, Micheline Kamber, 2014) gives the information needed to classify an object. Info (D) is also called Entropy.

$$\text{Info(D)} = \sum_{i=1}^{m} p_i \log_2(p_i)$$

Assume we used feature 'A' to partition D into v partitions. Information needed for further classification can be expressed as follows:

$$\text{Info}_A(D) = \sum_{j=1}^{v} \frac{D_j}{D} \text{Info}(D_j)$$

Based on the above equation, information gained from using feature 'A' for classification can be expressed as follows:

$$\text{Gain(A)} = \text{Info(D)} - \text{Info}_A(D)$$

From the given set of features {A ... K}, select the feature that lends the maximum Information Gain. Use the selected feature for classification. Repeat the process with the rest of the features until optimal classification is achieved.

Gini Index

Consider a classification problem where each object is characterized by a set of features {A ... K}. Assume that there are 'm' distinct classes. C_i (i = 1, ..., m). Let D be a dataset of labeled objects to be classified.

The probability 'p_i' that an object in D belongs to class Ci.
= Number of objects of Class-i in D / Number of objects in D

$$p_i = \frac{|C_i, D|}{|D|}$$

The Gini index measures the impurity of D as expressed below (Jiawei Han, Micheline Kamber, 2014):

$$ini(D) = 1 - \sum_{i=1}^{m} (p_i)2$$

Consider that feature A splits D, the set of objects, into two partitions D1 and D2. After the split using feature A, the Gini index of D can be expressed as follows:

$$Gini_A(D) = \frac{|D_1|}{|D|} Gini(D_1) + \frac{|D_2|}{|D|} Gini(D_2)$$

After the partitioning of D using feature A, the reduction in impurity can be expressed as follows:

$$\Delta Gini(A) = Gini(D) - Gini_A(D)$$

From the given set of features {A ... K}, select the feature that provides the largest reduction in impurity. Use the selected feature for classification. Repeat the process with the rest of the features until optimal classification is achieved.

6.5.3 Overfitting and Tree Pruning

Overfitting: A decision tree may grow into too many branches. This may be attributable to noise or outliers, among other reasons. Such overfitting causes poor accuracy for unseen samples. Moreover, exhaustive tree construction or tree access is computationally prohibitive for big datasets. This necessitates tree pruning, which can be done during or after tree construction. Two approaches to avoid overfitting are pre-pruning and post-pruning.

Pre-pruning: Halt tree construction early. Do not split a node if this would result in the goodness of fit measure falling below a threshold. The challenge here is to determine an appropriate threshold.

Post-pruning: Remove branches from a 'fully grown' tree and get a progressively pruned tree. We may set apart some data from the training process and use it for testing and identifying the best-pruned tree.

6.5.4 Various Decision Tree Algorithms

There are several decision tree algorithms used in machine learning and data mining. Some of them are mentioned below.

ID3 is one of the earliest decision tree algorithms. It uses information gain to select the best attributes for splitting a node. However, it prefers attributes with many values and uses only categorical attributes for node split. The algorithm 'C4.5' is an improved version of ID3 that uses gain ratio as a criterion for attribute selection. This algorithm can also use continuously valued attributes and handle both classification and regression tasks.

CART (Classification and Regression Trees) is another widely used decision tree algorithm. It constructs binary trees using the Gini impurity for classification and mean squared error (MSE) for regression as the node splitting criteria. CART is known for its flexibility and ability to handle various data types.

CHAID (Chi-squared Automatic Interaction Detector) is primarily used for categorical data and uses chi-squared tests to find significant associations between attributes and the target variable. It can handle both classification and regression tasks.

Random Forrest is an ensemble method (discussed in Sect. 11.6.5) that combines multiple decision trees to improve predictive accuracy and reduce overfitting. XGBoost (Extreme Gradient Boosting) and LightGBM are some of the other gradient-boosting frameworks that use decision trees as base learners. They are optimized for speed and are scalable (work on large datasets efficiently). CatBoost: CatBoost is a gradient-boosting algorithm focusing on categorical feature handling.

Tutorial 6.5   Decision Tree Classification

In this tutorial, we are using the decision tree method for classifying iris flowers

Tutorial 6.5.1 Decision Tree Classification Report

```
import pandas as pd
import seaborn as sb
from sklearn.tree import DecisionTreeClassifier
from sklearn.metrics import classification_report, confusion_matrix

d = sb.load_dataset('iris')
d.info()
```

```python
X = d[['sepal_length', 'sepal_width', 'petal_length', 'petal_width']]
y = d.species

model = DecisionTreeClassifier()
model.fit(X, y)

model.classes_        # ['setosa', 'versicolor', 'virginica']

confusion_matrix(y, model.predict(X))
array([[50,  0,  0],
       [ 0, 50,  0],
       [ 0,  0, 50]])

print(classification_report(y, model.predict(X)))
              precision    recall  f1-score   support
      setosa       1.00      1.00      1.00        50
  versicolor       1.00      1.00      1.00        50
   virginica       1.00      1.00      1.00        50
    accuracy                           1.00       150

model.criterion  # 'gini'index measure for node splitting

model.feature_importances_.round(3)
([0.013, 0.013, 0.051, 0.923])
'petal_width'(1), 'petal_length'(2), 'sepal_length'(3), 'sepal_width'(3)

model.score(X,y) # 1.0 (100%)
```

Tutorial 6.5.2 Decision Tree Plot (simple)

```python
import pandas as pd
import seaborn as sb
from sklearn.tree import DecisionTreeClassifier

d = sb.load_dataset('iris')

X = d[['sepal_length', 'sepal_width', 'petal_length', 'petal_width']]
y = d.species

model = DecisionTreeClassifier()
model.fit(X, y)

from sklearn import tree
import matplotlib.pyplot as plt
plt.figure(figsize=(16,10),dpi=100, edgecolor='black', facecolor='white')

tree.plot_tree(model,fontsize=14)
```
See Figure 6-4: Complete Decision Tree Diagram

Tutorial 6.5.3 Decision Tree Textual Description

```
from sklearn.tree import export_text
```

The model generated in Tutorial 6.5.2 Decision Tree Plot is used here
```
dtctext = export_text(model, feature_names=['sl', 'sw', 'pl', 'pw'])
print(dtctext)
```
See Figure 6-5 (b): Decision Tree Diagram

Tutorial 6.5.4 Decision Tree elegant Graph

Install Graphviz
You may refer to: https://scikit-learn.org/stable/modules/tree.html
And https://graphviz.org/

```
!pip install graphviz

import graphviz
```

The model generated in the earlier Tutorial on Decision Tree is used here
```
tree_data =  tree.export_graphviz (model, out_file=None, max_depth=2,
               feature_names=(['sl','sw','pl','pw']),
               class_names=(['setosa', 'versicolor', 'virginica']),
               filled=True, rounded=True,special_characters=True)
tree_graph = graphviz.Source(tree_data)
tree_graph    # See  Figure 6-5 (a): Decision Tree Diagram
tree_graph.render("iris") # Tree written out to disk as pdf
```

6.6   Support Vector Machines

Support vector machines employ nonlinear mapping to convert the initial training data into a higher dimensional space. In this augmented dimensionality, SVM
seeks an optimal linear separating hyperplane known as the 'decision boundary'.
Through an effective nonlinear mapping to a suitably high dimension, the data from
two classes can consistently be separated by a hyperplane, as noted by (Deisenroth
2020; Géron 2019). SVM identifies and defines this hyperplane using support vectors and margins. Support vectors are data points on the decision boundary, crucial in determining the optimal separation. The margins these support vectors define
represent the spatial gaps or distances between the different classes (Fig. 6.6).

In Fig. 6.6, the dotted lines represent the margins. The blue and green points
that touch the dotted line on the left and the orange points that touch the dotted
line on the right are called support vectors. The thick line at the center is the optimal hyperplane. The distance between the dotted lines is the 'margin'. As we can
imagine, infinite hyperplanes separate any two classes. SVM searches for the
hyperplane with the largest margin, i.e., maximum marginal hyperplane (MMH).

Fig. 6.6 Support vectors, margins, and Kernel trick

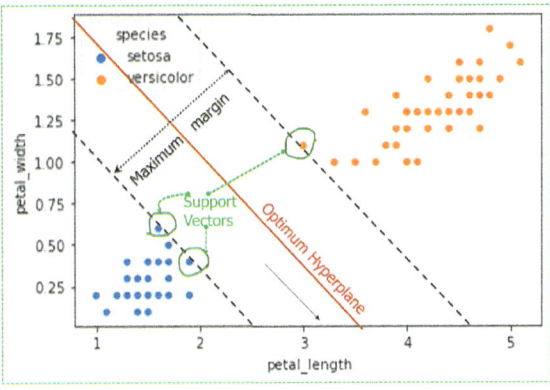

The data transformation functions used by SVM are called Kernel functions, which include a variety of functions such as polynomial, radial basis function (RBF), and sigmoid. The decision hyperplanes used for nonlinear SVMs are similar to some of the popular neural network classifiers. For example, an SVM with a Gaussian radial basis function (RBF) gives the same decision hyperplane as an RBF neural network (Jiawei Han, Micheline Kamber, 2014). An SVM with a sigmoid Kernel is equivalent to a two-layer neural network called a multilayer perceptron. Our tutorial will use the most popular Kernel function, RBF.

Transforming the data to higher dimensions can be costly if dimensions are high. The Kernel trick provides a practical solution for this. In this case, the Kernel function returns the inner product between two data points in a higher dimensional feature space.

Tutorial 6.6 Support Vector Classification

In this tutorial, we are using the support vector method for classifying iris flowers

```
import pandas as pd
import seaborn as sb
from sklearn.svm import SVC    # SVM Model Import
from sklearn.metrics import classification_report, confusion_matrix

d = sb.load_dataset('iris')

d.info()

X = d[['sepal_length', 'sepal_width', 'petal_length', 'petal_width']]
y = d.species

model = SVC(kernel = 'rbf')
model.fit(X, y)

model.classes_       # ['setosa', 'versicolor', 'virginica']
model.intercept_.round(3) # [0.126, -0.064, -0.107]
```

```
confusion_matrix(y, model.predict(X))
array([[50,  0,  0],
       [ 0, 48,  2],
       [ 0,  2, 48]])

print(classification_report(y, model.predict(X)))
              precision    recall  f1-score   support

      setosa       1.00      1.00      1.00        50
  versicolor       0.96      0.96      0.96        50
   virginica       0.96      0.96      0.96        50
    accuracy                          0.97       150
```

6.7   Other Classification Methods

This section will look at other popular classification methods such as Bayes classification, Bayesian belief network, rule-based classification, K nearest neighbors (KNN), backpropagation, and genetic algorithms.

Naive Bayes Classification

Bayesian classifiers are comparable in performance with decision trees and rule-based classification. There are many classifiers modeled based on the Bayes theorem. These classifiers predict class membership probabilities. Given training data X, the posterior probability of a hypothesis H, $P(H|X)$, follows the Bayes' theorem (Han et al. 2014)

$$P(H|E) = P(E|H)P(H)/P(E)$$

where

P(H) is Prior probability, the probability before getting the evidence,

$P(H|E)$ is Posterior probability, the probability after getting evidence.

Bayesian classifiers assume that the effect of a feature on classification is independent of the other features, which helps to simplify the computations. This assumption is called class conditional independence, leading to the 'Naïve' Bayes classification method.

Bayesian Belief Network

As we mentioned, Bayesian classifiers assume independence of the features. Multi-collinearity of features is a major challenge in multivariate analysis. Bayesian belief networks do not assume class conditional independence. Bayesian belief networks are probabilistic graphical models that represent dependencies among subsets of features and are modeled on joint conditional probability distributions.

Fig. 6.7 A Bayesian Belief
Network for CRM

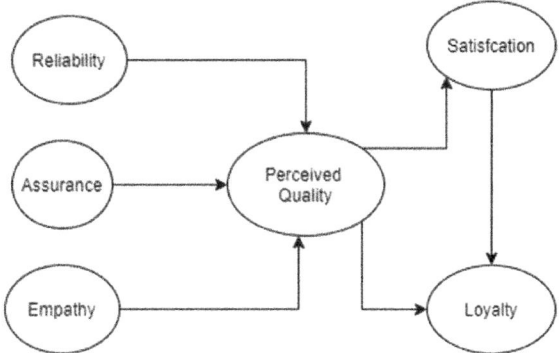

A belief network has two components—a directed acyclic graph (see Fig. 6.7) and a set of conditional probability tables. Each node represents a feature (a random variable)—discrete or continuous valued. The edges build relationships among the features.

Rule-Based Classification

If the conditions in the intermediate nodes of a decision tree are written as {if … then … else} rules, it is called rule-based classification. One rule is created for each path from the root to a leaf. Rules must be mutually exclusive and collectively exhaustive. Figure 6.5b gives an example for the rule-based classification of iris flowers. A simple way to rank the rules is to order them based on the decreasing order of misclassification. A leaf node holds the predicted data (ideally belonging to one class). A rule is assessed based on coverage and accuracy:

- Coverage: the number of tuples covered by the rule/sample size.
- Accuracy: the number of tuples correctly classified by rule/coverage.

KNN (K Nearest Neighbors)

KNN maintains a set of training data objects described by n features or dimensions. Given an unknown object, the classifier searches for the K nearest objects in the training dataset. Nearness may be measured using a distance metric, such as Euclidean distance. After examining the class labels of the nearest neighbors, the label with the highest frequency will be assigned to the new object. Initially, a small set of data objects is selected; all other data objects are classified based on the procedure mentioned above. The feature variables may be scaled to speed up the process.

During the training time, KNN stores the complete dataset. Computations (for classifying a test object) are deferred to testing time. Therefore, KNN is called a lazy learner.

The KNN algorithm can also make use of the regression method. In this method, KNN identifies the K nearest neighbors and predicts a numerical value (class label) for the new data point based on the average (or weighted average) of the target values of those neighbors.

Backpropagation (Neural Networks)

A neural network consists of multiple layers composed of neurons. Neurons take multiple inputs and generate one output using an activation function. A connection between two neurons has a weight associated with it. A neural network learns by adjusting the weights to optimize its ability to predict the class labels correctly. Backpropagation is the most popular neural network learning algorithm. See Chap. 12, 'Artificial Intelligence', for a detailed description.

Genetic Algorithms

Genetic algorithms, inspired by the principles of natural evolution, offer a powerful approach to problem-solving and optimization. In genetic learning, the process initiates with the creation of an initial population. This population comprises randomly generated rules, represented as bit patterns ranging from {0000 to 1111}. This binary coding extends to features and classes, creating a framework for systematically representing various elements.

The essence of genetic algorithms lies in their ability to create offspring by applying genetic operators, specifically crossover and mutation. In the crossover process, substrings from pairs of rules are exchanged, forming new pairs of rules. This mimics the idea of genetic recombination seen in natural evolution. On the other hand, the mutation operation involves randomly selecting bits in a rule's string and inverting them. This introduces variability and randomness into the population, simulating the mutation process observed in biological organisms.

The iterative nature of genetic algorithms involves generating a new population based on the existing rules. This process continues until a population, denoted as P, evolves, where each rule within P meets a predetermined fitness threshold. The fitness threshold represents the specific criteria or performance level that rules must achieve to be considered suitable for the given problem or task.

One notable advantage of genetic algorithms is their inherent parallelizability. This means the algorithm can efficiently leverage parallel processing capabilities, enhancing its computational efficiency and speed. Genetic algorithms find diverse applications, including classification problems and general optimization tasks. In data mining, these algorithms are not only employed for problem-solving. However, they can also serve as evaluative tools for assessing the fitness and performance of other algorithms, contributing to the broader landscape of algorithmic exploration and refinement.

Data Analytics in Action

Risk factors for diabetic peripheral sensory neuropathy (Adler et al. 1997).
Peripheral neuropathy is a case of numbness and pain, usually in the hands and feet, resulting from damage to the nerves outside the brain and spinal cord (peripheral nerves). In peripheral motor neuropathy, damage happens to the nerves that control movement. In peripheral sensory neuropathy, damage happens to the sensory nerves, the nerves that carry touch, temperature, pain, and other sensations to the brain. See Fig. 6.8.

A study on risk factors affecting diabetic peripheral sensory neuropathy is presented here. A cohort of 288 diabetic veterans who did not have neuropathy was selected for the study. Over several years, 20% developed insensitivity to the foot resulting from diabetic peripheral sensory neuropathy. The feature variables in the study included (a) factors such as gender and ethnicity, age, height, duration of diabetes, glycohemoglobin (glucose-bound hemoglobin) level, history of lower extremity ulceration, callus, edema, alcohol score, smoking, and albumin level.

Multivariate logistic regression analysis was conducted, controlling for gender and ethnicity. The result showed independent and significant associations with age, height, duration of diabetes, glycohemoglobin level, history of lower extremity ulceration, callus, and edema. The conclusions of the study are listed below:

- Poor glycaemic control increases the risk of neuropathy.
- Height and age increase the risk of neuropathy.
- Neuropathy in diabetic veterans may be worsened by alcohol ingestion.
- Early identification of risk factors and risk-prone subjects can be done. This might provide a means for interventions or treatments.

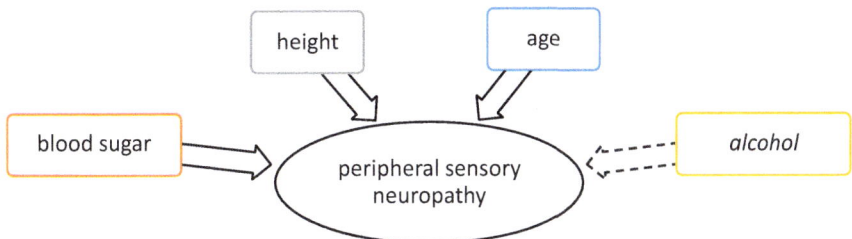

Fig. 6.8 Factors affecting peripheral sensory neuropathy

Summary

Classification is an extensively used statistical technique in data mining. It is used for the prediction of categorical class labels.

A regression model uses continuous-valued functions for numeric prediction—e.g., sales. A classification model predicts categorical (discrete or unordered) class labels—e.g., credit/loan approval, fraud detection, and medical diagnosis. The typical performance measures for a classifier include—precision, recall, and F1-score.

Binary logistic regression uses a set of numeric features for a two-way classification. This can be extended for multi-class classification; if we have c classes, we need a maximum of c-1 classifiers.

Linear discriminant analysis is a technique for analyzing data when the target variable is categorical and the features are numeric. If there are c target classes, we need a maximum of c-1 discriminant functions using the feature variables, which will best discriminate between the target categories. In stepwise discriminant analysis, the features are entered (or removed) one after the other based on their ability to discriminate between the groups.

A decision tree is a collection of nodes arranged as a binary tree. The interior nodes evaluate a logical condition for classification. The leaf nodes carry the actual data. Exhaustive tree construction or tree access is computationally prohibitive for big datasets, leading to overfitting. This necessitates tree pruning during the process of tree construction or after that. In practice, we come across data with numerous feature variables, and there is a need to identify, rank, and select the vital ones. Information Gain and Gini index are popular measures for selecting the best features. A decision tree may grow into too many branches. This may be attributable to noise or outliers, among other reasons. Such overfitting causes poor accuracy for unseen samples. There are two approaches to avoid overfitting. In pre-pruning, we halt the branch construction early if the goodness of fit falls below a threshold. In post-pruning, we remove branches from a 'fully grown' tree.

Support vector machines utilize nonlinear mapping to elevate the original training data into a higher dimensional space. When appropriately mapped to a sufficiently high dimension, it becomes possible to separate data from two classes using a hyperplane consistently. SVM accomplishes this by identifying the hyperplane through support vectors and margins. Support vectors, the data points touching the decision boundary, play a pivotal role in determining the optimal separation, and these support vectors define the margins. The functions responsible for transforming the data in SVM are called Kernel functions, encompassing various types such as polynomial, radial basis function (RBF), sigmoid, and more. Notably, the decision hyperplanes employed in nonlinear SVMs align with those used in popular neural network classifiers.

Bayesian classifiers predict class membership probabilities. Bayesian classifiers assume that the effect of a feature on classification is independent of the other features, which helps to simplify the computations. This assumption is called class conditional independence, leading to the 'Naïve' Bayes classification method. Multi-collinearity of features is a major challenge in multivariate analysis.

Bayesian belief networks are probabilistic graphical models that represent dependencies among subsets of features and are modeled on joint conditional probability distributions. If the conditions in the intermediate nodes of a decision tree are written as {if … then … else} rules, it is called rule-based classification. One rule is created for each path from the root to a leaf. Rules must be mutually exclusive and collectively exhaustive. The nearest neighbor maintains a set of training data points described by n features or dimensions. Given an unknown tuple, a K nearest neighbor classifier searches for K nearest training data points. Nearness is measured using a distance metric, such as Euclidean distance. A neural network learns by adjusting the weights to optimize its ability to predict the class labels correctly. Backpropagation is the most popular neural network algorithm. Genetic algorithms are easily parallelizable. They find applications in classification and diverse optimization problems. They can be utilized to evaluate the fitness of other algorithms.

Questions

Comprehension

1. Compare and contrast classification with regression.
2. Mention some practical applications of classification.
3. Mention some methods that are undertaken to improve prediction accuracy.
4. State and explain the relationship between logit function and the sigmoid function.
5. Write a brief note on the method to develop a multi-class classification technique using logistic regression.
6. State and explain the linear discriminant analysis model.
7. Write a brief note on feature selection in linear discriminant analysis.
8. Define information Gain, entropy, and Gini index.
9. Write a note on overfitting and tree pruning.
10. Define support vector and Kernel.
11. Write a short note on SVM.
12. Discuss various metrics for model performance.
13. State and explain binary logistic regression.
14. Write a brief note on the Decision Tree building procedure.
15. Explain how Decision Trees can be used for feature selection.

Analysis

16. Explain the concept of precision in the context of classification performance measurement. How does it relate to false positives and true positives?
17. In binary logistic regression, what is the role of the sigmoid function (inverse logit) in mapping the linear combination of features and coefficients to class probabilities? How does it impact the decision boundary?
18. Compare and contrast linear discriminant analysis with linear regression.
19. Explain the concept of discriminant functions in LDA. How are these functions used to discriminate between different target categories?

20. Describe the basic structure of a decision tree. How are nodes, leaves, and interior nodes used in decision-making?
21. What is overfitting in the context of decision trees, and why is it a concern? How can pre-pruning and post-pruning be used to avoid overfitting?
22. The following topic is not covered in the book; however, we encourage you to explore it: How does SVM use kernel functions to transform data into higher dimensions? Explain the role of support vectors and margins in finding the optimal hyperplane.
23. Compare and contrast the Naive Bayes classification approach with decision trees and rule-based classification. What are the key assumptions of Naive Bayes?
24. How does the concept of feature selection relate to classification accuracy? Explain Information Gain and the Gini index as methods for feature selection.
25. Choose a classification problem where data separation is not linearly achievable. Describe how SVM, particularly with the RBF Kernel, can be applied to solve this problem.

Exercises

The questions in this section are based on the diamonds dataset accessible from the Seaborn Library. The diamonds dataset consists of features that determine the diamond price. We will categorize the price into three categories—0, 1, and 2—low, medium, and high. Develop a classifier based on the numeric features to classify an object to one of the above price categories. Apply the various methods shown in the Tutorials. The data preprocessing step is shown below. The exercises are listed after that.

Diamonds Dataset preprocessing

```
import seaborn as sb
d  = sb.load_dataset('diamonds')
d  = d.dropna()
d.info()
```

Creating a categorical variable pCat with values 0,1,2 based on price
```
import numpy as np
import pandas as pd
d['pCat'] = np.zeros(d.shape[0])
for i in d.index:
    if   d.loc[i, 'price'] < 2000: d.loc[i, 'pCat'] = 0
    elif d.loc[i, 'price'] < 8000: d.loc[i, 'pCat'] = 1
    else: d.loc[i, 'pCat'] = 2
```

Create DataFrame X that consists of predictor variables
```
    X = d[['carat', 'depth', 'table', 'x', 'y', 'z']]
```
Create DataFrame y for class labels (pCat =0,1,2)
```
    y = d.pCat
```

Exercise 6.1: Classify Diamonds

Predict diamond price category based on the given features using various classification methods such as

(a) Binary Logistic Regression.
(b) Multi-Class Logistic Regression.
(c) Linear Discriminant Analysis.
(d) Decision Tree.
(e) Support Vector.

Exercise 6.2: Compare Model Performance

Compare the performance of the above models and interpret the results.

Exercise 6.3 Decision Tree Classification

Figure 6.9 represents the decision tree classification of iris flowers. Explain the feature selection based on gini index. Build a rule-based classification (if–then–else rules). Discuss bias-variance tradeoff in the context of the diagram.

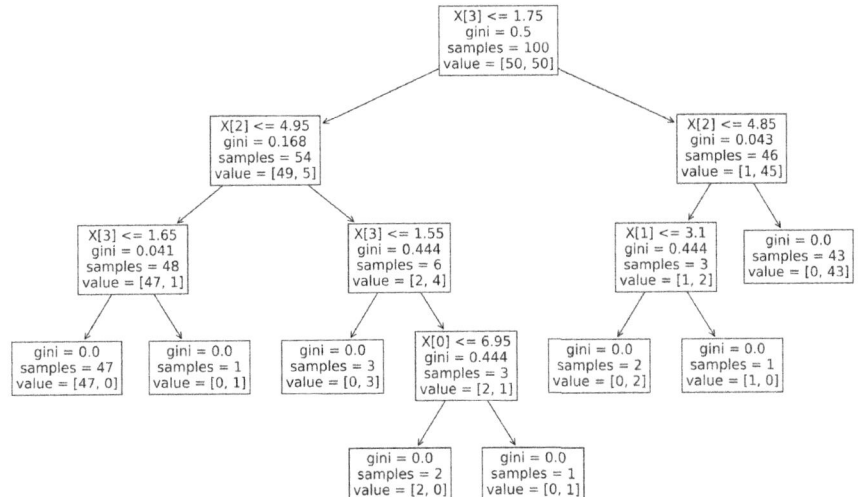

Fig. 6.9 Decision tree for two-way classification of Virginica and Versicolor flowers

Exercise 6.4 Classification Model Performance

In the classification of 150 iris flowers, the following results were obtained. Explain the results and underlying computation.

```
a) Categories: setosa, versicolor, virginica

Confusion matrix

array([[50,  0,  0],
       [ 0, 45,  5],
       [ 0,  1, 49]])
b) Classification_Report
```

	precision	recall	f1-score	support
setosa	1.00	1.00	1.00	50
versicolor	0.98	0.90	0.94	50
virginica	0.91	0.98	0.94	50
accuracy			0.96	150
macro avg	0.96	0.96	0.96	150
weighted avg	0.96	0.96	0.96	150

References

Adler AI, Boyko EJ, Ahroni JH, Stensel VX, Forsberg RC, Smith DG (1997) Risk factors for diabetic peripheral sensory neuropathy: results of the seattle prospective diabetic foot study. Diabet Care

Deisenroth MP, FAA, OCS (2020) Mathematics for machine learning. Cambridge University Press. https://mml-book.github.io

Géron A (2019) Hands-on machine learning with scikit-Learn, keras, and tensor flow (2019, O'reilly). In: Hands-on machine learning with R

Han J, Kamber M, Pei J (2014) Data mining. Concepts and techniques, 3rd edn. (The Morgan Kaufmann series in data management systems). In: Proceedings 2013 international conference on machine intelligence research and advancement, ICMIRA 2013

Malhotra NK (2020) Marketing research an applied prientation, 7th edn. Pearson Education

Chapter 7
Factor Analysis

Learning Objectives

- Understand the basic principles of factor analysis.
- Examine the application of eigenvalue analysis for factor extraction.
- Explain principal component analysis.
- Discuss the common terms used in the factor analysis procedure.
- Discuss the critical aspects of measurements and data collection.
- Describe the basic data validation done in factor analysis.
- Describe the steps involved in the computational procedure for factor analysis.
- Determine the number of factors that emerge from factor analysis.
- Illustrate confirmatory factor analysis.
- Demonstrate factor analysis using Python.

Overview

In business and industrial practices, we may encounter numerous measured features of a phenomenon or a problem. We may need to derive a smaller set of valuable factors that define the phenomena from this. Factor analysis is a broad term encompassing a set of techniques primarily employed for the purposes of data reduction and summarization.

The chapter explains the theoretical background of factor analysis and introduces two methods—principal component analysis (PCA) and common factor

---

Supplementary Information The online version contains supplementary material available at https://doi.org/10.1007/978-981-99-0353-5_7.

S. Sundararajan, *Multivariate Analysis and Machine Learning Techniques*,
Transactions on Computer Systems and Networks,
https://doi.org/10.1007/978-981-99-0353-5_7

analysis (CFA). This is followed by a discussion of eigenvalue analysis and the extraction of principal components. General concepts associated with factor analysis, such as instrument design, validity, and rotation, are introduced. After that, the computational procedure for factor analysis is described in detail. This includes the iterative procedure for removing trivial variables, the criteria for narrowing down on a desired number of factors, and naming/conceptualizing the factors. The section describes confirmatory factor analysis using structural equation modeling. This is followed by a detailed tutorial on factor analysis using Python based on a case study of IT project performance.

Definitions

Bartlett's test of sphericity: Inspects whether sufficient correlations exist among variables for them to group into factors. Bartlett's test statistic must be significant (p-value $<= 0.05$).

Communality: This is the amount of variance a variable shares with all the other variables.

Confirmatory Factor Analysis (CFA): CFA considers only common variance. This is usually done following principal component analysis, i.e., after we have gained some insight into the variables and their correlations.

Cronbach's alpha: This statistic is a good measure of internal consistency reliability. The value is expected to be above 0.6.

Eigenvalue: This represents the total variance explained by each factor. A value of 1 indicates that the factor can explain the variance of 1 variable, and so on.

Factor Analysis: Factor analysis is a broad term encompassing a set of techniques primarily employed for the purposes of data reduction and summarization.

Factor loading matrix: This matrix contains the factor loadings of all the variables on all the factors extracted.

Factor loading: These are simple correlations between the variables and the factors.

Kaiser–Meyer–Olkin (KMO): A measure of sampling adequacy that indicates the proportion of variance in the variables that the underlying factors might cause.

Orthogonal rotation: Orthogonal rotation generates a set of uncorrelated factors from the factor loading matrix. This includes methods like Varimax, Equimax, or Quartimax.

Principal Component Analysis: Principal component analysis is a popular technique for data summarization and reduction. PCA is an exploratory data analysis method that considers total variance. PCA reduces a set of measured features to a few conceptual/latent factors.

Structural Equation Modelling (SEM): This multivariate statistical methodology uses a confirmatory factor analysis approach to analyze a structural theory. SEM can accommodate multiple interrelated dependence relationships in a single model.

Variance: There are three types of variance considering a set of features. Specific variance is the variance associated with the specific variable. Common variance is the variance shared with all other variables. Error variance is the unexplained variance that results from some data-gathering error or random error.

Varimax rotation: This orthogonal rotation method usually results in factors with high factor loadings by fewer variables. At the same time, these variables are expected to have low factor loadings for the other variables. This enhances the interpretability of the factors.

7.1   Factor Analysis—Introduction and Overview

We frequently encounter situations where the feature dimensions are high. From this multitude of features, discerning the significant ones that genuinely impact the problem under investigation can be challenging. Furthermore, the presence of correlations among these features can lead to inflated estimations. As a result, it becomes vital to summarize and reduce the number of features to construct a meaningful and computationally effective model.

7.1.1   Factor Analysis

Factor analysis is a broad term encompassing a set of techniques primarily employed for the purposes of data reduction and summarization. In this method, we examine a set of observed variables for interrelationship and attempt to explain their variance in terms of a few concepts called constructs or factors. Each factor encompasses a set of highly correlated variables, which are simultaneously anticipated to exhibit minimal correlation with variables belonging to other factors. As these variables are mapped into a smaller set of factors, the data volume gets reduced.

Assume that there are 'k' variables $\{X_1 \ldots X_k\}$ and they map to 'n' underlying factors $\{F_1 \ldots F_n\}$ (where $n < k$). The following equation represents the relationship of the variables to with the factors F_i:

$$F_i = W_{i1}X_1 + W_{i2}X_2 + \ldots + W_{ik}X_k \tag{7.1}$$

where:

F_i estimate of the ith factor.
W_{ij} loading of variable X_j on factor F_i.

The total variance of a variable can be expressed in terms of the following (Sundararajan et al. 2019):

(a)   Specific variance—the variance associated with the specific variable.
(b)   Common variance—the variance shared with all other variables.
(c)   Error variance—the unexplained variance that results from some data-gathering error or random error.

There are many methods for factor analysis, such as Principal Component Analysis (PCA) and Common Factor Analysis (CFA). PCA is an exploratory data analysis method that considers total variance. It is used where prior knowledge indicates that specific variance and error variance may represent a relatively small portion of the total variance, and the primary objective of the researcher is data reduction. PCA is the most popular technique used in Factor Analysis. CFA is a confirmatory factor analysis method that considers only common variance.

7.1.2 PCA—Theoretical Basis

Principal Component Analysis (PCA) is a dimensionality reduction technique commonly used in data analysis and machine learning. It is a mathematical method that aims to transform a dataset into a new coordinate system of fewer dimensions in such a way that it captures as much variance in the data as possible. In other words, PCA reduces the dimensionality of the dataset while retaining as much information as possible in the selected principal components. The main method used for PCA is Eigenvalue analysis. Let us look at the mathematical basis and the corresponding computer algorithm.

Mathematical Basis for PCA Using Eigen Value Analysis

Consider a matrix A of size n × n. Then there is a non-zero vector 'v', such that,

$$Av = \lambda v, \quad \text{where } \lambda \text{ is an eigenvalue of A}$$

Consider an identity matrix I. The diagonals of an identity matrix are 1, and all other cells are zero. The above equation can be rewritten as,

$$Av = \lambda Iv$$

This implies that, $Av - \lambda Iv = 0$

Or

$$(A - \lambda I)v = 0$$

The above equation can be solved using the following method (Strang 2022). Compute the determinant of $(A - \lambda I)$. The result will be a polynomial in λ, of degree n (as A is an n × n matrix). Set the determinant (det) to zero and find roots. The n roots are the 'n' *eigenvalues* of A.

$$\det (A - \lambda I)v = 0$$

For each eigenvalue λ, solve $(A - \lambda I)v = 0$, to find the corresponding eigenvector v.

These eigenvectors represent the factor loadings, which convey the strength of the relation between each observed variable and the underlying factor. Factor loadings can be positive or negative, representing the direction of the relationship.

Algorithm for PCA Using Eigen Value Analysis

The steps involved in performing PCA using eigenvalues and eigenvectors are shown below:

1. Standardize the data: This step ensures that variables with different units and scales don't disproportionately influence the results. Data standardization is discussed in Sect. 2.8.2. For example, consider z-score standardization. An observation x_i can be transformed into z-score as follows:

$$z_i = (x_i - x_{mean})/sd, \text{ where sd is the standard deviation of the sample}$$

2. Compute the covariance matrix: The covariance matrix summarizes the relationships between variables in the dataset. It is a square matrix where the diagonal elements represent the variances of individual variables, and the off-diagonal elements represent the covariances between pairs of variables.
3. Calculate the eigenvalues and eigenvectors: Here we transform the data into a different coordinate system. The eigenvalues represent the variances explained by each principal component axis, while the eigenvectors are the directions of this component axis.
4. Sort the eigenvalues: Arrange the eigenvalues in descending order. We aim to retain the principal components with the highest eigenvalues. (The corresponding principal components capture most of the variance in the data.)
5. Select the principal components: Decide how many principal components we need for dimensionality reduction. Typically, this is based on the explained variance (discussed later).
6. Create the projection matrix: Form a matrix of the selected eigenvectors (corresponding to the chosen principal components).
7. Project the data: Multiply the standardized data by the projection matrix to obtain the transformed data in the reduced-dimensional space.

7.1.3   A Case Study

Let us consider the case study discussed in Sect. 5.4. Example. A study was conducted to explore the variables that influence the performance of a software project (Sundararajan 2021). In this exercise, we consider 11 variables measured on a scale of 1–7 and an additional variable, 'project_type', with values $\{0,1,2\}$. Note that 'case number' is not considered in the analysis. The outcome variable, project performance (project_perf), was measured in terms of deviation from the estimated effort. Data were collected from 100 software projects. The frequency distributions of 11 variables are shown in Fig. 5.3.

Tutorial 7.1   Factor Analysis Case Study - Basics

The code in this tutorial strictly follows six steps in the Algorithm for PCA using Eigen Value Analysis. The training data having 12 feature variables is reduced to 2 factors!

Refer: Section 1.3 for dataset or {(Sundararajan, 2023)}

```python
import pandas as pd
import seaborn as sb
import numpy as np
import matplotlib.pyplot as plt

d=pd.read_csv('itprojects.csv')

d.columns
# 'project_perf' is the outcome; drop it. also drop 'Case_No'

X=d[['change_mgmt', 'project_plan', 'tech_mentoring', 'pm_tools',
 'dev_process', 'system_arch', 'design_think', 'team_skills',
 'core_team', 'prior_exp', 'rewards_recog', 'project_type']]
```

Plot the frequency distributions of the variables
```python
plt.title("Features affecting software project performance")
sb.kdeplot(data=X)
```
See Figure 7.1: Variable Reduction Using Factor Analysis

```python
training_data = X.copy()
training_data.columns
['change_mgmt', 'project_plan', 'tech_mentoring', 'pm_tools',
 'dev_process', 'system_arch', 'design_think', 'team_skills',
 'core_team', 'prior_exp', 'rewards_recog', 'project_type']
```

1. The data is already standardised. All the features are on 1-7 scale.

2. Compute the covariance matrix: 12 features: 12x12 matrix
```python
covariance_matrix = np.cov(training_data, rowvar=False)
print("Covariance Matrix:", np.round(covariance_matrix,1))
Covariance Matrix:
[[ 0.9  0.5 -0.1 -0.1  0.1  0.1  0.2  0.2  0.1  0.1  0.1  0.2]
 [ 0.5  0.9 -0.2 -0.1  0.1  0.1  0.1  0.2 -0.   0.1  0.2  0.2]
 [-0.1 -0.2  0.7  0.2 -0.2 -0.2 -0.2 -0.2 -0.1 -0.2 -0.3  0.2]
 [-0.1 -0.1  0.2  0.5 -0.1 -0.2 -0.1 -0.1 -0.1 -0.2 -0.2  0.2]
 [ 0.1  0.1 -0.2 -0.1  0.5  0.3  0.3  0.1  0.1  0.2  0.2 -0.2]
 [ 0.1  0.1 -0.2 -0.2  0.3  0.5  0.3  0.2  0.2  0.1  0.3 -0.2]
 [ 0.2  0.1 -0.2 -0.1  0.3  0.3  0.5  0.1  0.2  0.2  0.2 -0.2]
 [ 0.2  0.2 -0.2 -0.1  0.1  0.2  0.1  0.6  0.2  0.2  0.3 -0.1]
 [ 0.1 -0.  -0.1 -0.1  0.1  0.2  0.2  0.2  0.5  0.2  0.3 -0.2]
 [ 0.1  0.1 -0.2 -0.2  0.2  0.1  0.2  0.2  0.2  0.4  0.2 -0.2]
 [ 0.1  0.2 -0.3 -0.2  0.2  0.3  0.2  0.3  0.3  0.2  0.9 -0.3]
 [ 0.2  0.2  0.2  0.2 -0.2 -0.2 -0.2 -0.1 -0.2 -0.2 -0.3  0.5]]
```

3. Calculate the eigenvalues and eigenvectors

```
eigenvalues, eigenvectors = np.linalg.eig(covariance_matrix)
print("Eigen Values:", np.round(eigenvalues,2))
Eigen Values:
[2.47 1.36 0.66 0.11 0.16 0.19 0.23 0.28 0.51 0.38 0.42 0.47]
```

4. Sort the eigenvalues: the values are in sorted order

5. Select the principal components:

If eigen value = n, the respective principal component can represent the
variation due to n variables. Select the components with eigen values > 1

```
selected_eigenvalues = eigenvalues[:2]
print("Eigen Values of Selected Principal Components:",
np.round(selected_eigenvalues,2))
Eigen Values of Selected Principal Components: [2.47 1.36]
```

Show the eigen vectors corresponding to selected principal components

```
selected_eigenvectors = eigenvectors[:2, :]
print("Eigen Vectors of Selected Principal Components:",
np.round(selected_eigenvectors,2))
[-0.26 -0.62 -0.23 -0.1  -0.23  0.13  0.15 -0.12 -0.38 -0.45 -0.2   0.02]
[-0.25 -0.59  0.22  0.12 -0.16  0.16 -0.18  0.13  0.36  0.42  0.26 -0.23]]
```

Take a look at the original data

```
print("Original Data in 12 Dimensions:", training_data)
```

6. Create the projection matrix

Project the data to the selected principal component axes

```
projected_data = np.dot(training_data, selected_eigenvectors.T)
print("Data Projected to 2 Principal Components:", np.round(projected_
data,2))
Data Projected to 2 Principal Components:
[[  1.68     1.39]
 [-11.12     0.09]
 [-10.72     1.43]
 [-10.19     0.98]
 [ -9.67     2.18]
    ...      ...
 [-10.42     1.2 ]
 [-10.41     1.43]
 [ -9.5      1.41]
 [-11.96     0.96]
 [-10.81     1.54]]
```

Table 7.1 shows data dimension reduction from 12 features to 2 factors

Table 7.1 Factors/eigen values that emerged from the factor analysis

Training Data												Projected Data	
5	5	2	1	5	5	5	5	5	5	5	1	1.68	1.39
6	6	1	2	6	6	6	6	5	5	5	1	- 11.12	0.09
5	5	2	2	5	5	5	6	6	5	4	1	- 10.72	1.43
5	5	1	2	5	5	5	5	5	5	5	1	- 10.19	0.98
2	5	2	2	5	5	5	5	5	5	5	0	- 9.67	2.18
...	...	...	...	...	...	...	...	...	...	...	...	...	...
5	5	2	2	5	5	5	5	5	5	5	1	- 10.42	1.2
5	5	2	3	4	5	4	5	5	5	5	2	- 10.41	1.43
5	4	3	2	4	4	4	4	5	4	5	2	- 9.5	1.41
6	6	2	2	5	5	6	5	6	6	5	1	- 11.96	0.96
5	5	1	3	6	6	6	5	6	5	6	1	- 10.81	1.54

7.2 Important Concepts Associated with Factor Analysis

Here, we are going to discuss important concepts associated with factor analysis.

7.2.1 The Measurement Instrument

Due diligence is needed in the instrument design (the variables and their measurement) for a successful factor analysis technique (Malhotra 2020). The instrument must be well designed to ensure clarity and comprehensiveness of the measured phenomena. A checklist is provided below.

1. Identify the objectives of factor analysis.
2. Decide the variables to be measured. This should be based on past research, theory, and the researcher's judgment. Include only the variables relevant to the concept being explored.
3. Measure the variables using an appropriate scale—e.g., interval or ratio scale.
4. Select the appropriate sample size to reflect the population. As a rough guideline, there should be at least five times as many observations as variables.
5. Exclude instances of variables with missing values. Otherwise, attempt to impute missing values if the researcher is certain that the data imputations preserve the properties of the variables under consideration to an acceptable level.

7.2.2 Data Validation

In statistical data analysis, validity and reliability must be inspected as the first step in the analysis of results. The data collected must be representative and of adequate quality. The following criteria apply to factor analysis (Malhotra 2020).

- Bartlett's test of sphericity: Sufficient correlations exist among variables for them to group into factors. Bartlett's test statistic must be significant (p-value <= 0.05). (Note: The null hypothesis in this test is that significant correlations do not exist among the variables. This is rejected when p <= the significance level assumed, e.g., 0.05.)
- Kaiser–Meyer–Olkin's (KMO) measure of sampling adequacy indicates the proportion of variance in the variables that the underlying factors might cause. High values (0.6–0.9) generally indicate that factor analysis may be helpful with the data.
- Cronbach's alpha (α) statistic is a good measure of internal consistency reliability—its value needs to be above 0.6. Check Cronbach's alpha over all the variables taken together and then for each factor separately.

7.2.3 Common Terms Associated with Factor Analysis

Factor loadings represent straightforward correlations between variables and factors (principal components). Within the factor matrix, you find the factor loadings for all variables across all extracted factors. Communality indicates the extent of variance a variable shares with other variables, as indicated by the derived factors. Eigenvalues, on the other hand, signify the overall variance explained by each factor. The values are expected to be 1 or more. A value of 1 indicates that the factor can explain the variance of 1 variable.

7.2.4 Rotation

Several procedures have been suggested for factor extraction and determining the number of factors in Principal Component Analysis (Hair et al. 2010). The factor loading matrix (see Table 7.4) represents the correlation between factors and variables. A large value indicates a closer relationship between the variable with the factor. Ideally, each variable should exhibit a maximum correlation (1.0) with a single factor and a minimum correlation (0.0) with all other factors. However, practical scenarios seldom conform to this ideal, leading to challenges in interpretation. To address this, the factor matrix is rotated—a process designed to simplify the structure and improve interpretability. It's important to note that this rotation doesn't improve the overall variance explained by the principal components;

rather, its objective is to present a more comprehensible and meaningful depiction of the relationships among variables.

If our objective is to generate uncorrelated factors, we use orthogonal rotation. The rotation is called orthogonal if the axes are maintained at right angles. This includes methods like Varimax, Equimax, or Quartimax.

The commonly used method for orthogonal rotation is the varimax procedure. This method usually results in factors with high factor loadings by fewer variables. At the same time, these variables are expected to have low factor loadings for the other variables. This enhances the interpretability of the factors. The first factor provides the best summary of linear relationships in the data. The second factor, orthogonal to the first, is derived from the variance remaining after the one explained by the first factor. This process is repeated to extract an optimal number of factors.

There are many other ways of rotation. The rotation is called oblique rotation when the axes are not maintained at right angles, and the factors are correlated. Sometimes, allowing for correlations among factors may simplify the factor pattern matrix. Oblique rotation should be used when factors in the population are likely to be strongly correlated.

To understand orthogonal rotation for variable reduction, see Fig. 7.2, which shows data distribution in two dimensions (X-Axis and Y-Axis). If we need to characterize the data variance using one dimension only, which one will we choose? X-axis? Y-axis? From Fig. 7.2, it is intuitive to note that the diagonal 'orange' axis represents the data variations more than the X-axis or Y-axis, taken individually. This is an example of the application of axis rotation. Assuming we have several dimensions (hyperspace), we will choose the principal component of the new coordinate system that explains the maximum data variation. Then, we will look for a second dimension (second principal component) for further characterization of the variance in data. Having chosen the 'orange' diagonal as the first principle component that represents most of the data variance, we would now cut an orthogonal line, as represented by the 'blue' diagonal, to supplement the characterization. The blue line is the second principal component in orthogonal rotation. The above process is repeated till we capture a reasonably good amount of data variance (using a relatively fewer number of factors.

Fig. 7.1 Variable reduction using factor analysis

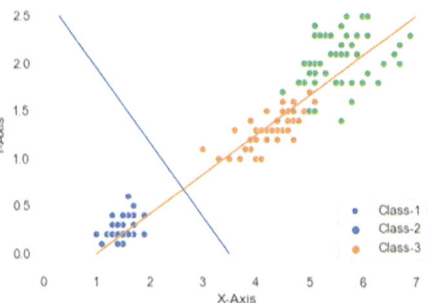

Fig. 7.2 Scree plot showing two prominent factors

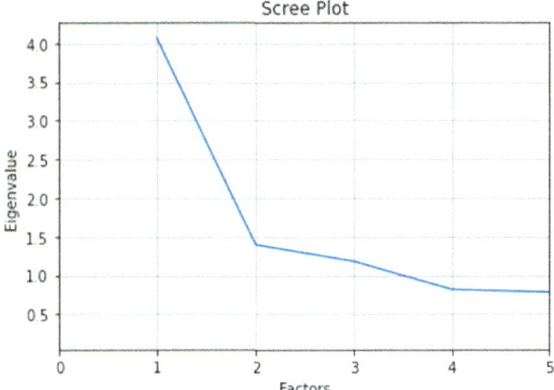

7.2.5   The Steps Involved in Factor Analysis

1. Perform factor analysis.
2. Check the factor analysis results and respecify the structure iteratively by removing the variables that load less or cross-load, one variable at a time.
3. The correlation of a variable with the factors is called factor loadings. The variables that load less than |0.4| may be removed. A correlation of $r = 0.4$ implies an explained variance of $r^2 = 0.16$ or 16%. Note that the researcher may choose to retain variables that load less than |0.4| if they deem to do so due to theoretical considerations or prior history. Check the absolute value of the loading, discarding the sign.
4. If a variable loads multiple factors with high loadings, it is called cross-loading. Remove the variable if the difference in cross-loadings is less than a threshold (e.g., 0.2).
5. Repeat steps 1–3 until the criteria for factor selection are met.
6. The factors that emerge from factor analysis can be interpreted and named in terms of the variables that load high on them. Look at the factor loading matrix and identify the variables that load a factor high.

7.2.6   How to Determine the Number of Factors

The number of factors that emerge from factor analysis is the same as the number of variables. The first factor explains the maximum variance in the data, followed by the second, and so on. The idea of factor analysis is a reduction in the number of variables. Therefore, the researcher is left with the question of how many factors will be selected. The method for factor selection is discussed below.

Fig. 7.3 Reduction of 9
variables to 3 factors

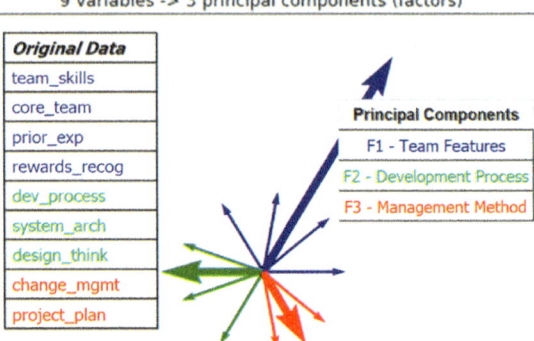

1. Select only factors with Eigenvalues greater than 1.0.
2. Scree Plot: A scree plot is a plot of the Eigenvalues against the number of factors in order of extraction. Experimental evidence indicates that the point at which the screen begins denotes the true number of factors. See Fig. 7.3.
3. Percentage of Variance: The number of factors extracted may be determined by the cumulative percentage of variance extracted by the factors. Check whether it is satisfactory (e.g., 60% or more).
4. A predetermined number of factors are based on the researcher's intent.
5. A combination of the above methods.

7.3   The Project Performance Case Study

This section is a continuation of the case study described in Sect. 7.2.2. A set of tutorials is given at the end of this section to demonstrate the detailed steps associated with principal component analysis for variable reduction. An abstract of the detailed analysis is provided here.

7.3.1   Factor Analysis Procedure

Instead of 12 variables in Tutorial 7.1, here we started with 11 variables (we dropped project_type). We used a package for factor analysis. It has several modules that help us in detailed factor analysis. After preliminary data analysis, nine variables were selected for factor analysis—team_skills, core_team, prior_exp, rewards_recog, dev_process, system_arch, design_think, change_mgmt, and project_plan. Three factors (principal components) with eigenvalue>= 1 emerged during the principal component analysis. (See Tutorials 7.3.3–7.3.5) (Table 7.2).

Table 7.2   Factors/eigen values that emerged from the factor analysis

Factor1	Factor2	Factor3	Factor4	Factor5	Factor6	Factor7	Factor8	Factor9
3.52	1.39	1.18	0.70	0.64	0.53	0.41	0.37	0.25

Table 7.3   Factor loading matrix for 3-factor solution

Variable (X_j)	F_1	F_2	F_3	Factor (principal component)
team_skills	0.57	0.16	0.25	F1 (team)
core_team	0.69	0.2	−0.04	
prior_exp	0.76	0.23	0.09	
rewards_recog	0.5	0.24	0.14	
dev_process	0.24	0.64	0.07	F2 (development process)
system_arch	0.26	0.77	0.12	
design_think	0.22	0.77	0.11	
change_mgmt	0.11	0.15	0.54	F3 (management process)
project_plan	0.08	0.02	0.92	

The eigenvectors represent the factor loadings, which convey the strength of the relation between each observed variable and the underlying factor. Factor loadings can be positive or negative, representing the direction of the relationship. As we discussed earlier, Factors are linear sums of variables that can expressed as

$$F_i = W_{i1}X_1 + W_{i2}X_2 + \dots + W_{ik}X_k \quad (\text{See Eq. 7.1})$$

The factors may be named based on the variables that load a factor high. Table 7.3 shows that the variables prior_exp, core_team, team_skills, and rewards_recog constitute factor F1, which may be named the 'team' factor. The variables system_arch and dev_process constitute factor F2, which may be named the 'development process' factor. The variables project_plan and change_mgmt constitute factor F3, which may be named the 'management process' factor.

7.3.2   Data Transformation

As discussed in the algorithm given in Sect. 7.2.1, the last two steps are—(a) create the projection matrix of the selected eigenvectors (corresponding to the chosen principal components) and (b) multiply the input data by the projection matrix to obtain the transformed data in the reduced-dimensional space. This transformation can be achieved by the dot product of the input data with the projection matrix, as demonstrated at the end of Tutorial 7.1. Another method is shown under Tutorial 7.3.7. Each of the 100 observations will be transformed from 9 variables to 3 factors: -

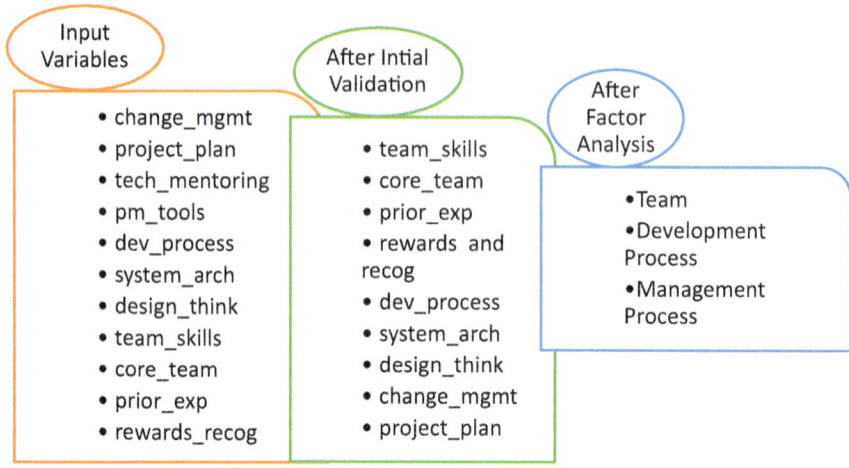

Fig. 7.4   Variable reduction using factor analysis

Factor 1 (Team) =
    $0.57 \times$ team_skills $+0.69 \times$ core_team $+0.76 \times$ prior_exp $+$
    $0.5 \times$ rewards_recog $+0.24 \times$ dev_process $+0.26 \times$ system_arch $+$
    $0.22 \times$ design_think $+0.11 \times$ change_mgmt $+0.08 \times$ project_plan

Factor 2 (Development) =
    $0.16 \times$ team_skills $+0.20 \times$ core_team $+0.23 \times$ prior_exp $+$
    $0.24 \times$ rewards_recog $+0.64 \times$ dev_process $+0.77 \times$ system_arch $+$
    $0.77 \times$ design_think $+0.15 \times$ change_mgmt $+0.02 \times$ project_plan

Factor 3 (Management) =
    $0.25 \times$ team_skills - $0.04 \times$ core_team $+0.09 \times$ prior_exp $+$
    $0.14 \times$ rewards_recog $+0.07 \times$ dev_process $+0.12 \times$ system_arch $+$
    $0.11 \times$ design_think $+0.54 \times$ change_mgmt $+0.92 \times$ project_plan
    Figure 7.4 depicts the 3-factor solution pictorially.

7.3.3   Case Study—Conclusion

The summary of the factor analysis case study is shown in Fig. 7.5. Note that there were initially 11 variables, but we discarded two after initial data validation. From the factor analysis of 9 variables a three-factor solution has emerged. The total variance explained by the model is 54%. This is not a good model, as the explanatory power is just above 50%.

Table 7.4 A data snap shot after factor analysis

Original Data (2 Rows x 9 Columns)										3 more variables
Row	CM	PP	DP	SA	DT	TS	CT	PE	RR	-
1	5.00	5.00	5.00	5.00	5.00	5.00	5.00	5.00	5.00	-
2	6.00	6.00	6.00	6.00	6.00	6.00	5.00	5.00	5.00	-

Reduced Data after Variable to Factor Transformation (2 Rows x 3 Columns)

Row	Factor 2 (Management)	Factor 1 (Development)	Factor 3 (Team)	-
1	0.26	-0.05	0.09	-
2	1.40	1.35	0.05	-

Let us look at the data before and after variables to factors transformation. Let's use abbreviations for variable names due to space considerations—team_skills (ts), core_team (ct), prior_exp (pe), rewards_recog (rr), dev_process (dp), system_arch (sa), design_think (dt), change_mgmt (cm), and project_plan (pp). Table 7.4 shows the transformation of the first two observations from nine variables to three factors.

Tutorial 7.3   Factor Analysis Case Study - Detailed

This is a continuation of Tutorial 7.1. Instead of 12 variables in that tutorial, here we consider 11 variables. We are using a package for factor analysis. It has several modules that help us in detailed factor analysis.

Tutorial 7.3.1 Environment Setup

```
import pandas as pd
import seaborn as sb
import numpy as np
import matplotlib.pyplot as plt
```

For data description, refer to Tutorial 7.1 or {(Sundararajan, 2023)}
Note:- if the data has missing values, they must be removed/imputed

```
d=pd.read_csv('itprojects.csv')
d.columns
```

Discard 'Case_No','project_type', 'project_perf'. Select all the other 11 features

```
X = d[['change_mgmt', 'project_plan', 'tech_mentoring',
       'pm_tools', 'dev_process', 'system_arch', 'design_think',
       'team_skills', 'core_team', 'prior_exp', 'rewards_recog']]
```

The frequency distribution of 12 variables is shown in Figure 5-3. Note that the distribution of the variables - tech_mentoring and pm_tools differ from the other variables. These variables get dropped during the factor analysis procedure.

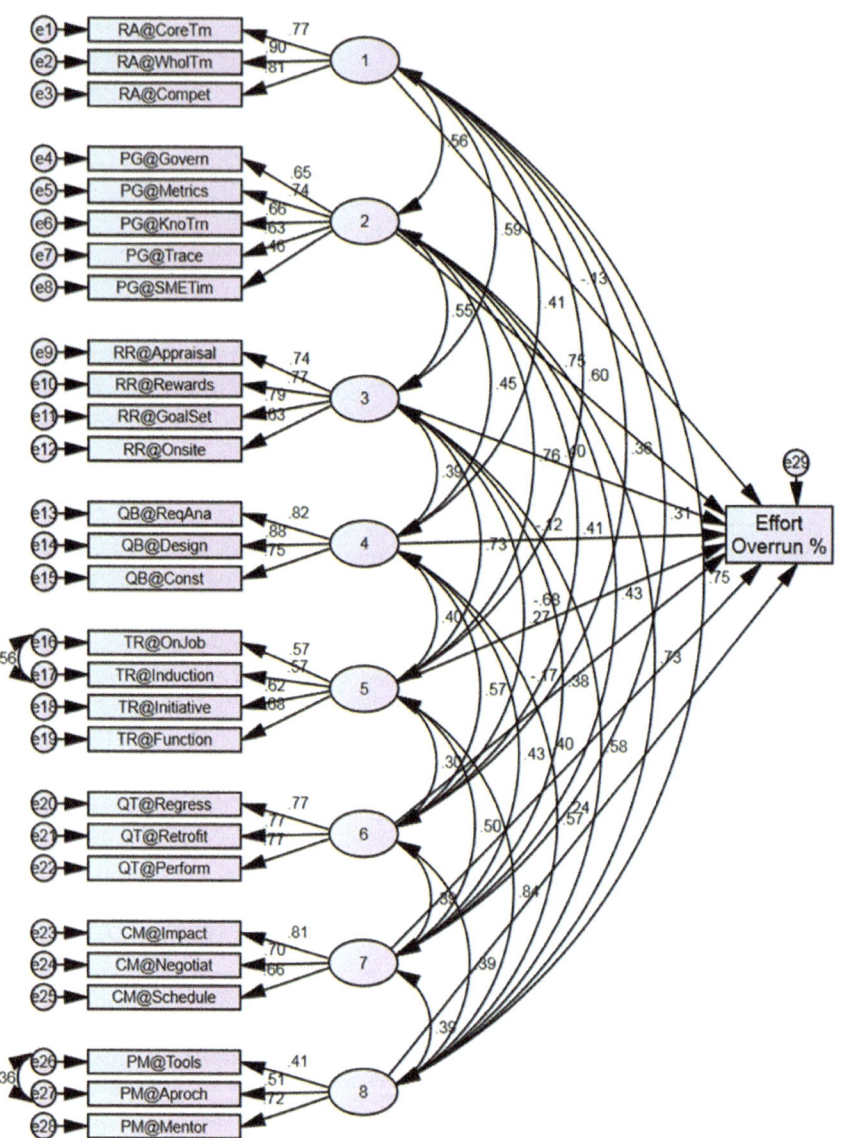

Fig. 7.5 Structural equation model representing software project characteristics

Tutorial 7.3.2 Data Validation

You may refer to: https://factor-analyzer.readthedocs.io/en/latest/
```
!pip install scipy==1.10.1
# quick fix; install old scipy library, as the factor-analyzer module is
deprecated
!pip install factor-analyzer
```

Import factor analysis main library
```
from factor_analyzer import FactorAnalyzer
```
Import routines for data validation
```
from factor_analyzer.factor_analyzer import calculate_bartlett_sphericity
from factor_analyzer.factor_analyzer import calculate_kmo
```

Bartlett's test of sphericity: for the existence of correlations
```
chi_square_value,p_value=calculate_bartlett_sphericity(X)
round(p_value,3) # 0.000
```
p-value of BTS is < 0.05. Therefore, it is inferred that significant correlations exist for factor formation

Kaiser-Meyer-Olkin (KMO) measure
```
kmo_item,kmo_model=calculate_kmo(X)
round(kmo_model,2)     # 0.77
np.round(kmo_item,2)
```
```
(0.6 , 0.59, 0.87, 0.86, 0.8 , 0.75, 0.79, 0.86, 0.74, 0.74, 0.87)
```
Kaiser-Meyer-Olkin (KMO) measure of sampling adequacy statistic indicates the proportion of variance in the variables that the underlying factors might cause. A value> 0.6 is generally acceptable. Here, KMO is 0.66. Therefore, the sample is good for factor analysis

Tutorial 7.3.3 Factor analysis: Iteration - 1

Perform factor analysis iteratively, removing low-loading or high cross-loading variables one at a time

```
fa = FactorAnalyzer(rotation='varimax')
fa.fit(X)
```

Check the Eigenvalues
```
ev, v = fa.get_eigenvalues()
np.round(ev,2)
[3.52, 1.39, 1.18, 0.7 , 0.64, 0.53, 0.41, 0.37, 0.25]
```

The eigenvalues of the first three factors are above 1. other factors must be discarded, as their eigenvalues are < 1.

Scree plot to select the number of factors

```
x_axis = range(1,12,1)  # to accommodate 11 variables
plt.plot(x_axis,ev)
plt.title  ('Scree Plot')
plt.xlabel ('Factors')
plt.ylabel ('Eigenvalue')
plt.xlim(0,5)
plt.grid()
plt.show()
```

See Figure 7-2: Scree plot showing Two prominent factors. The scree plot bends sharply at 2, suggesting a two-factor solution. However, all the first 3 factors have an eigenvalue of more than 1. Therefore, we will consider a 3-factor solution. Note that 11 variables are reduced to 3 factors: 11 dimensions to 3.

Explore factor loading - rotated component matrix; See Table 7-4

```
fa.loadings_.shape  # (9,3)
nFactorsSelected = fa.loadings_.shape[1]

fa.loadings_round = np.round(fa.loadings_,2)
fa.loadings_round

flmatrix  = pd.DataFrame()
flmatrix['var'] = list(X.columns)
for i in range(nFactorsSelected):
    flmatrix[('Factor'+str(i+1))]  = fa.loadings_round[:,i]
flmatrix
Factor Loading Matrix
0      change_mgmt      0.11     0.15     0.54
1      project_plan     0.1      0.03     0.89
2      tech_mentoring  -0.4     -0.3     -0.23
3      pm_tools        -0.32    -0.29    -0.18
4      dev_process      0.25     0.64     0.07
5      system_arch      0.26     0.78     0.11
6      design_think     0.22     0.74     0.1
7      team_skills      0.59     0.15     0.24
8      core_team        0.66     0.2     -0.06
9      prior_exp        0.75     0.23     0.07
10     rewards_recog    0.53     0.24     0.14
```

Discard variables that cross load / load low. In the factor loading matrix, pm_tools loads less than 0.4. and it also cross loads f1(0.32), f2(0.29), with a difference of < 0.2. So remove pm_tools and re-run factor analysis

Tutorial 7.3.4 Factor analysis: Iteration - 2

```
X = d[['change_mgmt', 'project_plan', 'tech_mentoring',
       'dev_process', 'system_arch', 'design_think',
       'team_skills', 'core_team', 'prior_exp', 'rewards_recog']]
X.columns
```

```
fa = FactorAnalyzer(rotation='varimax')
fa.fit(X)

fa.loadings_.shape  # (9,3)
nFactorsSelected = fa.loadings_.shape[1]

fa.loadings_round = np.round(fa.loadings_,2)
fa.loadings_round

flmatrix  = pd.DataFrame()
flmatrix['var'] = list(X.columns)
for i in range(nFactorsSelected):
    flmatrix[('Factor'+str(i+1))]  = fa.loadings_round[:,0]
flmatrix
```

In the new factor loading matrix, tech_mentoring loads less than 0.4 and it also cross loads f1(0.39), f2(0.29), with a difference of < 0.2. So, remove tech_mentoring and re-run factor analysis

Tutorial 7.3.5 Factor analysis: Iteration - 3

```
X = d[['change_mgmt', 'project_plan',
        'dev_process', 'system_arch', 'design_think',
        'team_skills', 'core_team', 'prior_exp', 'rewards_recog']]

fa = FactorAnalyzer(rotation='varimax')
fa.fit(X)
ev, v = fa.get_eigenvalues();np.round(ev,2)
(0   3.52   1.39   1.18   0.70   0.64   0.53   0.41   0.37   0.25)

fa.loadings_.shape  # (9,3)
nFactorsSelected = fa.loadings_.shape[1]

fa.loadings_round = np.round(fa.loadings_,2)
fa.loadings_round

flmatrix  = pd.DataFrame()
flmatrix['var'] = list(X.columns)
for i in range(nFactorsSelected):
    flmatrix[('Factor'+str(i+1))]  = fa.loadings_round[:,0]
flmatrix
```

In the new factor loading matrix, variables load more than 0.4. There are no 'big' cross-loadings. So, the factor analysis procedure is complete.

Tutorial 7.3.6 Final Check

The following were tested.
a) Bartlett's test of sphericity for the existence of correlations; p-value of BTS must be < 0.05.
b) Kaiser-Meyer-Olkin (KMO) measure of sampling adequacy; value > 0.6 is generally acceptable.
c) Checked the scree plot around the knee; the 3-factor solution is fine

d) factor loading matrix shows that every factor has eigenvalues > 1. So, each factor is capable of explaining at least one variable

```
fa.get_factor_variance()
```

The sum of Squares Loadings   [1.81161258, 1.78944568, 1.25508526]
Proportion of Variance   [0.20129029, 0.1988273 , 0.13945392]
Cumulative Variance Explained   [0.20129029, 0.40011758, 0.5395715]
Total Variance Explained is   **54%**

Tutorial 7.3.7 Compute the Factor Scores

```
F = fa.fit_transform(X)
F[0:2,]
array([[ 0.08786772,  -0.05484817,   0.26012858],
       [ 0.04872067,   1.35416323,   1.3952079 ]])
```

Add the factor score columns to the data frame d
```
d['team'] = F[:,0]
d['build'] = F[:,1]
d['management'] = F[:,2]
```
Write dataframe d with factor scores to a CSV data file, without index
```
d.to_csv('itprojectsWithfactors.csv',index = False)
```

Tutorial 7.3.8   Using Factor Scores for Regression

Consider Tutorial 7.1, where we conducted a Principal Component Analysis to reduce the 11 variables to a more manageable number of factors. In this tutorial, we use the factors that emerged from the factor analysis to build a regression model to predict project_perf. The regression model using all the variables gives a better accuracy than the regression model with the factors. However, variable reduction is helpful when the number of variables is too high and their interactions (multi-collinearity) are high.

```
import pandas as pd
import matplotlib.pyplot as plt

dt=pd.read_csv('itprojectsWithfactors.csv') # saved in earlier tutorial
dt.columns
'Case_No',
'change_mgmt', 'project_plan',
'tech_mentoring', 'pm_tools', 'dev_process', 'system_arch',
'design_think', 'team_skills', 'core_team', 'prior_exp',
'rewards_recog',
'project_type', 'project_perf'
'team', 'build', 'management'

import statsmodels.api as sm  # for regression
y = dt.project_perf
X = dt[['team', 'build', 'management']]
X['intercept'] = 1
model = sm.OLS(y,X,missing='drop').fit()        # build regression model
```

```
print(model.summary())
prob(F-statistic)   <= 0.05 significant
team                0.002   significant
build               0.545 > 0.05 / not significant
management          0.000   significant
drop the factor build, as it is not significant

Drop the factor 'build' as it is not significant (p > 0.05)
X.drop('build',inplace=True,axis=1)
model = sm.OLS(y,X,missing='drop').fit()        # build regression model
print(model.summary())
prob(F-statistic) <= 0.05 overall model     is significant
team             0.002   sig. factor team     is significant
management       0.000   sig. factor management is significant

Validations: -
Durbin-Watson: 1.854 (~2 observations do not have autocorrelation)
Prob(Omnibus) / Prob(JB) < 0.05 (the outcome variable is normal)
percentage of variance explained: -
Adj. R-squared:         0.505
Intercept       -0.0000 ~ round(float(-5.551e-17),4)
team            0.2687
management      0.7234
regression model: -
project_perf = 0.2687 * team + 0.7234 * management
check the relative importance of the factors: -
The coefficient 'management' (0.7234) > that of 'team' (0.2687). Therefore,
the variance in the factor 'management' influences 'project_perf', more than
the factor 'team'
Note: -
1) Both team, and management are measured using variables on a scale of 1 ...
7; otherwise, we need to standardize them before regression.
2) The model performance with the 3-factor model (Adj. R-squared = 0.505)
is less than that of the 9-variable model. See the note at the beginning of
the tutorial
```

7.4   Confirmatory Factor Analysis

Confirmatory Factor Analysis (CFA) aims to uncover hidden factors that influence the observed variables without assuming anything about specific variations or measurement errors. CFA focuses only on the shared variance among variables. CFA is a good choice when a researcher already understands the underlying factor relationships and wants to confirm if their model is valid. The researchers use prior knowledge and theories to predict how the observed measurements are related to these underlying factors and then test these predictions with statistical methods.

Structural Equation Modeling (SEM) is widely used in various fields, including psychology, sociology, economics, and other social sciences, as well as in biology, marketing, and other disciplines where complex relationships among variables must be examined. It offers a flexible and powerful tool for hypothesis testing and model validation.

The SEM is a statistical tool that validates structural theories. SEM permits the inclusion of numerous interconnected relationships within a single model. In contrast to other statistical methods, SEM can simultaneously estimate and connect multiple equations. This means that the results from one equation can serve as the input for other equations. This versatility empowers researchers to analyze complex relationships that might be challenging with other approaches. Additionally, researchers can use the factor structure derived from Principal Component Analysis as a foundation for creating their SEM model (Hair et al. 2010; Malhotra 2020). The components of SEM are listed below:

- The structural model represents the relationships among latent variables.
- The measurement model describes the relationships between latent variables and their observed indicators.
- Error terms account for unobserved influences on the observed variables.
- Various fit indices, such as chi-square, comparative fit index (CFI), tucker-lewis index (TLI), root mean square error of approximation (RMSEA), etc., to assess how well the model fits the observed data.
- The path diagrams visually depict the relationships between observed and latent variables through arrows and paths.

Let us consider the software project discussed in Sect. 7.2 and demonstrated in Sect. 7.10, the project performance case study. We considered 11 input variables and 100 software projects. However, in the current example, we consider 31 variables influencing software project performance and 145 software projects to demonstrate SEM modeling. For detailed information about the method used, refer to Sundararajan et al. (2019).

Initially, the data were subjected to Principal Component Analysis with Varimax rotation. Three variables were found trivial and discarded. Nine latent constructs (factors) emerged from the PCA, underlying 28 variables. The theoretical model proposed above, with specified relationships between latent and observed variables, was presented to SEM to test whether the data (observations from 145 projects) support the proposed model.

Seven factors emerged from the SEM model as shown in Fig. 7.6. It comprises of eight latent factors—resource competency (1), project planning and governance (2), resource motivation (3), quality of build (4), training (5) quality of test (6), change management (7), and project methodology (8). The variable-factor loadings show the covariance of the variables. The factors, in turn, predict effort variance, using a regression model. Note that effort variance is a commonly accepted measure of software project performance.

Table 7.5 Statistics for goodness of fit in the CFA case study

Statistic	Fit index	Measures	Observed value	Acceptable values for "good" model fit
1	CMIN/df	χ^2 value and the associated degree of freedom	1.93	Less than 5 indicates a good fit; less than 3 very good fit
2	CFI	Incremental fit index; the badness of fit index	0.83	Above 0.90 indicates a good fit
3	RMSEA	Absolute fit index	0.08	Less than 0.08 indicates a good fit

As we mentioned earlier, the overall fit of an SEM model is assessed using several measures of fit. Typically, we may consider a set of statistics such as the χ^2 value and the associated degree of freedom, one of the absolute fit indices, one of the incremental fit indices, and one of the badness of the fit index. Some of the important statistics observed (Sundararajan et al. 2019) are shown in Table 7.5.

Summary

Factor analysis is a broad term encompassing a set of procedures primarily employed to condense and summarize data. Principal component analysis (PCA), a widely used technique within factor analysis, examines the overall variance exhibited by variables. Through orthogonal rotation, uncorrelated factors are generated from the factor loading matrix, employing methods such as Varimax, Equimax, or Quartimax. Varimax rotation typically yields factors characterized by high loadings from a smaller set of variables, with low loadings expected for other variables, enhancing factor interpretability.

The factor loading matrix contains simple correlations between all variables and the extracted factors. Communality denotes the variance shared by a variable with all other variables, while eigenvalues signify the total variance explained by each factor. Before conducting factor analysis, assessing the data's suitability is crucial. Bartlett's test of sphericity examines whether there are adequate correlations among variables for grouping into factors. The Kaiser–Meyer–Olkin (KMO) statistic gauges sampling adequacy by indicating the proportion of variance in variables potentially caused by underlying factors. Cronbach's alpha is a reliable measure of internal consistency.

Confirmatory Factor Analysis (CFA) does not make any assumptions about specific variance and error variance. The CFA focuses solely on common variance. Structural Equation Modeling (SEM), a multivariate statistical methodology, adopts a confirmatory approach to analyze structural theories. SEM enables researchers to incorporate multiple interrelated dependence relationships within a single model.

Questions

Comprehension

1. Describe the use of Eigen Value Analysis in Principal Component Analysis.
2. Write a note on orthogonal rotation and the principal components that emerge from orthogonal rotation.
3. List the tests for validity and reliability in Factor Analysis.
4. How do we determine the number of factors extracted in Principal Component Analysis?
5. Describe the steps in Factor Analysis.
6. Describe Confirmatory Factor Analysis using SE.

Analysis

7. Compare and contrast Exploratory Factor Analysis with Confirmatory Factor Analysis.
8. How does factor analysis help in data reduction and summarization? Can you provide a real-world example where this technique is beneficial?
9. How is data standardization achieved in PCA, and why is it important for the analysis?
10. Describe the role of factor loadings in factor analysis. How do positive and negative factor loadings impact the interpretation of the results?
11. How does the choice of orthogonal rotation method (e.g., Varimax, Equimax, Quartimax) in factor analysis impact the interpretability of the derived factors?
12. Explain the concept of Cronbach's alpha and its relevance in factor analysis. How can it be used to assess the reliability of a set of variables?
13. In factor analysis, what is meant by the Kaiser–Meyer–Olkin (KMO) measure of sampling adequacy, and how does it affect the analysis?
14. Can you illustrate the differences between specific, common, and error variance in the context of factor analysis? How are these types of variance related?
15. In the context of factor analysis, why is the design of the measurement instrument crucial? How does the selection of variables and measurement scales impact the success of factor analysis?

16. Explain the significance of data validation in the factor analysis process. What are the key criteria, such as Bartlett's test of sphericity, Kaiser–Meyer–Olkin's measure, and Cronbach's alpha, in ensuring data quality and representativeness?

17. What is the role of factor loadings in factor analysis, and how do they help understand the relationships between variables and underlying factors? Provide an example from the case study.

18. In the factor analysis procedure, how does the iterative process of removing variables with low factor loadings contribute to refining the factor structure?

19. Discuss the methods for determining the number of factors in factor analysis.

20. Evaluate the success of the factor analysis in the case study based on the total variance explained. Is a 54% explanatory power considered good, and what factors might influence this result?

Application

21. Imagine you are working with a dataset from a marketing research study. How could factor analysis be applied to identify key customer segments or factors influencing purchase decisions?

22. Suppose you are a financial analyst working with stock market data. How might PCA be used to summarize and analyze the relationships between different stocks and identify latent factors affecting stock prices?

23. You are tasked with analyzing employee satisfaction data in a large organization. How could factor analysis be applied to uncover key factors contributing to employee well-being and engagement?

24. Imagine you are working on customer feedback data for a restaurant chain. How could factor analysis help identify areas for improvement, such as service quality, food quality, or ambiance, based on customer responses?

25. You are a researcher in the field of psychology studying personality traits. Explain how factor analysis can be applied to identify underlying personality factors and assess their correlations with behaviors and life outcomes.

26. In a customer churn analysis for a telecommunications company, how could factor analysis help identify factors contributing to customer attrition and inform customer retention strategy?

27. See the results below from the Principal Component Analysis. Interpret the observations in each table. Write the equations for computing Factor Scores (Table 7.6).

Table 7.6 Principal component analysis exercise

KMO and Bartlett's test			
Kaiser–Meyer–Olkin measure of sampling adequacy			0.69
Bartlett's test of sphericity significance			0.04
Factor loading matrix			
Factor->	1	2	3
Var_a	0.49	0.16	0.46
Var_b	−0.10	0.23	0.80
Var_c	0.40	0.57	0.37
Var_d	0.51	0.48	0.32
Var_e	−0.06	0.80	0.08
Var_f	0.28	0.79	0.09
Var_g	0.76	0.10	0.32
Var_h	0.77	0.17	0.30
Var_i	0.86	0.06	0.02
Var_j	0.63	0.44	0.08
Var_k	0.84	0.17	−0.03
Var_l	0.45	0.00	0.65

Total variance explained			
Factor	Eigen value	% of variance explained	Cumulative %
1.00	3.97	33.06	33.06
2.00	2.15	17.93	50.99
3.00	1.72	14.31	65.30

Overall reliability statistics	
Cronbach's alpha	Number of variables
0.87	12.00

Exercises

Exercise 7.1 Factor Analysis (Principal Component Analysis)
Do Factor Analysis of Diamonds Dataset, using PCA. Consider all variables other than 'price' for PCA. See Sect. 1.6 for a brief description of the diamonds dataset.

Exercise 7.2 Use Factor Scores for Predicting Diamond Prices
Perform regression analysis to predict diamond prices based on factor scores. Compare the result with regression analysis using the observed variables as inputs. Explain the finding.

References

Hair JF, Black WC, Babin BJ, Anderson RE (2010) Multivariate data analysis. In: Vectors. https://doi.org/10.1016/j.ijpharm.2011.02.019

Malhotra NK (2020) Marketing research an applied orientation, 7th edn. Pearson Education

Strang G (2022) Introduction to linear algebra. Wellesley-Cambridge Press

Sundararajan S (2021) Business analytics-overview, curriculum, opportunities and skills. Researchgate.Net

Sundararajan S (2023) MVA-ML. https://github.com/sun-sri/MVA-ML

Sundararajan S, Marath B, Vijayaraghavan PK (2019) Variation of risk profile across software life cycle in IS outsourcing. Softw Qual J 27(4). https://doi.org/10.1007/s11219-019-09451-8

Chapter 8
Cluster Analysis

Learning Objectives

- Examine the basic concepts of cluster analysis.
- Discuss and demonstrate hierarchical cluster analysis.
- Discuss and demonstrate partitioning methods.
- Discuss and demonstrate the performance evaluation of cluster analysis.
- Acquire familiarity with advanced cluster analysis methods.

Overview

Cluster Analysis has important applications in almost all areas, such as business (e.g., customer segmentation), social network analysis, recommendation systems, information retrieval, information security, computational biology, and climatology.

The chapter starts with an overview of cluster analysis and the methods/procedures for cluster analysis. An exposition of hierarchical methods, partitioning methods, and distance measurements follows this. We then explore the performance evaluation of clustering methods in detail and discuss advanced cluster analysis methods.

Supplementary Information The online version contains supplementary material available at https://doi.org/10.1007/978-981-99-0353-5_8.

Definitions

Agglomerative clustering: Agglomerative hierarchical clustering uses a bottom-up strategy. It starts by placing each object as an individual cluster. Then, it merges these atomic clusters into larger and larger clusters, iteratively, until all the objects are in a single cluster or until certain termination conditions are satisfied. Termination conditions can be a pre-specified number of clusters, a specific number of iterations, or the diameter of clusters reaching a certain threshold.

Calinski-Harabasz index: The Calinski-Harabasz index is the ratio of the sum of between-clusters dispersion to the sum of inter-cluster dispersion for all clusters. It is also called the Variance Ratio Criterion. A higher score implies better-defined clusters.

CLARANS method: The k-medoids method is expensive with large datasets and numerous features. We can use data samples rather than the entire dataset. CLARANS (Clustering Large Applications based upon RANdomized Search) is a trade-off between the cost and the effectiveness of using samples.

Cluster analysis: Cluster analysis is a class of techniques used to group objects or cases into relatively homogeneous groups called clusters. Objects in each cluster tend to be similar to each other and dissimilar to objects in other clusters.

Dendrogram: A tree structure called a dendrogram is commonly used to represent the process of hierarchical clustering. It shows how objects are grouped step by step. Dendrogram helps us to have an idea of cluster division. However, it does not give us an idea of the spatial distribution of clusters or the number of optimal clusters.

Density-based clustering: Density-based clustering identifies clusters of arbitrary shapes in a dataset, assuming that a cluster in a data space is a contiguous region of high data-point density, separated from other clusters by contiguous regions of low data-point density. Examples: - DBSCAN, OPTICS, and DENCLUE.

Divisive clustering: Divisive clustering starts with the entire dataset that is progressively split into multiple groups iteratively until each object is in a separate cluster of its own or until it satisfies certain termination conditions. See agglomerative clustering for termination condition.

Fowlkes-Mallow's Index: The Fowlkes-Mallows Index (FMI) is the geometric mean of the pairwise precision and recall. The score ranges from 0 to 1. A high value indicates a good similarity between a pair of clusters.

Hierarchical clustering: In hierarchical clustering, we group objects into a tree of clusters. Generally, there are two hierarchical clustering methods——agglomerative and divisive.

K-medoids method: In the k-means algorithm, the cluster center is computed as the arithmetic mean of the coordinates of all objects in a cluster. It is, therefore, sensitive to outliers. An almost centrally located actual object may be considered instead. Such a point is called medoid, and the method is called the k-medoids method.

Normalized mutual information (NMI): Mutual Information or Normalized Mutual Information measures the agreement of clusters with ground truth labels with the predicted cluster solution.

Partitioning method: In the partitioning method, k-objects selected randomly represent the initial cluster centers. All other objects are assigned to the nearest clusters. The cluster center is recomputed, and objects are reassigned to new clusters iteratively.

Probability model-based clustering: The probability model-based clustering method assumes a cluster is a parameterized distribution.

Rand index: The Rand index, or the adjusted Rand index, measures the similarity of clusters with ground truth labels with the predicted cluster solution.

Silhouette coefficient: The silhouette coefficient is calculated using the mean intra-cluster distance and the mean nearest-cluster distance. The score varies from -1 to $+1$. High positive scores indicate dense and well-separated clusters.

STING method: STING is a grid-based multi-resolution clustering technique. In this method, there are several levels of cells with increasing levels of resolution from top to bottom. Statistical information of each cell is calculated and stored beforehand and is used to answer queries.

V-measure: The V-measure is equivalent to normalized mutual information (NMI). V-measure is the harmonic mean of homogeneity and completeness.

Ward's method: Ward's method is a commonly used, variance-based method for agglomeration clustering. The squared Euclidean distance to the respective cluster means is calculated for each object. At each stage, the two clusters with the smallest increase in the sum of squares within-cluster distances are combined.

8.1   An Overview of Cluster Analysis

Cluster analysis is a class of techniques used to group objects or cases into relatively homogeneous groups called clusters. Objects in each cluster tend to be similar to one another and dissimilar to objects in other clusters. Similarities are computed in terms of proximity between the feature values ($X_1 \ldots X_k$).

8.1.1   Proximity Measures

8.1.1.1   Common Measures

Some of the proximity measures used for clustering are listed below (Han and Micheline Kamber, 2014).

- Euclidean, Manhattan, and Minkowski are common distance measures for numeric data.
- Cosine similarity helps compare rankings (e.g., recommendation systems, document ranking based on term frequency, etc.). The cosine similarity between vectors X and Y is computed as the ratio of the dot product of X and Y divided

by the L2 norms of X and Y. See Fig. 8.1b. The use of cosine similarity is demonstrated in Sect. 10.4, recommendation systems. Cosine distance is equal to 1 minus cosine similarity.

$$\text{Cosine similarity}(X, Y) = X.Y/||X||.||Y||$$

- Jaccard distance can be used for comparing asymmetric binary vectors, showing the presence or absence of features as bit strings. Jaccard similarity (A, B)=(A and B)/(A. or. B)
- Edit distance can be used to compare two strings.
- TF.IDF (Term Frequency times Inverse Document Frequency) is used to compare the similarity of documents based on the relative frequency of the terms used and the rarity of the terms.
- Hamming distance can be used for error correction in data communication.

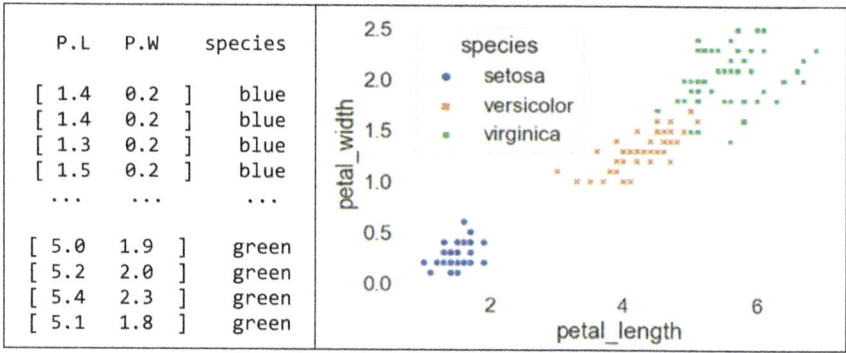

(a) Iris Flower Clusters in 2D Euclidean Distance

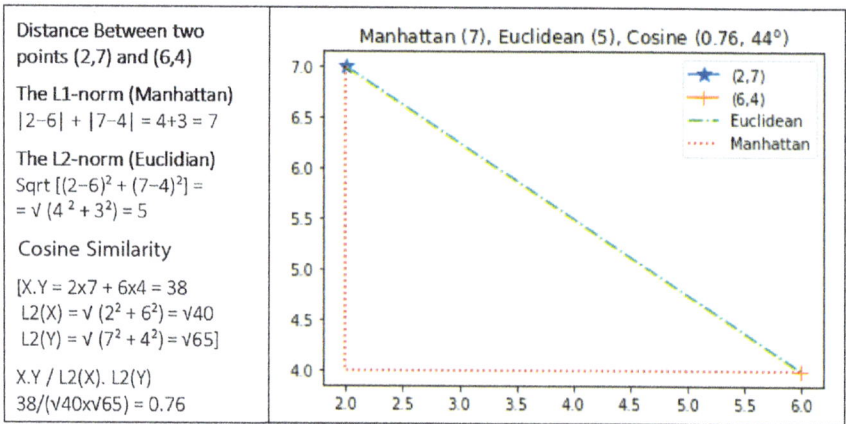

(b) Comparison of Three Different Distance Measures

Fig. 8.1 Popular distance measures

L1 and L2 Norms

L1 norm, also known as the Manhattan distance, measures the sum of the absolute differences between the corresponding components of two vectors. An example is the distance we travel from one point to another in a city, where we can only move along the streets in a grid-like fashion (horizontally and vertically), not through buildings and walls. L2 norm or the Euclidean Distance calculates the square root of the sum of the squared differences between the corresponding components of two vectors. It represents the straight-line distance between two points in a Cartesian coordinate system. The Minkowski distance is a generalized distance metric that measures the dissimilarity between two data points in a multi-dimensional space. It incorporates both the L1 (Manhattan distance) and L2 (Euclidean distance) norms as special cases, making it a flexible distance metric.

Euclidean distance is the most popular distance measure (the straight-line distance between two data objects). Figure 8.1a shows Iris Flower Clusters in Euclidean space, based on two dimensions——petal length and width. Figure 8.1b compares Manhattan, Euclidean, and Cosine distance measures.

Let X_1 and X_2 be two data objects having k numeric features. The Euclidean distance between X_1, X_2 can be expressed as follows

$$\text{Distance}(X_1, X_2) = \sqrt{\left[\left(X_1^1 - X_2^1\right)^2 + \left(X_1^2 - X_2^2\right)^2 + \cdots + \left(X_1^k - X_2^k\right)^2\right]}$$

Examples Euclidean Distance between two points (0,1) and (2,3)

$= (0-2)^2 + (1-3)^2 = 4 + 4 = 8$
Euclidean Distance between two points (1, 3, 5) and (4, 8, 16)
$= (1-4)^2 + (3-8)^2 + (5-16)^2 = 9 + 25 + 121 = 155$

Various Methods for Classification

Numerous methods are available for cluster analysis. Some of these methods are mentioned below. In this chapter, we will focus on Hierarchical and Partitioning Methods. The other methods are given a passing mention in Sect. 8.6

- Hierarchical Methods
- Partitioning Methods
- Density-based Methods
- Grid-based Methods
- Probability-model-based clustering
- Clustering high dimensional data
- Clustering Large Datasets
- Clustering Graphs and Network Data

8.1.2   Challenges in Cluster Analysis

Let us briefly examine the challenges in cluster analysis (Han and Micheline Kamber, 2014). Many clustering algorithms are effective with small datasets and a few features. In practice, we face datasets of a million rows and/or high dimensionality (several attributes). Therefore, we need scalable clustering algorithms. Many algorithms rely on numeric measurements and Euclidian distance. However, we must work with binary data, categories, densities, and network graphs. Common clustering algorithms tend to cluster in hypersphere; however, clusters may have irregular shapes.

Real-world datasets contain outliers or missing/unknown/noisy data, which can skew the results (see Sect. 2.8.2). The clustering results can be sensitive to input parameters and the order of data. Domain knowledge is necessary to determine the number of desirable clusters and for feature selection (especially for objects with several attributes). The clustering results must be interpretable and usable in practice.

In the case of fuzzy datasets, the data points may belong to multiple clusters to varying degrees. Various algorithms exist for fuzzy clustering, such as fuzzy C-Means (FCM) and probabilistic C-Means (PCM), though choosing the right algorithm for meaningful clustering is a challenge. Interpretability and scalability (as the data size increases) are some of the other major challenges faced.

8.1.3   General Procedure for Cluster Analysis

Feature Selection

The researcher must exercise care in selecting relevant features, for which she/he must properly understand the theoretical basis of the problem (Malhotra, 2020). Irrelevant features must be avoided, as they may distort the clustering solution. To the extent possible, we need a comprehensive set of features to define the phenomenon under investigation. For example, we need to measure financial, functional, service, and psychological features to study how a product can be positioned in a competitive market. The selected features must be able to describe the similarity and dissimilarity between the data objects.

The Right Number of Clusters

Hierarchical and nonhierarchical methods must be used together to arrive at the right number of clusters. Initially, hierarchical clustering analysis is done to understand the underlying structure. Note that the solution may depend on the order of cases in the dataset.

Determining the optimal number of clusters lacks quick fixes. Both theoretical and practical considerations should guide this process. In hierarchical clustering, the distances at which clusters merge serve as criteria for determining cluster count, extractable from the agglomeration schedule or dendrogram. Conversely, nonhierarchical clustering involves plotting the ratio of total within-group variance to between-group variance against cluster numbers, allowing identification of the suitable cluster count, as detailed in the discussion on the scree plot. Additionally, assessing the relative sizes of clusters from various solutions and employing diverse performance measures is crucial. The cluster shape is also important——it varies with the addressed phenomena.

8.2 Hierarchical Methods

In hierarchical clustering, we group objects into a tree of clusters. Generally, there are two hierarchical clustering methods——agglomerative and divisive.

8.2.1 Agglomerative/Divisive Clustering

Agglomerative hierarchical clustering employs a bottom-up approach, commencing by assigning each object as a distinct cluster. Subsequently, it systematically merges these initial clusters into increasingly larger ones until all objects are part of a single cluster or specific termination conditions are met. In contrast, divisive clustering initiates with the entire dataset. It iteratively divides it into multiple groups until each object resides in a separate cluster or until pre-determined termination conditions are fulfilled. These termination conditions may include a pre-determined cluster count, a specified number of iterations, and the diameter of clusters reaching a certain threshold, among other criteria.

8.2.2 Dendrogram

A tree structure called a dendrogram is commonly used to represent the process of hierarchical clustering. It shows how objects are grouped step by step. For example, Figs. 8.2 and 8.3 shows the dendrogram generated during the clustering of iris flowers. The X-axis represents the leaf nodes and the number of objects in the leaf. The horizontal lines of the grid divide the data into a specific number of clusters. The grid line through ($Y = 15$) divides the data into two clusters—the yellow ones and the green ones. Figures 8.2 and 8.3a and b show the agglomeration of data points into bigger and bigger clusters.

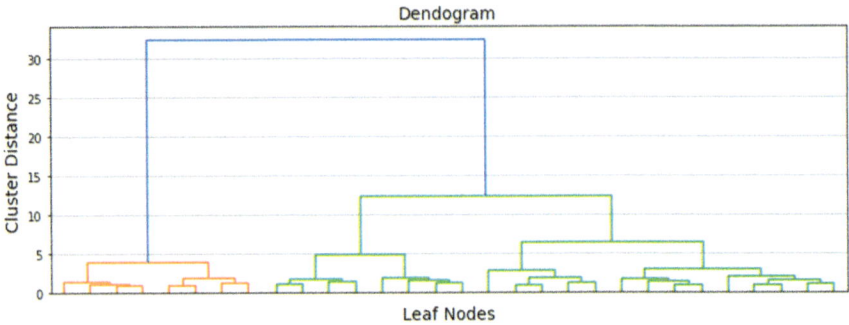

Fig. 8.2   Dendrogram

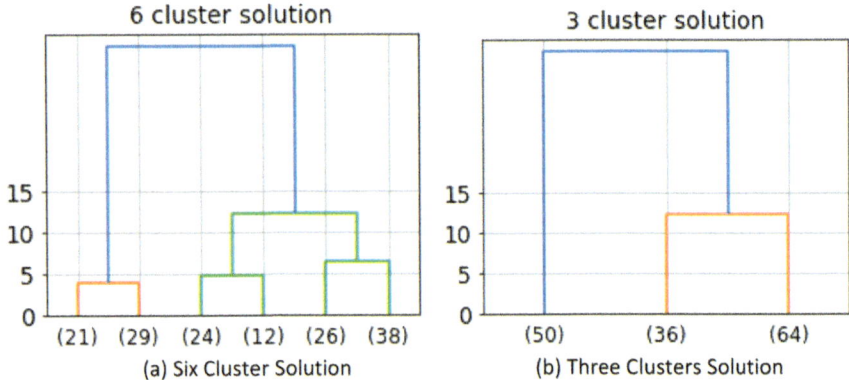

Fig. 8.3   Three clusters solution and six clusters solution

The Y-axis measure helps to quantify the relative distance between a pair of child clusters. From a visual check of Fig. 8.3a, we find that the inter-cluster distance between the first cluster of 21 flowers and the second cluster of 29 flowers is 4. Similarly, the inter-cluster distance between the third cluster of 24 flowers and the fourth cluster of 12 flowers is 5.

Dendrogram helps us to have an idea of cluster division. However, it does not give us an idea of the spatial distribution of clusters or the number of optimal clusters. Usually, the ideas gained from hierarchical clustering (e.g., the distribution of objects in each cluster) are used along with k-means clustering analysis to get a better picture.

8.2.3   Distance Measures in Hierarchical Clustering

The distance measures used in hierarchical and k-means clustering are different. We will see some popular hierarchical clustering methods—centroid, linkage, and variance method (Malhotra, 2020).

Centroid Method

In the centroid method, the distance between two clusters is the distance between their centroids. The cluster centroid is a k-dimensional vector formed by the means of each of the 'k' feature variables. During the agglomeration process, the centroid is recomputed after each re-grouping of objects.

Linkage Methods

The **single linkage method** is based on minimum distance or the nearest neighbor rule—the distance between two clusters is the distance between the two closest objects (each object chosen from each cluster). The steps in the single linkage method are shown below.

1. Start with one cluster for each point (sum of squares $= 0$).
2. Merge two clusters with the smallest gap (distance between the two closest points) between them.

Repeat steps 1 and 2 until k clusters emerge.

Single-link clustering can handle any cluster shape. However, it does not give due diligence to the emergence of cluster shape or balance of cluster sizes.

In **complete linkage**, the distance between two clusters is the distance between the two furthest objects (each object chosen from each cluster). In the **average linkage** method between two clusters, we make pairs of objects taken one from each cluster; the average distance between all such pairs is computed.

Variance Methods

The variance methods attempt to generate clusters that minimize the within-cluster variance. A commonly used variance method is Ward's method. The actions involved are:-

- Computation of Cluster Means: The mean of the feature variables vector is calculated for each cluster. This mean represents the central tendency of the data points within that cluster.
- Calculation of Squared Euclidean Distances: For each object (data point), the squared Euclidean distance between the object and the mean of the cluster to which it belongs is computed. Euclidean distance measures the straight-line distance between two points in space.
- Summation of Distances: The squared Euclidean distances calculated for all objects in all clusters are summed. This provides a measure of the total within-in-cluster variance, representing how spread out the data points are within each cluster.
- Cluster Merging: At each stage of the clustering process, the two clusters that, when combined, result in the smallest increase in the overall sum of squares

of within-cluster distances are merged. This helps in gradually forming clusters that minimize the total within-cluster variance.

The three-step procedure for merging clusters to arrive at optimal clusters is shown below.

1. Start with one cluster for each point (sum of squares $= 0$).
2. Merge two clusters, which results in the smallest increase in the sum of squares (of the distance between objects and the merged cluster mean). The sum of squares can be called merge cost. We must consider all possible pairs of clusters for merging.
3. Repeat steps 1 and 2 until k clusters emerge or the merging cost is beyond a limit.

In the sum of squares method, distances are measured equally in all directions. Therefore, the clusters that emerge will be round or spheroid. However, in practice, clusters do not come in regular spheroid shape. Therefore, the solution may not be accurate.

Tutorial 8.2 Hierarchical Clustering

Tutorial 8.2.1 Data Setup and Visualization

```
import pandas as pd
import seaborn as sb
pdf = sb.load_dataset('iris')
X = pdf.drop('species',axis=1)
```

Scatter Plot
```
sb.scatterplot(x='petal_length',y='petal_width', data=pdf, hue='species')
```

frequency distribution of all four measures
```
sb.kdeplot(data=pdf)
```

Tutorial 8.2.2 Dendrogram

```
import scipy.cluster.hierarchy as shc
import numpy as np
import matplotlib.pyplot as plt

plt.figure(figsize=(4,4))  # default 6.4 width, 4.8 height
yTicks=np.arange(0,20,5)
plt.yticks(yTicks, fontsize=14)
plt.grid()
plt.title('3 cluster solution', fontsize=14)
dend = shc.dendrogram((shc.linkage(X, method ='ward')),
                      p = 3, truncate_mode='lastp')
```
See Figure 8-3 (b): 3 cluster solution

```
plt.title('6 cluster solution', fontsize=14)
dend = shc.dendrogram((shc.linkage(X, method ='ward')),
                    p = 6, truncate_mode='lastp')
```
See Figure 8-3 (c): 6 cluster solution

```
plt.figure(figsize=(12,4))  # default 6.4 width, 4.8 height
plt.grid()
plt.title('Dendogram', fontsize=14)
plt.ylabel('Cluster Distance', fontsize=14)
plt.xlabel('Leaf Nodes', fontsize=14)
dend = shc.dendrogram((shc.linkage(X, method ='ward')),
                    truncate_mode='lastp')
xTicks=np.arange(0,1,1)
plt.xticks(xTicks)
plt.show()
```
See Figure 8-2: Dendrogram

8.3 Partitioning Methods

Partitioning Methods are about dividing the dataset into a specific number of clusters based on some criterion. There are various methods for partitioning, such as k-means, k-medoids, and CLARANS. We will briefly discuss these methods.

8.3.1 K-Means Method

The k-means algorithm iteratively refines the clusters by adjusting the cluster centers based on the mean of the objects within each cluster, aiming to optimize the overall similarity or distance criterion.

Given a dataset D with n objects and a specified number of clusters, k, a partitioning algorithm arranges the objects into k partitions (where $k < n$), each representing a cluster. The formation of these clusters is guided by the optimization of an objective function, often based on similarity or distance. The aim is to ensure that objects within a cluster are considered 'similar,' while those in different clusters are deemed 'dissimilar' based on the dataset's attributes.

Let C_i be a cluster with $\hat{c}_i$ as the center. Let 'o' be an arbitrary object in that cluster. The objective function can be expressed as:-

$$\text{Minimize } E = \Sigma_{i=1}^{k} \Sigma_{o \in C_i} (o - \hat{c}_i)^2$$

The algorithm is outlined:

1. Randomly choose k-objects, which serve as the initial cluster centers for the k clusters.
2. Assign all remaining objects to the cluster that is closest to them, determined by the distance between the object and the cluster mean.

3. Recalculate the mean for each cluster.
4. Iterate through Steps 2 and 3 until the criterion function converges. The criteria for convergence can be set as one of the following: (a) a fixed number of iterations, (b) partitions remaining unchanged, or (c) cluster centers not changing.

8.3.2   K-Medoids Method

The k-means algorithm is sensitive to outliers. The cluster center is computed as the arithmetic mean of the coordinates of all objects in a cluster. Outliers can skew the computation of the mean value. Therefore, an almost centrally located *actual object* may be considered the central reference point of a cluster. Such a point is called medoid, and the method is called the k-medoids method. The k-medoids algorithm is described.

1. Initialization: Randomly select 'k' objects from the dataset as initial medoids.
2. Assignment: Assign each remaining object in the dataset to the cluster whose current medoid is the closest (based on a chosen distance metric).
3. Update Medoids: Calculate the total cost (or distance) of all objects to the current medoid for each cluster. For each object in the cluster, swap it with the medoid and calculate the total cost. If the swap reduces the total cost, update the medoid to be the object with the minimum total cost within the cluster.
4. Repeat the Assignment and Update Medoids steps until convergence criteria are met (e.g., no further changes in medoids or a specified number of iterations have been reached).

The key difference between k-means and k-medoids lies in how the cluster center is updated. In k-means, the center is the mean of the points in the cluster, while in k-medoids, the center is one of the actual data points. This makes k-medoids more robust to outliers, as extreme values affect the medoid less.

8.3.3   CLARANS Method

The k-medoid method is expensive with large datasets and/or numerous features. We can use data samples rather than the entire dataset, but the effectiveness will depend on the sample size. CLARANS (Clustering Large Applications based upon RANdomized Search) is a trade-off between the cost and the effectiveness of using samples.

This method randomly selects 'k' objects as the medoids. Then we randomly select a medoid 'm_i,' and an object 'o' that is not designated as a medoid. Check whether replacing 'm_i' with 'o' improves the optimizing criterion. If yes, replace m_i with 'o'. Repeat this process a specific number of times. The result is the best local optimum solution.

The key idea behind CLARANS is to explore different potential cluster config-urations by iteratively considering random medoid-object pairs. This randomiza-tion allows the algorithm to escape local optima and find a solution that might be more globally optimal. By operating on a subset of the data through random sam-pling, CLARANS achieves computational efficiency, making it more feasible for large datasets. The effectiveness of CLARANS depends on the balance between the number of iterations performed and the quality of the random sampling. A larger number of iterations increases the likelihood of finding a better solution but also incurs more computational cost.

The algorithm offers a practical compromise for cases where the full dataset is too large to handle directly with traditional k-medoid methods.

8.3.4 Distance Measures in Partitioning Methods

As mentioned earlier, the distance measures used in hierarchical and k-means clustering differ. The distance measures used in k-means clustering include sequential threshold, parallel threshold, and optimizing partitioning.

In the sequential threshold method, the process begins by arbitrarily choos-ing an object as a cluster center. All objects within a pre-determined threshold distance from this initial object are grouped to create a cluster. Once clustered, these objects are considered fixed. Another object is selected from the remaining un-clustered objects, reiterating the process.

The parallel threshold method involves simultaneously selecting multiple clus-ter centers. Objects within a specified threshold distance from each chosen center are grouped with their respective nearest center.

Objects can be reassigned to different clusters to optimize a global criterion in the optimizing partitioning method. This method involves adjusting cluster assign-ments to enhance the overall performance of the clustering solution.

8.4 Performance Evaluation

There are several measures for determining the performance of cluster analysis. If the data labels are not known, we may use measures such as the silhouette coef-ficient (Jiawei Han, Micheline Kamber, 2014) and the Calinski-Harabasz index (Halkidi et al., 2001). If data labels are available for the training dataset, meas-ures such as Fowlkes-Mallow's score, rand index, normalized mutual information (NMI), adjusted mutual information (AMI), V-measure, etc. may be used. A scree plot can determine the effective number of clusters, built on the ratio of inter-clus-ter to intra-cluster variance.

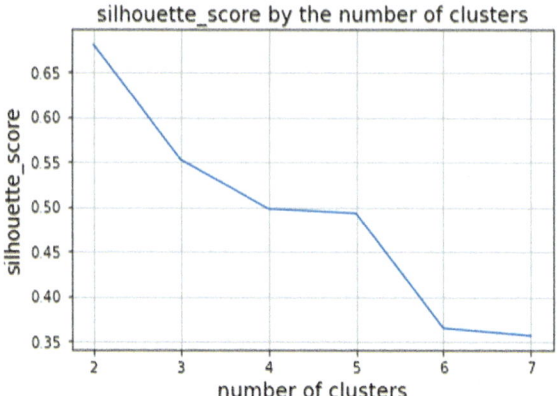

Fig. 8.4 K-Means silhouette score

8.4.1 Silhouette Coefficient

The silhouette coefficient of an object is calculated using the mean intra-cluster distance and the mean nearest-cluster distance. This can be expressed as,

$$S = \frac{b - a}{\max(a, b)}$$

where:

'a' is the Mean distance of an object with all other objects in the same cluster,

'b' is the mean distance of an object with all other objects in the next nearest cluster.

The average silhouette coefficient value of all the objects in a cluster indicates the goodness of fit of that cluster. The average of the silhouette coefficient values of all the objects over all the clusters taken together indicates the fitness of the clustering solution.

The score varies from -1 to $+1$. High positive scores indicate dense and well-separated clusters. Scores around zero indicate overlapping clusters. A negative value indicates incorrect clustering. Figure 8.4 shows that 2-cluster, 3-cluster, 4-cluster, and 5-cluster solutions are relatively better. Then, we find the score dropping steeply.

8.4.2 Calinski-Harabasz Index

The Calinski-Harabasz index (CH) is the ratio of the sum of between-clusters dispersion (SSB) to the sum of inter-cluster dispersion (SSW) for all clusters. It is also called the Variance Ratio Criterion. A higher score implies better-defined clusters. Assume a dataset with N objects with $\hat{c}$ as the overall cluster center. Let o_j be an arbitrary object.

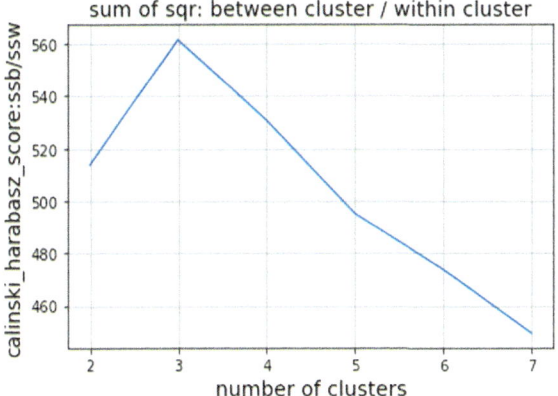

Fig. 8.5 K-means Calinski-Harabasz score

$$SSB(SS_{between}) = \Sigma_{j=1}^{N}(o_j - \widehat{c})^2 \text{ with } k - 1 \text{ degrees of freedom}$$

As we saw in the criterion function stated in the section on k-means, assume k clusters, let C_i be a cluster with $\widehat{c_i}$ as the center, and 'o' be an arbitrary object in that cluster. Then SSW can be expressed as follows:-

$$SSW(SS_{within}) = \Sigma_{i=1}^{k} \Sigma_{o \in C_i}(o - \widehat{c_i})^2 \text{ with } n - k \text{ degrees of freedom}$$

$$CH = \frac{SSB}{SSW} \frac{(n-k)}{(k-1)}$$

Figure 8.5 shows that 2-cluster, 3-cluster, 4-cluster, and 5-cluster solutions have relatively better scores, after which the scores drop steeply. The 3-cluster solution is observed to be the best.

8.4.3 Evaluation of the Quality of Clustering

If we know the target labels, we can evaluate the effectiveness of the clustering solution that has emerged from cluster analysis. The metrics include Fowlkes-Mallow's score, Rand index, Normalized mutual information (NMI), Adjusted mutual information (AMI), V-measure, etc. We will take a brief look at these measures. See Tutorial 8.5/Fig. 8.6.

Confusion Matrix

Let us recap the confusion matrix, covered in the chapter 'Classification'. See Table 8.1. Here TP is the count of True Positives, FP is the count of False Positives, FN is the count of False Negatives, and TN is the count of True Negatives. Common performance measures include Precision, Recall, Accuracy,

Fig. 8.6 Clustering scores
when ground truth is known

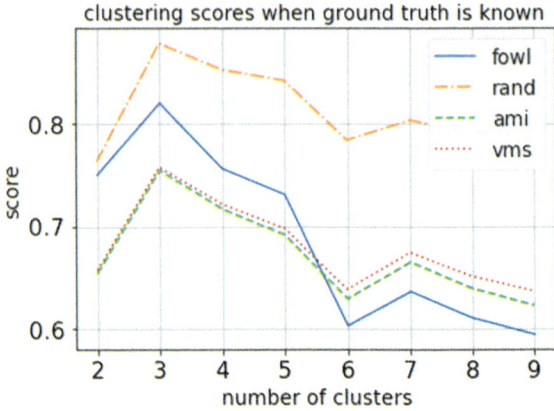

Table 8.1 Confusion matrix—when ground truth is known

		Predicted labels		
		1	0	
Observed labels	1	True Positives (TP)	False Negatives (FN)	R (Recall) = TP / (TP+FN)
	0	False Positives (FP)	True Negatives (FN)	FMI = √(PR) Geometric mean
		P (Precision) = TP / (TP+FP)	Arithmetic Mean = (P+R) / 2	F1 Score = 2PR / (P+R) Harmonic mean

F1 Score, and FMI. It is the prerogative of the researcher to choose the appropriate
score for performance rating.

Fowlkes-Mallow's Index (FMI)

The Fowlkes and Mallows Index (FMI) is defined as the geometric mean of the
pairwise precision and recall, as expressed below. The score ranges from 0 to 1. A
high value indicates a good similarity between a pair of clusters. Based on Table
8.1, FMI can be expressed as follows.

$$\text{the FMI} = \sqrt{\text{Precision} * \text{Recall}}$$

$$= \sqrt{\text{TP}/(\text{TP} + \text{FP}) * \text{TP}/(\text{TP} + \text{FN})}$$

$$= \text{TP}/\sqrt{(\text{TP} + \text{FP})(\text{TP} + \text{FN})}$$

Rand Index

Rand index or the adjusted rand index, measures the similarity of clusters with ground truth labels with the predicted cluster solution, ignoring permutations and with chance normalization.

For a sample size of n, the total number of pairs of data objects $= nC_2$.

Let CG be the actual clustering of data objects with known ground truth labels. Let CP be the predicted clustering solution. Rand Index is defined as follows.

$$\text{Rand Index} = (\text{known pairs} + \text{strange pairs})/nC_2$$

In the above equation, 'known pairs' is the count of object pairs that appear in the same clusters in CG, and CP and 'strange pairs' is the count of all the pairs that reside in different clusters in CG and CP. Adjusted Rand Index is obtained by scaling Rand Index to $\{0 \ldots 1\}$, so that it can be used for comparison.

Normalized Mutual Information (NMI)

Mutual Information or Normalized Mutual Information measures the agreement of clusters with ground truth labels with the predicted cluster solution, ignoring permutations.

V-measure

The V-measure is equivalent to normalized mutual information (NMI). V-measure is the harmonic mean of homogeneity and completeness. A homogeneous clustering solution is composed of pure clusters. Here, each cluster is pure——it consists of only objects of one class label. In a complete clustering solution, *all the objects* belonging to a specific cluster (as per ground truth) must be assigned to one cluster. The V-measure can be expressed as:-

$$V - \text{measure} = \frac{(1 + \beta)}{\beta} \frac{\text{homogeneity} * \text{completeness}}{\text{homogeneity} + \text{completeness}}$$

where: β is Given a Value of 2 by Default

8.4.4 Scree Plot

The methods mentioned above do not give an exact method to decide the number of clusters. Figure 8.7 shows a plot of the ratio of the mean squared intra-cluster distances to the mean of squared inter-cluster distances. The 'scree plot' plot offers a simple method to guess the optimal cluster count. The knee of the scree plot gives us a good guess at the cluster count. The scree plot in Fig. 8.7 offers a 3-cluster solution.

Fig. 8.7 Scree plot to
determine the optimum
number clusters

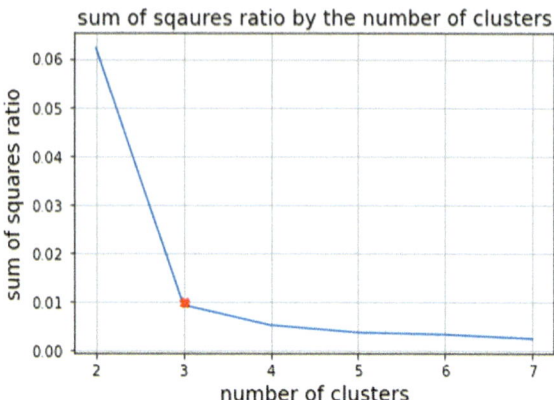

Tutorial 8.4.1 K-Means Clustering

Do Data Setup as in Tutorial 8.2.1 Data Setup and Visualization
```
    from sklearn.cluster import KMeans
```
k = 3; Three Cluster Solution
```
    cluster = KMeans(n_clusters=3, random_state=0).fit(X)
    print(np.round((cluster.cluster_centers_),1))
    [[5.9 2.7 4.4 1.4]
     [5.0 3.4 1.5 0.2]
     [6.8 3.1 5.7 2.1]]
```
Predict the cluster membership of each object
```
    cluster.fit_predict(X)
```
Let us inspect some of the parameters:-
```
    cluster.get_params() # 'max_iter': 300, 'tol': 0.0001,
```
Distance of data points from the cluster centers
```
    X2D   =  cluster.fit_transform(X)
    print(np.round(X2D,2))
    [[3.42 0.14 5.06]
     [3.4  0.45 5.11]
     [3.57 0.42 5.28]

     ..

     [1.18 4.41 0.65]
     [1.51 4.6  0.84]
     [0.83 4.08 1.18]]
```

Tutorial 8.4.2 Cluster Count by Silhouette and Calinski_harabasz scores

Objectives:
1. Plot silhouette_score and visually compare the clustering solutions
2. Plot calinski_harabasz_score and visually compare the clustering solutions

Do Data Setup as in Tutorial 8.2.1 Data Setup and Visualization

```
from sklearn.cluster import KMeans
from scipy.spatial import distance
from sklearn import metrics # silhouette, calinski_harabasz
import matplotlib.pyplot as plt
```

The lower the score, the lower the better. Score for 2,3,4,5,6,7 clusters:

```
kstop = 8
sil     = [0] * (kstop-2)
cal     = [0] * (kstop-2)
for k in range(2,kstop,1):
    cluster  = KMeans(n_clusters=k, random_state=0).fit(X)
    sil[k-2] = metrics.silhouette_score(X, cluster.labels_,
metric='euclidean')
    cal[k-2] = metrics.calinski_harabasz_score(X, cluster.labels_)
```

Plot the Silhouette Score for 2,3,4,5,6,7 cluster solutions
```
sslabels = list(range(2,kstop,1)); sslabels
plt.plot(sslabels, sil)
plt.xticks(sslabels)
plt.grid()
plt.title ('silhouette_score by the number of clusters', fontsize=14)
plt.xlabel('number of clusters', fontsize=14)
plt.ylabel('silhouette_score', fontsize=14)
```
See Figure 8-4: K-Means Silhouette Score

Plot the Calinski Harabasz Score for 2,3,4,5,6,7 cluster solutions
```
sslabels = list(range(2,kstop,1)); sslabels
plt.plot(sslabels, cal)
plt.xticks(sslabels)
plt.grid()
plt.title ('sum of sqr: between cluster / within cluster', fontsize=14)
plt.xlabel('number of clusters', fontsize=14)
plt.ylabel('calinski_harabasz_score:ssb/ssw', fontsize=14)
```
See Figure 8-5: K-Means Calinski-Harabasz Score

Tutorial 8.4.3 Scree Plot to Determine the Cluster Count

Objective: Generate Scree Plot to determine the optimum number clusters.

Do Data Setup as in Tutorial 8.2.1 Data Setup and Visualization

Define ssw

```
from sklearn.cluster import KMeans
def compute_ssw (X, centroids, predicted_class):
    ssw = 0
    for i in range(len(X)):
        xi = X.loc[i]
        xc = centroids[predicted_class[i]]
        ssw = ssw + distance.euclidean(xi,xc)**2
        return ssw
```

Define ssb

```
def compute_ssb (centroids):
    ssb = 0
    nclusters = len(centroids)
    for p in range(0,nclusters):
        for q in range(1, nclusters):
            if (p != q):
                ssb = ssb +
    distance.euclidean(centroids[p],centroids[q])**2
    return ssb
```

Compute ssw/ssb: -

Setup libraries and parameters

```
from sklearn.cluster import KMeans
from scipy.spatial import distance
import matplotlib.pyplot as plt
kstop    = 8  # to generate 7 clusters
ssratio  = [0] * (kstop-2) # 5 elements [0, 0, 0, 0, 0]
```

Evaluate 5 different cluster solutions, 2, 3, 4, 5, 6

```
for k in range(2,kstop,1):
    cluster = KMeans(n_clusters=k, random_state=0).fit(X)
```
Predict the cluster membership of each object
```
    predicted_class = cluster.fit_predict(X)
    centroids = cluster.cluster_centers_
    ssw = compute_ssw (X, centroids, predicted_class)
    ssb = compute_ssb (centroids)
    ssratio[k-2] = ssw / ssb

sslabels = list(range(2,kstop))
sslabels # [2, 3, 4, 5, 6, 7]
```

```
plt.plot(sslabels, ssratio)
plt.xticks(sslabels)
plt.grid()
plt.title ('sum of sqaures ratio by the number of clusters', fontsize=14)
plt.xlabel('number of clusters', fontsize=14)
plt.ylabel('sum of squares ratio', fontsize=14)
```
See Figure 8-7: Scree Plot to Determine the Optimum number Clusters

Inference: -
See Figure 8-7: Scree Plot to Determine the Optimum number Clusters. The
scree plot bends at X=3. (This is the knee of the scree plot). Therefore 3
cluster solution is optimal

Tutorial 8.4.4 When the labels known

Do Data Setup as in Tutorial 8.2.1 Data Setup and Visualization

```
from sklearn.cluster import KMeans
import numpy as np
import matplotlib.pyplot as plt

from sklearn.cluster import KMeans
from sklearn.metrics.cluster import fowlkes_mallows_score
from sklearn.metrics.cluster import rand_score
from sklearn.metrics.cluster import adjusted_mutual_info_score
from sklearn.metrics.cluster import v_measure_score
```

Compute Scores for 2 ... 9 clusters:

```
y = pdf.species
leg = ['fowl','rand', 'ami', 'vms']
kstop = 10
fowl  = [0] * (kstop-2)
rand  = [0] * (kstop-2)
ami   = [0] * (kstop-2)
vms   = [0] * (kstop-2)
```

The iris dataset consists of data labels - the 'species'. Predict the cluster
membership of each object (y_predicted) and check with the given labels for
accuracy of the model.

```
for k in range(2,kstop,1):
    y_predicted = KMeans(n_clusters=k,random_state=0).fit(X).fit_predict(X)
    fowl[k-2] = fowlkes_mallows_score(y, y_predicted)
    rand[k-2] = rand_score(y, y_predicted)
    ami[k-2]  = adjusted_mutual_info_score(y, y_predicted)
    vms[k-2]  = v_measure_score(y, y_predicted)
```

```
Plot Scores for 2 ... 9 clusters
   sslabels = list(range(2,kstop,1))
   plt.plot(sslabels, fowl,linestyle='solid')
   plt.plot(sslabels, rand,linestyle='dashdot')
   plt.plot(sslabels, ami, linestyle='dashed')
   plt.plot(sslabels, vms, linestyle='dotted')
   plt.xticks(sslabels)
   plt.grid()
   plt.title ('clustering scores when labels are known', fontsize=14)
   plt.xlabel('number of clusters', fontsize=14)
   plt.ylabel('score', fontsize=14)
   plt.legend(leg, fontsize=14)
See Figure 8-6: Clustering Scores When Ground Truth is Known
```

8.5   Other Clustering Methods

In this chapter, we explored two popular cluster analysis methods——partition-ing and hierarchical. There are many other methods in practice. We will have a cursory view of some of them (Han and Micheline Kamber, 2014).

8.5.1   The Choice of Algorithms

The underlying assumptions of the clustering algorithm and the data distribu-tion influence the cluster shape. For example, the K-means algorithm assumes that clusters are spherical and equally sized, as it minimizes the squared Euclidean dis-tance between points and cluster centroids. Hierarchical clustering builds a tree of clusters, and the shape of the clusters is not constrained. The DBSCAN algorithm identifies clusters based on the density of data points. It can find clusters of arbi-trary shapes and is not sensitive to the assumption of spherical clusters.

When dealing with non-spherical clusters, an algorithm that aligns with the underlying structure of the data is crucial. Experimenting with different algorithms and validating their performance on your specific dataset is often necessary to find the most suitable approach. Let us conclude the chapter with an overview of other important clustering algorithms.

- Density-based clustering
- Grid-based clustering
- Probability Model-Based Clustering
- Clustering high dimensional data
- Clustering Large Datasets
- Clustering Graphs and Network Data.

8.5.2   Density-Based Clustering   .

Density-based clustering identifies clusters of arbitrary shapes in a dataset, assuming that a cluster in a data space is a contiguous region of high data-point density, separated from other clusters by contiguous regions of low data-point density. Cluster density is passed to the algorithm as a parameter. The data objects in the cluster separation regions may be considered noise or outliers. The algorithm works in one data scan. Examples of Density-based clustering algorithms are.-

- DBSCAN—Density-Based Spatial Clustering of Applications with Noise
- OPTICS—Ordering Points to Identify the Clustering Structure
- DENCLUE—Uses Statistical Probability Density Functions.

8.5.3   Grid Based Clustering

To understand the general principle underlying grid-based clustering, let us consider the example of the 'STING' method. STING is a multi-resolution clustering technique based on a grid structure. The spatial area is initially divided into rectangular cells, organized into several levels with increasing resolution from top to bottom. At higher levels, each cell is further partitioned into numerous smaller cells at lower levels.

Before query processing, each cell's statistical information is computed and stored, facilitating efficient query responses. Parameters such as count, mean, standard deviation, min, max, and probability distribution type are pre-calculated for both high and low-level cells. The top-down approach of the grid allows for the straightforward computation of high-level cell parameters based on those of the lower-level cells.

Spatial data queries are addressed top-down, starting from a selected layer with a small cell count and progressively identifying relevant cells in lower layers to obtain detailed information. Notably, the method is well-suited for parallelization, enhancing computational efficiency.

8.5.4   Probability Model-Based Clustering

This method assumes that a cluster is a parameterized distribution. The given dataset is considered a data sample. Based on the dataset, we estimate the parameters of the clusters.

8.5.5   Clustering High Dimensional Data

When the dimensionality (the number of features) is high, conventional distance measures can be dominated by noise. This necessitates specialized algorithms for clustering high dimensional data. There are two major methods for
this—(a) search for clusters existing in subspaces, where a subspace is a space
woven by a subset of the attributes, e.g., Clique, Proclus, correlation-based clustering, bi-clustering, etc. (b) dimensionality reduction, e.g., feature selection and
summarization, principal component analysis method, etc.

8.5.6   Clustering Large Datasets

Clustering algorithms scale poorly in terms of computing time as the dataset
size increases. BIRCH and CHAMELEON are advanced hierarchical clustering
methods for large datasets.

BIRCH has a two-step process. In the first step, the dataset is scanned sequentially. The data points are added one after the other to a structure similar to the
B + tree. An incoming data point is added to the closest leaf entry. The intermedi-
ary nodes contain pointers to leave nodes and cluster features such as the number
of data points, the sum of data points, and the sum of the squares of data points
(the zeroth, first and second moments). With the addition of new data points,
cluster features are updated. Two parameters are in the tree construction——the
maximum number of children per non-leaf node and the maximum diameter of
sub-clusters at the leaf nodes. These parameters determine the merge/split of
nodes. In the second step, an appropriate clustering algorithm is used to cluster
the leaf nodes of the CF tree, removing the sparse clusters and grouping dense
clusters into larger ones. BIRCH algorithm is fast O(n). However, it is sensitive
to the order of insertion of data points, as is the case with B + trees. If the data is
unordered, building the tree takes longer. Since the size of the leaf nodes is fixed
as a parameter, the emerging clusters may not be natural. Since cluster diameter is
another parameter used, the emerging clusters tend to be spherical.

CHAMELEON is a hierarchical clustering algorithm designed to effectively
group data into clusters by incorporating dynamic modeling. First, small sub-clusters are created. Then, they are combined into larger clusters iteratively. The merging decisions are based on evaluating inter-cluster relationships and within-cluster
cohesion, promoting the formation of meaningful and well-balanced clusters. The
steps are elaborated below.

The objects are initially divided into many small sub-clusters using a
graph-partitioning algorithm. This step creates a set of initial clusters, each representing a subset of similar objects. The algorithm then agglomerates these small
sub-clusters to form larger clusters iteratively.

The decision to merge two clusters is based on careful consideration of inter-connectivity and proximity measures. Two clusters are merged if the interconnectivity (relationships between data points in different clusters) and closeness (proximity of data points within a cluster) between the clusters are notably higher than the within-cluster measures (relationships and proximity within each cluster). This criterion ensures that the merging process maintains a balance——clusters are combined when the connections between them are significant, indicating a meaningful similarity while also considering the closeness of data points within each cluster.

8.5.7   Clustering Graphs and Network Data

Data represented as graphs and networks have become quite common today, especially in online social networks, the World Wide Web (WWW), and digital libraries. These networks can represent connections between various entities, such as people in a social network or web pages.

Traditional clustering focuses on grouping similar items based on their attributes. In the context of network data, clustering and community detection are primarily concerned with grouping entities based on their connections within the network. Let us see some examples. In community detection, the entities are grouped based on their connections within the network. Citation networks are clustered based on relationships between research papers and citations. In social network analysis, individuals are based on their connections (friendship, communication, etc.). So, network-graph-based similarity and clustering methods necessitate very different clustering methods. See Fig. 8.8. Clustering network graphs are discussed further in the chapter on 'Computational Techniques'.

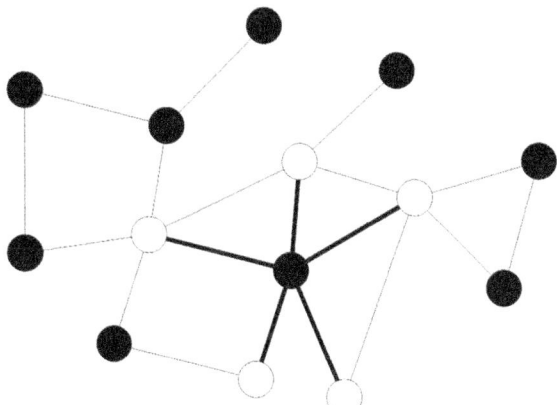

Fig. 8.8 Social network clusters

Data Analytics in Action

State-of-the-art clustering data streams. Big Data Analytics (Ghesmoune et al., 2016).
   Data stream clustering applications are deployed in numerous spheres, including network intrusion detection, transaction streams, phone records, web click-streams, social streams, weather monitoring, etc. In data streams, we assume that the data is voluminous and arrives so rapidly that storing it in a conventional database for processing later is not feasible. Therefore, if not processed in real-time, the data is lost. Extending our imagination, we assume we can make only a small number of passes on the data to generate clusters. Here, both the CPU time and memory pose challenges. The paper attempts a comprehensive survey of the algorithms for clustering over data streams. This is diagrammatically shown in Fig. 8.9. The paper discusses popular open-source streaming platforms such as Spark Streaming, Flink, MOA, and SAMOA. The paper also describes open challenges in data stream clustering——data privacy, data variety, data veracity, data summarization/data sampling, distributed streams, autonomous and self-diagnosis, and combining offline and online models.

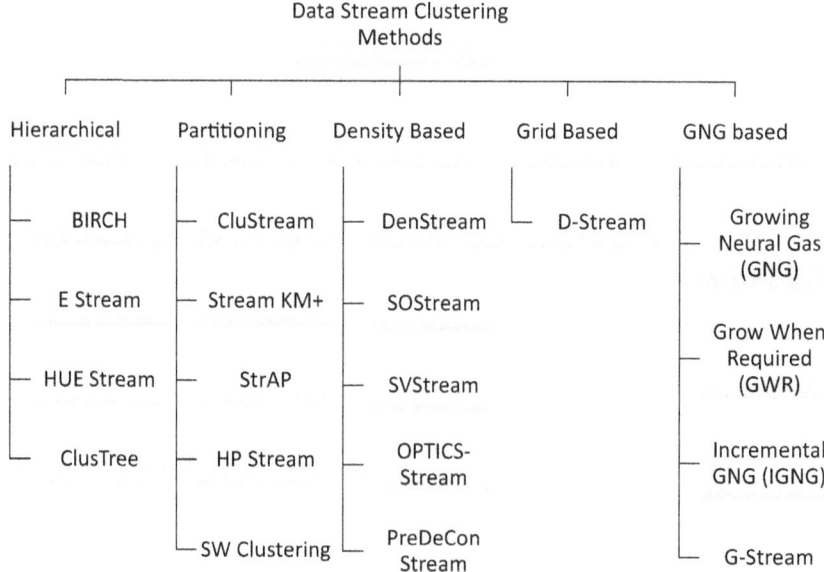

Fig. 8.9 Clustering streams

Summary

Cluster analysis is a class of techniques used to group objects or cases into relatively homogeneous groups called clusters. Objects in each cluster tend to be similar and dissimilar to objects in other clusters. Similarities are computed in terms of proximity between the feature values. The proximity measures used for clustering can be varied——euclidean for numeric data, Jaccard distance for similarity of documents, edit distance to compare strings, hamming distance for error correction in data communication, etc. Density-based clustering and clustering of network graphs need different measures.

There are many challenges in cluster analysis. For high volume/high dimensional data, we need scalable clustering algorithms. The distance measures can be varied. Common clustering algorithms tend to cluster in hypersphere; however, clusters come in irregular shapes. Real-world datasets contain outliers, missing, or noisy data. We need a comprehensive set of relevant features with a well-founded theoretical basis. The clustering results must be interpretable and usable in practice.

In hierarchical clustering, objects are grouped into a tree of clusters. Agglomerative hierarchical clustering employs a bottom-up approach, progressively combining individual objects into clusters. On the other hand, divisive clustering starts with the entire dataset. It iteratively divides it into multiple groups until each object is in a separate cluster or termination conditions are met. A dendrogram, a tree structure, visually represents the hierarchical clustering process, illustrating how objects are grouped step by step. Although the dendrogram provides insights into cluster division, it lacks information about the spatial distribution of clusters or the optimal number of clusters. Agglomeration clustering employs various methods such as the centroid, linkage, and variance. Variance methods aim to create clusters that minimize within-cluster variance.

The distance measures used in k-means clustering include sequential threshold, parallel threshold, and optimizing partitioning. There are various methods for partitioning, such as k-means, k-medoids, and clarans. In the k-means algorithm, the cluster center is computed as the arithmetic mean of the coordinates of all objects in a cluster. It is, therefore, sensitive to outliers. An almost centrally located actual object may be considered instead. Such a point is called medoid, and the method is called the k-medoids method. With large datasets and numerous features, we may use data samples rather than the entire dataset. CLARANS (Clustering Large Applications based upon RANdomized Search) is a trade-off between the cost and the effectiveness of using samples.

There are several measures for determining the performance of cluster analysis. This includes the silhouette coefficient, Calinski-Harabasz index, etc. If the ground truth is known, measures such as Fowlkes-mallow's index; normalized mutual information (NMI), adjusted mutual information (AMI), V-measure, etc.

may be used. A scree plot can be used to determine the effective number of clusters, which works on the ratio of intra-cluster to inter-cluster variance.

This chapter explored two popular cluster analysis methods——partitioning and hierarchical methods. There are many other methods in practice, such as density-based clustering, grid-based clustering, probability model-based clustering, high dimensional data, large datasets, clustering graphs, and network data.

Questions

Comprehension

1. Define Cluster Analysis
2. Write a note on the applications of Cluster Analysis
3. How do you tackle the problem of outliers in cluster analysis?
4. What is the rationale behind the CLARANS method?
5. Discuss the challenges in clustering high dimensional data. Suggest clustering methods to tackle them.

Write brief notes on

6. Hierarchical Methods
7. Partitioning Methods
8. Dendrogram
9. Fowlkes-Mallow's score
10. Density-based Methods
11. Probability-model-based clustering.
12. Grid-based clustering
13. Clustering graphs and network data
14. BIRCH algorithm for clustering large datasets
15. What are the challenges in cluster analysis?
16. How to determine the cluster proximity (the distance measures) in hierarchical and k-means clustering
17. How do we determine the optimal number of clusters?
18. Describe the k-means clustering algorithm with an example.
19. Describe the various performance measures used in cluster analysis.

Analysis

20. Compare and Contrast Cluster Analysis with Classification
21. How does the cosine distance measure work, and in what scenarios is it beneficial?
22. When would you use the Jaccard distance, and what types of data does it apply to?

23. Can you explain the difference between L1 and L2 distance norms and how they are applied in clustering?

24. Can you explain how domain knowledge is critical in addressing challenges like determining the right number of clusters and selecting relevant features for clustering?

25. What are the advantages of the k-medoid method over the traditional k-means approach, particularly in handling data with outliers and noise?

26. How does the CLARANS method balance cost and effectiveness when dealing with large datasets, and what are the trade-offs associated with using data samples?

27. What role does the dendrogram play in hierarchical clustering, and why is it often used in conjunction with k-means clustering for a more comprehensive understanding of the data's structure?

28. How do the challenges of cluster analysis in high dimensional data differ from those in low-dimensional data, and what advanced techniques can be applied to address these challenges?

29. What criteria can be considered when selecting the 'right' number of clusters, and how do hierarchical clustering and nonhierarchical methods contribute to this decision?

30. In your experience, how do outliers and noisy data impact the choice of clustering algorithms and the quality of clustering results, and can you provide strategies to handle these issues effectively?

31. What is the significance of the confusion matrix in evaluating cluster analysis results when the ground truth is known? How can precision, recall, accuracy, F1 Score, and the Fowlkes-Mallow's Index be derived from this matrix?

32. In the context of clustering, what is the Rand Index, and how does it measure the similarity of clusters with known ground truth labels? Can you explain the concept of 'known pairs' and 'strange pairs' in the Rand Index calculation?

33. What is a 'scree plot,' and how can it help determine the optimal number of clusters in cluster analysis? Are there specific guidelines for interpreting scree plots effectively?

34. Explain the importance of feature selection in cluster analysis. How can irrelevant features affect the quality of clustering solutions, and what strategies can be employed to address this issue?

35. Compare density-based clustering methods (e.g., DBSCAN) with grid-based clustering methods (e.g., STING). What are the key differences in their approaches, and in what situations would you choose one over the other for a clustering task?

36. BIRCH and CHAMELEON are hierarchical clustering methods designed for large datasets. Explain the fundamental differences between these two approaches and how they handle clustering in large-scale scenarios. What are the trade-offs associated with each method?

37. How does the choice of clustering algorithm impact the identification of clusters in a dataset? Specifically, how does the k-means algorithm influence cluster shapes, and how do K-means and DBSCAN handle different cluster shapes?

38. Clustering graphs and network data is distinct from traditional attribute-based clustering. Discuss the primary differences and challenges in clustering graph-based data. Share an example of a situation where network graph clustering led to meaningful insights or applications.

Application

39. Write a brief note on Euclidean Distance in k dimensions. Find the Manhattan, Euclidean, and cosine distance between the two points [1, 6, 9] and [2, 4, 8]
40. How can cluster analysis be applied in business, and what are some specific use cases for customer segmentation?
41. How does cluster analysis help suggest products or content to users in recommendation systems? Can you give an illustration of this process?
42. How is cluster analysis applied in information retrieval, and what benefits does it offer in terms of organizing and retrieving information?
43. How can density-based clustering algorithms like DBSCAN be applied in practice? Provide an example of a real-world scenario where density-based clustering is valuable and explain the benefits of using such an approach.

Exercises

The questions in this section are based on the penguins dataset accessible from the Seaborn Library. Select all the male penguins. Develop different clustering solutions based on all the numeric features.

Penguins Dataset Preprocessing

```
import seaborn as sb
d = sb.load_dataset('penguins')
d = d.dropna()
dmales = d[d.sex=='Male']
X = dmales[['bill_length_mm', 'bill_depth_mm', 'flipper_length_mm',
            'body_mass_g']]
y = ground truth on species
```

Exercise 8.1 Dendrogram

Draw a Dendrogram. Compare clustering solutions with 2–10 clusters.

Exercise 8.2 k-Means

Do K-means clustering, giving 2–10 cluster solutions. Compare the various performance scores (silhouette, calinski_harabasz) without ground truth.

Exercise 8.3 Ratio of Ssw/SSB; Scree Plot; Optimal Count of Clusters

Do a K-means scree plot for giving 2–10 cluster solutions. Determine the optimal number of clusters based on the exercises. Explain your rationale.

Exercise 8.4 Compare Performance Scores with Ground Truth

Do K-means clustering giving 2 to 10 cluster solutions. Compare the various performance scores with ground truth.

References

Ghesmoune M, Lebbah M, Azzag H (2016) State-of-the-art on clustering data streams. Big Data Analytics, 1(1). https://doi.org/10.1186/s41044-016-0011-3

Halkidi M, Batistakis Y, Vazirgiannis M (2001) On clustering validation techniques. J Intelligent Inform Syst 17(2–3). https://doi.org/10.1023/A:1012801612483

Han J, Micheline Kamber JP (2014) Data mining. concepts and techniques, 3rd edn (The Morgan Kaufmann Series in Data Management Systems). In: Proceedings—2013 international conference on machine intelligence research and advancement, ICMIRA 2013

Malhotra NK (2020) Marketing research an applied Prientation, 7th Edn. Pearson Education

Chapter 9
Survival Analysis

Learning Objectives

- Gain an understanding of the basic concepts of survival analysis.
- Discuss and demonstrate Kaplan–Meier survival estimation.
- Discuss and demonstrate the Log Rank Test for comparison of groups.
- Discuss and demonstrate the Cox Proportional Hazards Survival Model.
- Demonstrate Grid Search Technique for Building a Parsimonious Model.
- Demonstrate Random Forest Survival Analysis Model for Survival Analysis.
- Illustrate key performance metrics for evaluating survival models.

Overview

This chapter introduces the concepts and techniques of survival analysis. We also investigate a case study on the survival analysis of subjects with breast cancer, using the various techniques introduced in this Chapter.

The chapter begins by delving into censored data and survival models. Subsequently, it explores survival estimation through the Kaplan–Meier method and group comparisons using the log-rank test. The chapter then extensively examines multivariate survival analysis, focusing on the Cox proportional hazards survival model. Afterward, it covers the grid search method for constructing a concise survival model, transitioning to exploring the random forest survival analysis model. Lastly, the chapter investigates the range of performance metrics employed in survival analysis.

Supplementary Information The online version contains supplementary material available at https://doi.org/10.1007/978-981-99-0353-5_9.

271
S. Sundararajan, *Multivariate Analysis and Machine Learning Techniques*,
Transactions on Computer Systems and Networks,
https://doi.org/10.1007/978-981-99-0353-5_9

The chapter commences with a discussion on censored data and survival models. Subsequently, it explores the Kaplan–Meier method for survival estimation and the log-rank test for comparing groups. A detailed discussion of multivariate survival analysis using the Cox proportional hazards survival model ensues. Afterward, it covers the grid search technique for building a parsimonious survival model and explores the random forest survival analysis model. Lastly, the chapter investigates the various performance metrics used in survival analysis.

Definitions

AUC: AUC and concordance index are common performance metrics used in survival analysis. AUC is the area under the receiver operating characteristic (ROC) curve. The AUC value is expected to be between 0.5 (random prediction) and 1 (perfect prediction). Also, see ROC.

Brier score: The Brier score is used to evaluate the accuracy of a survival prediction at a given time t. It represents the *mean value of the squared distances between the observed survival status and the estimated survival probability.* It takes a value between 0 (best) and 1 (worst).

Censored Data. In survival analysis, an individual may not experience the event during follow-up, or we may lose contact with him/her, or he/she may withdraw from the study. Despite follow-up, we may still need to determine the exact survival time. This leads to the incompleteness of data, called censoring. Data is called right-censored when we are unaware of the exact time of occurrence of the event under study. Similarly, data is left censored when we do not know the exact start time. Data is interval-sensitive, or interval-censored when we have a time interval, rather than the exact time of an event.

Concordance index: The concordance index is the proportion of observations the model can order correctly in terms of survival time, i.e., concordant pairs divided by the total number of possible evaluation pairs. The concordance index is a generalization of AUC, which is useful in classification based on a single event to accommodate censored data.

Concordant pairs: Please see the definition with an example under the section on concordance index.

Cox proportional hazards model: Numerous features influence the likely course of a medical condition—e.g., Age, Gender, Genomics, months since occurrence/diagnosis, prior therapy, and current treatment. The Cox proportional hazards model is a multivariate regression analysis model used for studying the effects of multiple numeric predictors, and covariates on survival.

Kaplan–Meier survival estimate: Kaplan–Meier survival estimate is a simple survival probability estimate computed from "survival time" data.

Log-rank test: The log-rank test can be used to compare the survival probability distribution across stratified covariates.

Parsimonious model: A parsimonious model has a minimum number of parameters that offer good explanatory power. To build a parsimonious model, we need to identify the vital features.

Random Forest Survival Analysis: A Random Forest is a collection of decision trees with a single and aggregated result. Multiple trees in the random forest reduce the chances of overfitting. Random Forest Survival Analysis (RSF) is used for risk prediction of right-censored outcomes. RSF uses multiple split criteria and methods to form an ensemble of survival trees.

ROC: ROC (receiver operating characteristic curve) is a probability distribution obtained by plotting the True Positive Rate (y-axis) against the False Positive Rate (x-axis) depicting the performance of a classifier at all classification thresholds. Also, see AUC.

Survival analysis: Survival analysis is a statistical study of time and events. The event of interest may be death, recovery, or any specific experience of interest. Time is the duration from the beginning of the follow-up of an individual until an event occurs. We may not know the exact survival time (censored data), posing a challenge to the analysis.

Survivor function: The survivor function, also called the survival function or reliability function, gives the probability that a person survives longer than some specified time.

9.1 Introduction and Overview

Survival analysis is a statistical study of time and events. The event of interest may be death, recovery, or any specific experience of interest. Let us take a medical study. Time is the duration from the beginning of the follow-up of an individual until an event occurs. Numerous features influence the likely course of a medical condition—e.g., Age, Gender, Grade, Genomics (a comprehensive set of genetic information), months since occurrence/diagnosis, prior therapy, current treatment, etc. An individual may not experience the event during follow-up, or we may lose contact with him/her, or he/she may withdraw from the study. Despite follow-up, we may not know the exact survival time. These are the premises under which a survival analysis study is undertaken (David and Mitchel 2012).

9.1.1 What is Censored Data?

Consider a hypothetical example—a study of cancer subjects over twelve weeks, depicted in Fig. 9.1.

The x-axis shows the time in weeks. The horizontal lines represent the subjects who volunteered for the study. The subjects joined the study at different times during this period. We can track a subject who has consented to the study during

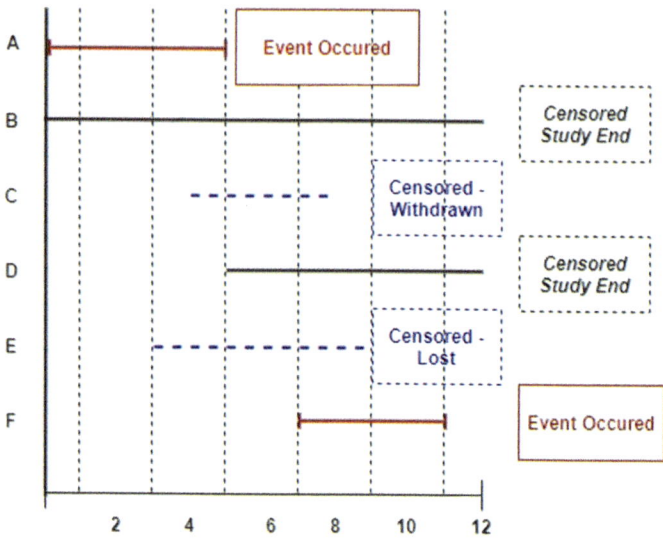

Fig. 9.1 A survival study

the study period. At the end of the study over 12 weeks, we know that A and F have met with the event under study. Their survival period is known (A—5 weeks, F—4 weeks) However, the survival periods of others (B, C, D, E) are not known.

In some cases, we may end up with only partial information about survival time. This problem is called censoring. Censoring occurs when we have some information about individual survival time, but we do not know the exact survival time (David and Mitchel 2012). Let us look at the cases shown in Fig. 9.1.

1. A and F have met with the event (e.g., death). We know their survival time (A—5 weeks, F—4 weeks)
2. C withdrew from the study, and we lost contact with E. This implies that C and E have become unavailable for follow-up.
3. B and D did not encounter the event throughout the study. We do not know what happens to them after the period of study.
4. B, C, D, and E are called right-censored data.

Note that data can also be left-censored. For example, let us say that a subject tests positive for a particular disease. However, we are unaware of the exact time when the subject contracted the disease, so we miss out on a time period before our reckoning. Such data is called left censored. Let us consider another variation when two tests were administered. The first one proved negative, and the second proved positive. In this case, we have a time interval within which the subject contracted the disease. Such data are called interval-sensitive data.

In our tutorials, we deal with only the right censored data. Primarily our data will consist of a set of observations (X, Y). X is a set of features that influence the

Table 9.1   Veterans lungs cancer survival data

X: 6 Features (DataFrame)				Y: survival data (structured array)
Age_yrs Celltype		Prior_therapy	Treatment	(Status, Survival-days)
69.0 squamous	...	no	standard	(True, 72.)
64.0 squamous	...	yes	standard	(True, 411.)
38.0 squamous	...	no	standard	(True, 228.)
... ...	...	...	...	...
67.0 large	...	yes	test	(True, 231.)
65.0 large	...	no	test	(False, 182.)
37.0 large	...	no	test	(True, 49.)

prognosis—e.g., Age, Gender, Grade, Genomics; and Y is a tuple (Died—True/False, Survival Period). For example, see Table 9.1, and Fig. 9.2.

9.1.2   The Goals of Survival Analysis

The primary goals of survival analysis are:-

1. To estimate and interpret survivor and/or hazard functions from survival data.
2. To assess the relationship of explanatory variables to survival time.
3. To build survival models.

Survivor Function

The survivor function S(t), also called the survival function or reliability function, gives the probability that a person survives longer than some specified time t. See Fig. 9.2a.

$$S(t) = P(T > t)$$

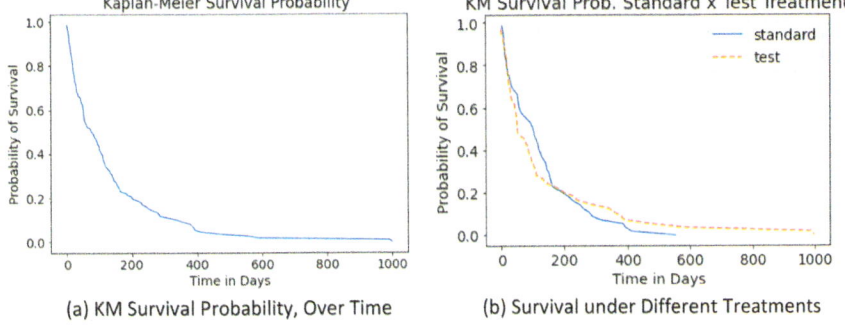

(a) KM Survival Probability, Over Time   (b) Survival under Different Treatments

Fig. 9.2   Kaplan–Meier survival probability—veterans lung cancer

Assume that 't' is a continuous variable and f(t) the probability that an event occurs at time t. The cumulative distribution function F(t) gives the probability that the event has occurred by duration t.

$$F(t) = P\{T \le t\}$$

$$S(t) = P(T > t)$$

$$= 1 - F(t)$$

$$= \int_t^\infty f(x)dx$$

Hazard Function

The hazard function h(t) gives the instantaneous **potential per unit time for the event to occur**, given that the individual has survived up to time 't' (David and Mitchel 2012).

$$h(t) = \lim_{dt \to 0} \left(P\left(t \le T < t + \frac{dt}{T} \ge t \right) \right) \Big/ dt$$

Here the numerator is the conditional probability that the event will occur in the interval $[t, t+dt]$, given that it has not occurred before. The denominator is the width of the interval 'dt'. The hazard function is also known as conditional failure rate, conditional mortality rate, or instantaneous failure rate. It provides the probability of occurrence of an event.

The cumulative hazard function is computed as follows:-

$$H(t) = \int_0^t h(u)du$$

The relationship between survival function and hazard function can be expressed as:-

$$S(t) = \exp\left(-\int_0^t h(u)du \right)$$

Survival Analysis Models

The Kaplan–Meier estimator is a non-parametric method used to estimate the survival function from observed data. It is employed to visualize and analyze the cumulative survival probability over time, offering insights into how the probability of survival changes throughout the study period.

The log-rank test is a non-parametric test used to compare the survival curves of two or more groups. It helps determine whether there are significant differences in survival probabilities among different groups, making it a valuable tool for assessing the efficacy of different treatments or interventions. For example, to compare the survival probability of subjects undergoing 'standard treatment' and subjects undergoing 'test treatment'—see Fig. 9.2b.

The Cox proportional hazards regression model is a semi-parametric method used to assess multiple covariates' impact on survival. It allows the simultaneous examination of numeric predictors and known covariates, providing hazard ratios to quantify the change in hazard with variations in predictors. For example, when researchers want to assess the impact of numeric predictors (e.g., blood pressure) and known covariates (e.g., age, gender, cancer grade) on survival simultaneously, the Cox proportional hazards regression model is commonly employed. It provides hazard ratios, indicating how the hazard (risk of an event) changes with a one-unit change in the predictor.

The RSF is an ensemble learning method that combines multiple decision trees to provide a comprehensive and aggregated prediction. It is beneficial when dealing with datasets with both numeric and categorical predictors. RSF is versatile and capable of incorporating classification and regression models into the survival analysis framework.

Artificial neural network models, such as Cox-NNET, have found applications in biomedical fields, especially in tasks like imaging analysis and patient prognosis prediction. Cox-NNET is specifically designed to predict patient prognosis using high-throughput transcriptomics data. It has demonstrated comparable or superior performance over traditional methods like the Cox-proportional hazards regression method and the Random Forest survival method. The source code is available on GitHub (Ching et al. 2018).

9.1.3 A Survival Analysis Case Study

Survival analysis software package installation is a simple one-step process. Check (Sundararajan 2023) for instructions for package installation, if necessary. In addition, the software package ELI5 is used in the random forest Tutorial—check the URL mentioned above for installation instructions.

A survival analysis study was conducted on breast cancer patients to predict distant metastases (dm). Distant metastasis refers to cancer that has spread from the primary organ to distant organs/lymph nodes. The sampling frame was breast cancer patients, who have reported lymph node-negative (N-) condition. The sampling size was 198 cases. The measurements included 76 gene signatures (prognostic) and four additional parameters: age, estrogen receptor, grade, and size. For more details, see Sect. 1.6. In this study, the endpoint was the presence of distance metastases, which occurred for 51 patients (25.8%). We will use the above dataset for all the tutorials in this chapter.

9.2   Kaplan–Meier Survival Estimate

A simple survival probability estimate can be computed from 'survival time' data, considering subjects with known survival times, such as individuals who died within the study period, as illustrated in the 'Breast Cancer Metastases Survival Data' in Table 9.2. The column "*Y: survival data (days to metastases)*" is in the form of a set of tuples (e.tdm, t.tdm), where each tuple represents one subject. Let us focus on all the subjects with status (e.tdm)=True.

A simple survival estimate beyond time t, $\hat{S}(t)$, is arrived below.

Let $S(t) = P(T > t)$

= number of patients surviving beyond (t)/total number of patients.

Kaplan–Meier survival estimate can be used to compute and plot the observed survival probability. The general formula for Kaplan–Meier survival probability at failure time t_f is shown below (David and Mitchel 2012).

$$\hat{S}(t_f) = \hat{S}(t_{f-1})\hat{P}(T > t_f | T \geq t_f)$$

Here, $\hat{P}(T > t_f | T \geq t_f)$ is the conditional probability of surviving past time t_f and $\hat{S}(t_{f-1})$ represents the product of all conditional survival probabilities, through the previous f–1 failures. This can be expressed as,

$$\hat{S}(t_{f-1}) = \prod_{i=1}^{f-1} \hat{P}(T > t_i | T \geq t_i)$$

Table 9.2   Breast cancer metastases survival—data sample

X: 76 gene features + 4 other Features (DataFrame)					Y: survival data (days to metastases)
X200726_at	X200965_s_at	...	grade	size	(e.tdm, t.tdm)
10.926361	8.962608	...	poorly differentiated	3.0	(True, 723.)
				...	
11.939616	9.615587	...	well differentiated	2.5	(False, 2225.)
11.848449	10.528911	...	intermediate	1.2	(False, 2722.)
11.425778	9.901486	...	poorly differentiated	2.5	(False, 1781.)

Tutorial 9.1   BC Case - Kaplan-Meier Survival Estimate

Model Kaplan-Meier survival estimate of breast cancer patients to predict distant metastases (dm). We can install the scikit-survival package and re-solve package installation issues, if any, using the following command
!pip install scikit-survival Refer: {(Sundararajan, 2023)}

Tutorial 9.1.1 Install Package / Resolve Problems / Setup Data

```
!pip install scikit-survival
import matplotlib.pyplot as plt
```
Load survival analysis libraries
```
from sksurv.datasets import load_breast_cancer
from sksurv.nonparametric import kaplan_meier_estimator
```

Load Breast Cancer Data
```
X, Y = load_breast_cancer()
```
X is a set of 80 features, and Y is survival time

```
Y.dtype    # [('e.tdm', '?'), ('t.tdm', '<f8')]
```
e.tdm: Boolean; True: indicates that metastasis has occurred
t.tdm: time to distant metastasis in days (See the data description)

Tutorial 9.1.2 Build KM model; Plot KM Model

Compute the survival time and the associated probability
```
time, prob = kaplan_meier_estimator(Y['e.tdm'], Y['t.tdm'])

plt.rcParams['font.size'] = '14'
plt.plot(time, prob)
plt.title ('Kaplan-Meier Survival Probability ')
plt.ylabel('Probability of Survival S(t) ->')
plt.xlabel('Time in Days')
```
See Figure 9-3: Kaplan Meier Survival Probability...

Tutorial 9.1.3 Compare Survival Under Different 'er' Types

er is estrogen-receptor: category labels (positive, negative)
```
X['er'].value_counts() # positive cases: 134; negative cases: 64
```

Plot the estimate for er positive cases first
```
Filter = (X['er'] =='positive')
time, prob = kaplan_meier_estimator(Y['e.tdm'][Filter],
Y['t.tdm'][Filter])
```

Fig. 9.3 Kaplan Meier survival probability of breast cancer patients

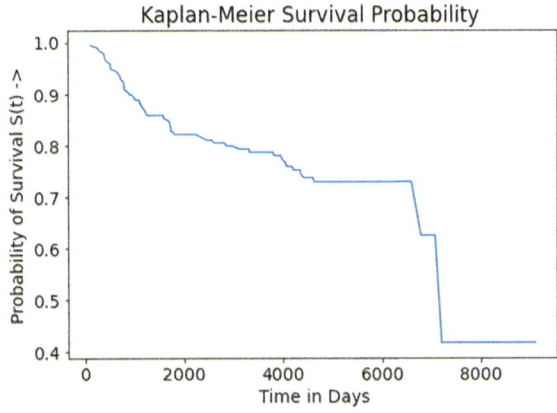

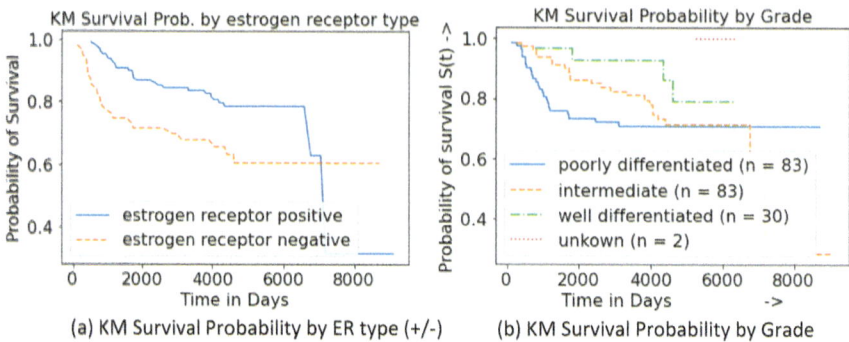

Fig. 9.4   KM survival probability by categories (ER type and Grade)

```
plt.plot(time, prob)
Plot the estimate for er negative cases
Filter = (X['er'] =='negative')
time, prob = kaplan_meier_estimator(Y['e.tdm'][Filter],
Y['t.tdm'][Filter])
plt.plot(time, prob)

plt.rcParams['font.size'] = '16'
leg = ['estrogen receptor positive','estrogen receptor negative']
plt.legend(leg, loc='best')
plt.title ('KM Survival Prob. by estrogen receptor type', fontsize=16)
plt.ylabel('Probability of Survival')
plt.xlabel('Time in Days')
See Figure 9-4: KM Survival Probability by Categories
```

Tutorial 9.1.4 Compare Survival Under Different 'grades'

```
X['grade'].value_counts()
Counts: intermediate: 83, poorly differentiated: 83, well differentiated:
30, unknown: 2

ls=['solid','dashed','dashdot','dotted']
i=0
for grade_type in X['grade'].unique():
    grade_filter = (X['grade'] == grade_type)
    time, prob = kaplan_meier_estimator(Y['e.tdm'][grade_filter],
                 Y['t.tdm'][grade_filter])
    plt.step(time, prob, where='post',linestyle=ls[i],
             label='%s (n = %d)' % (grade_type, grade_filter.sum()))
    i+=1
plt.rcParams['font.size'] = '16'
plt.legend(loc='best')
plt.title ('KM Survival Probability by Grade',fontsize=16)
plt.ylabel('Probability of survival S(t) ->')
plt.xlabel('Time in Days                        ->')
See Figure 9-4 (b): KM Survival Probability by Categories
```

9.3 Log-Rank Test and Cox Proportional Hazards Model

In this section, we take a look at the log-rank test for comparison of two or more independent groups (e.g., subjects with cancer grade = I, grade = II, or grade = III). This is followed by a detailed discussion of multivariate survival analysis using the Cox proportional hazards survival model.

9.3.1 The Log-Rank Test for Comparing Groups

Figure 9.5 depicts the stratification of the categorical variables in our case study. The ER estrogen-receptor feature can be classified into two categories: ER-positive and ER-negative. Similarly, the grade of cancer can be classified into three classes—intermediate, poorly differentiated, and well-differentiated.

Several tests are available to compare survival among independent groups, with the log-rank test being a particularly popular choice. This test is employed to compare the distribution of survival probabilities across different groups of participants. For instance, we might want to compare groups stratified by one of the covariates depicted in Fig. 9.5. Another scenario could be in a clinical trial with a survival outcome, where the goal is to compare the survival times of participants receiving a new drug to those receiving standard therapy.

Figure 9.2b illustrates that survival curves are estimated for each group separately using the Kaplan–Meier method and then compared using the log-rank test. A commonly used approach to compute the log-rank test is closely related to the chi-square test statistic (χ^2). This method compares the observed to expected numbers of events at each time point over the follow-up period. While the detailed derivations are not provided here (David and Mitchel 2012), the null hypothesis and the formulae are summarized below.

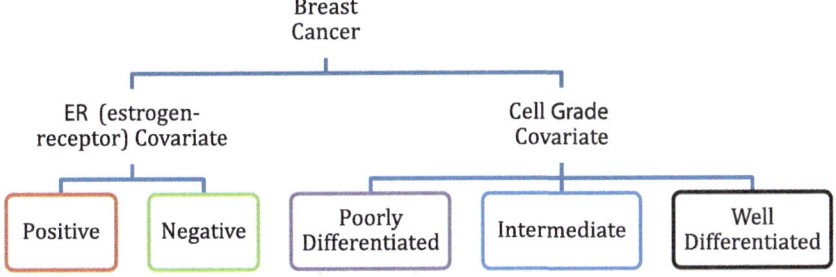

Fig. 9.5 Stratification of sample data

Null hypothesis (H_0): There is no difference in survival between two or more independent groups.

$$\chi^2 = \sum \frac{\left(\sum O - \sum E\right)^2}{\sum E}$$

Here,

$\sum O$ *is the observed number of events, over time t*
$\sum E$ *is the Expected Number of Events, Over Time t*

9.3.2   Cox Proportional Hazards Survival Model

The accuracy of survival estimates improves when we consider the influence of categorical covariates such as Age, Gender, Grade, Genomics, and numeric predictors such as blood pressure. The Cox proportional hazards model or Cox regression model is a multivariate parametric model that provides a way to estimate survival in the presence of multiple numeric predictors and covariates.

The model evaluates the influence of several features (X_1 ... X_k) on surviving a hazard. It allows us to examine how the features influence the rate of a particular event happening (e.g., death) at a particular point in time. This rate is commonly referred to as the hazard rate. The Cox PH model formula can be expressed as (David and Mitchel 2012):-

$$h(t) = h_0(t) * \exp\left(B_1 X_1 + B_2 X_2 + \cdots + B_k X_k\right)$$

where:

- *t represents the survival time,*
- *h(t) is the hazard function determined by a set of k features (X_1, X_2, ... X_k)*
- *coefficients (B_1, B_2, ..., B_k) measure the influence of the features.*
- *h_0 is called the baseline hazard. It corresponds to the hazard rate if the influence of the co-variates X_i is set to zero*

The exponent $B_1 X_1 + B_2 X_2 + \cdots + B_k X_k$ is a linear regression in k features.

The exponential quantities exp (B_i) are called hazard ratios (HR). A value of $B_i > 0$ implies $\exp(B_i) > 1$, which in turn means HR > 1. The following rules ensure.

- HR = 1 implies no effect on hazard h(t)
- HR > 1 implies an increase in hazard h(t)
- HR < 1 implies a decrease in hazard h(t)

Tutorial 9.3   BC Case - Cox Proportional Hazards Model

Consider our case study of breast cancer patients to predict distant metastases (dm). Do the following: -
1. Model the survival time estimate, considering the influence of the covariates.
2. Compute the log hazard ratios (HR), and interpret them
3. Compute the bi-variate influence of covariates with survival time and interpret them

Tutorial 9.3.1 BC Case - Data Setup

```
!pip install scikit-survival
import pandas as pd
import numpy as np
```

Load survival analysis libraries
```
from sksurv.linear_model import CoxPHSurvivalAnalysis
from sksurv.preprocessing import OneHotEncoder
from sksurv.datasets import load_breast_cancer
```

Load breast cancer data
```
X, Y = load_breast_cancer()
```

'er' and 'grade' are categories. Transform them using OHE to numeric. See
Table 9-3: One-Hot Encoding of the variable 'grade'
```
Xn = OneHotEncoder().fit_transform(X)
```

Tutorial 9.3.2 BC Case - Fit Cox PH Survival Model

```
estimator = CoxPHSurvivalAnalysis()
estimator.fit(Xn, Y)
Check Model Score
estimator.score(Xn, Y) # 0.950
```

Tutorial 9.3.3 Log Hazard Ratios

```
  B = pd.DataFrame()
B['variable'] = Xn.columns
B['log_hazard_ratio'] = estimator.coef_ # Coefficients
print(B)
```

	variable	log_hazard_ratio	Interpretation
0	X200726_at	-4.034343	decreases the HR
1	X200965_s_at	0.702673	decreases the HR
...	...	...	
77	er=positive	-2.429159	decreases the HR
78	grade=intermediate	4.932327	increases the HR
79	grade=poorly differentiated	3.362598	increases the HR
80	grade=unkown	-16.889889	not interpreted
81	size	0.804356	not interpreted

[82 rows x 2 columns]

Interpret the Log Hazard Ratios: -
LHR = 1 implies no effect on hazard h(t)
LHR > 1 implies an increase in hazard h(t)
LHR < 1 implies a decrease in hazard h(t)

The researcher must interpret the results based on theoretical and practical insights. grade=unknown has a negative HR. It is left to the researcher to interpret, as its impact is unknown. Similarly, it may be incorrect to say that size decreases the impact. Therefore, we skip the interpretation, as it is beyond the scope of this tutorial. Log Hazard Ratios are indicative of the influence of features. However, this measure is not unbiased as the features may exhibit multicollinearity.

Tutorial 9.3.4 Rank the Features

Define a Function to Fit and Score Survival, considering only one feature at a time

```python
def RankFeatures(X, Y):
    FeatureCount = X.shape[1]
    scores = np.empty(FeatureCount)
    survival_model = CoxPHSurvivalAnalysis()
    for j in range(FeatureCount):
        Xⱼ = X[:, j:j+1] # consider jth column (one feature)
        survival_model.fit(Xⱼ, Y)
        scores[j] = survival_model.score(Xⱼ, Y)
    return scores
```

Using OHE, transform the categorical variables to numeric variables
```python
    Xn = OneHotEncoder().fit_transform(X)
```

Compute feature scores using the function RankFeatures that we defined
```python
    feature_scores = RankFeatures(Xn.values, Y)
```

Display the Scores
```python
    pd.Series(feature_scores, index=Xn.columns).sort_values(ascending =
    False)
```

The features are ranked: -
```
X202240_at            0.659134
X218883_s_at          0.647393
X203306_s_at          0.644363
X204014_at            0.633506
size                  0.626373
...
X218914_at            0.501704
X217815_at            0.493498
X221344_at            0.492110
X205848_at            0.491983
grade=intermediate    0.478412
Length: 82, dtype: float64
```

```
Interpretation of the Feature Ranking: -
A higher feature score (bi-variate score of a feature with survival time)
shows a stronger influence of a feature or covariate. Therefore, the features/
covariates at the top are expected to be more important. Log hazard ratio
and the bi-variate score of features/covariates with survival time indicate
the influence of features/covariates. However, both measures are not unbiased
indicators, as the features may exhibit multicollinearity
```

9.4 Parsimonious Model for Survival Analysis

A parsimonious model has a minimum number of parameters that offer good explanatory power. Naturally, the reduction in parameters results in dimension reduction and data reduction. To build a parsimonious model, we need to identify the vital features. However, it will be daunting when the feature count is high. There are multiple methods for feature selection (Table 9.3):-

- Rank and select features using Cox Proportional Hazards Model.
- Grid Search.
- Random Forest Survival Analysis Model.

The previous section discussed feature selection using the Cox proportional hazards model. We will explore grid search and Random Forest Survival Analysis models subsequently.

9.4.1 Grid Search to Build a Parsimonious Model

The previous section discussed feature selection using the Cox proportional hazards model. How many variables would we select from the feature rank list to build a parsimonious model? The scikit-learn software library provides a solution through 'grid search'.

Table 9.3 One-hot encoding of the variable 'grade'

Original Variable	OHE	Recoded as 3 Variables		
grade		grade= intermediate	grade= poorly differentiated	grade= unkown
Well-differentiated	000	0	0	0
unknown	001	0	0	1
poorly differentiated	010	0	1	0
intermediate	100	1	0	0

SelectKBest Parameter

The parameter 'SelectKBest' returns k best features and discards all other features, based on univariate statistical tests. Note that for feature selection, different methods can be used in SelectKBest. For example, SelectKBest (chi2, k = 20) implies the Chi-square test is used for feature ranking and 20 features will be selected.

In this tutorial, we define the function 'RankFeatures'. This function uses the Cox regression test to score each feature against the survival time, one feature at a time. Before estimation, we use 'SelectKBest' in a pipeline, a pre-processing step.

Tutorial 9.4.1   BC Case - Grid Search for a Parsimonious Model

Setup Data and Libraries

```
from sksurv.datasets import load_breast_cancer
X, Y = load_breast_cancer()
from    sklearn.model_selection import GridSearchCV, KFold
from    sklearn.pipeline import Pipeline
from    sksurv.preprocessing import OneHotEncoder
from    sksurv.linear_model import CoxPHSurvivalAnalysis
from    sklearn.feature_selection import SelectKBest
import numpy  as np
import pandas as pd
```

Grid Search for Best Features

Cox Regression is performed for each feature X_i, taken one at a time, against the survival time. The concordance index that emerges (survival_model.score) indicates the ranking of the feature

Define a function to rank a given set of features {X_i | i=1 to n}

```
def RankFeatures(X, Y):
    FeatureCount = X.shape[1]          # 82 features in this tutorial
    scores = np.empty(FeatureCount)    # placeholder for 82 C-index

# Cox proportional hazards model
    survival_model = CoxPHSurvivalAnalysis()
    for j in range(FeatureCount):
        Xj = X[:, j:j+1] # select one feature (column) at a time

# Cox regression performed for the selected feature against survival time
        survival_model.fit(Xj, Y)

# Save the C-index computed above (selected feature vs survival time)
        scores[j] = survival_model.score(Xj, Y)
    return scores
```

```
# Setup Processing Pipeline
# The parameter SelectKBest is described in an earlier section
pipe = Pipeline([('encode', OneHotEncoder()),
                 ('select', SelectKBest(RankFeatures)),
                 ('model',  CoxPHSurvivalAnalysis())])
```

*** Very Important ***
It is computationally expensive if we decide to explore all 82 features and
covariates. For demonstration, let us consider a maximum of 61 features, from
which we will narrow down to a fewer number of vital features

```
feature_set = np.arange(1, 61) # Consider 1 to 60 features
param_grid = {'select__k': feature_set}
```

Setup 5-fold cross-validation with shuffle
```
KF  = KFold(n_splits=5, random_state=1, shuffle=True)
```

Setup Grid Search Parameters
```
gcv = GridSearchCV(pipe,
                   param_grid,
                   cv = KF,
                   return_train_score = True)
```

Perform Grid Search, varying the set of features from 1 to 60
```
gcv.fit(X, Y)
```

Display Grid Search Results

Check how many features are selected.
```
gcv.best_params_ #  {'select__k': 47}  # 47 features/covariates selected
gcv.best_score_  #  0.705             # test score
```

Sort the Results based on test data
```
Test_Results = pd.DataFrame(gcv.cv_results_).sort_values(by='mean_test_score',
                      ascending=False)
```

Print-Test Results
```
Test_Results[['param_select__k','params',
'mean_test_score','mean_train_score']]
```

The best test score is (**0.705492**) with 47 features selected

	param_select__k	params	mean_test_score	mean_train_score
46	47	{'select__k': 47}	0.705492	0.908115
47	48	{'select__k': 48}	0.698546	0.911145
45	46	{'select__k': 46}	0.697641	0.905013
42	43	{'select__k': 43}	0.694297	0.895746
..				
0	1	{'select__k': 1}	0.610806	0.667295
56	57	{'select__k': 57}	NaN	
57	58	{'select__k': 58}	NaN	

Sort TRAIN Results

```
Train_Results = pd.DataFrame(gcv.cv_results_).sort_values(by='mean_train_
score', ascending=False)
```

Print TRAIN Results

```
Train_Results[['param_select__k', 'params',
'mean_test_score','mean_train_score']]
```

The best train score is (0.953682) with 59 features selected

	param_select__k		params	mean_test_score	mean_train_score
58	59	{'select__k': 59}		0.659351	0.953682
59	60	{'select__k': 60}		0.650626	0.952504
...					
0	1	{'select__k': 1}		0.610806	0.667295
56	57	{'select__k': 57}		NaN	
57	58	{'select__k': 58}		NaN	

Run the Model with the Selected Model parameters

Run the model with the selected parameters, over the entire dataset

```
pipe.set_params(**gcv.best_params_)
pipe.fit(X, Y)
```

Extract the features and their score from 'pipe'

```
OHE, KBest, CoxPH = [s[1] for s in pipe.steps]
```

Check the contents of the variables OHE, KBest, and CoxPH

```
OHE          # OneHotEncoder()
KBest        # SelectKBest(k=47, score_func=<function RankFeatures)
CoxPH        # CoxPHSurvivalAnalysis()
```

Format and print the features and their score

```
pd.Series(CoxPH.coef_, index=OHE.encoded_columns_[KBest.get_support()])
```

features (47)	Score
-------------	---------
X200965_s_at	0.400214
X201091_s_at	-0.311317
X201288_at	-0.025255
X201368_at	-0.274931
X201663_s_at	1.520212
X201664_at	-0.798827
X202239_at	-0.534744
X202240_at	1.097430
X202418_at	0.045413
X202687_s_at	-0.396493
..	
er=positive	-2.742841
grade=poorly differentiated	-0.050478
size	0.112297

```
Interpretation of the Model Parameters from Grid Search:
1. The number of features selected in the model: 47
2. Selected Features and their Coefficients are listed above
3. Best Training Score: 0.908
```

9.4.2   Random Forest Survival Analysis Model

As we discussed in Sect. 9.5, a parsimonious model is a model with a minimum number of parameters that offers good explanatory power. For this, the following methods may be used—Rank and select features using the Cox Proportional Hazards Model, Grid Search, and Random Forest Survival Analysis Model. In this section, we will discuss the Random Forest Survival Analysis Model. This section introduces the basic concepts, followed by a tutorial.

Random Survival Forrest

A random survival forest (RSF) is an ensemble of decision trees that provide a single aggregated result (James et al. 2021). Random forests are discussed in Sect. 11.6.5—Ensemble Methods. Multiple trees in the random forest reduce the chances of overfitting. RFS is used for risk prediction of right-censored outcomes in various areas such as biomedical research, and finance.

RSF could use classification and regression models. Several split criteria can be used for splitting a node into subtrees, the most popular one being the log-rank test (see Sect. 9.4). To evaluate the prediction accuracy of the RFS model, we can use Harrell's concordance index (or C-index). As we have discussed, the concordance index is defined as the proportion of all comparable pairs in which the predictions and outcomes are concordant (please check the section on the concordance index for further description). A random forest survival analysis tries to ensure that individual trees are not correlated by employing the following measures:

- building each tree on a different bootstrap sample of the original training data.
- at each node, evaluate the split criterion only for a randomly selected subset of features and thresholds.

Predictions are formed by aggregating predictions of individual trees in the ensemble. The criterion for split is based on the log-rank test.

Permutation Importance

To rank features, we may use permutation-based methods. This involves removing features, one at a time, and measuring the reduction in the model score (R-squared, F1). After removing each feature, should we train the algorithm on the entire dataset? This is costly. Instead, we may consider only the test data set, as mentioned below. Take the test data set. Replace the selected feature with random noise. Use the original model to compute the score.

The Software Package

The default values for the parameters controlling the size of tree growth and pruning are critical while dealing with large data sets with numerous features, as they determine the processing complexity. The RandomSurvivalForest module is currently under development, and more features are expected. For example, parameters are currently unavailable to limit the tree's depth or rank the features.

Some of the parameters that may be important to a researcher are listed below: -

- n_estimators (integer, default: 100)—The number of trees in the forest.
- max_depth (int default: None)—The maximum depth of the tree. If None, then nodes are expanded until all leaves are pure (all samples belong to the same class) or until all leaves have samples less than 'min_samples_split'.
- min_samples_split (default: 6): A node with more samples than this parameter becomes a candidate for the split. The value can be also given as a fraction of the number of samples.
- minimum sample size (default: 3): A node split is considered only if it leaves at least 'min_samples_leaf' samples in the left and right branches. This can affect the regularization of the model, especially in regression. The value can be also given as a fraction of the number of samples.
- Setting the parameter random_state to a specific integer helps reproducibility results on repeated trials (Fig. 9.7).

Tutorial 9.4.2   BC Case - Random Forest Survival Analysis (RFS) Model

Note: - This Random Forest Survival Analysis tutorial was run in Google Colab. For RFS, install scikit-survival (!pip install scikit-survival). Install eli5 (!pip install eli5) for feature ranking. If you face a problem "restart and run" Colab

Tutorial 9.4.2.1 Data Setup

```
!pip install scikit-survival
load dataset
    from sksurv.datasets import load_breast_cancer
    X, Y = load_breast_cancer()

import libraries
    from sksurv.preprocessing import OneHotEncoder
    from sksurv.ensemble import RandomSurvivalForest
    import matplotlib.pyplot as plt

Recode Covariate Categories using OHE
    X = OneHotEncoder().fit_transform(X)
    random_state_chosen = 20
```

Tutorial 9.4.2.2 Build and Test RFS Model

```
Design the Model Random Forrest Model
  RFS = RandomSurvivalForest(n_estimators=256,
                              min_samples_split=4,
                              min_samples_leaf=2,
                              max_features="sqrt",
                              n_jobs=-1,
                              random_state = random_state_chosen)
```

```
Fit the Random Forrest Model
  RFS.fit(X, Y)
```

```
Check the Random Forrest Model Score on the entire Data
  print(RFS.score(X, Y)) #  0.9760
```

Tutorial 9.4.2.3 Survival Prediction - Plot

```
Select the first 4 subjects (rows 0,1,2,3)
  X_test_sel = X[0:4]
```

```
Predict the survival probability distribution of the selected subjects
  surv = RFS.predict_survival_function(X_test_sel, return_array=True)

  Plot the survival probability distribution of the 4 selected subjects
  ls=['solid','dashed','dashdot','dotted']
  i=0
  for i, s in enumerate(surv):
      plt.step(RFS.event_times_, s, where="post",
  label=str(i),linestyle=ls[i])
      i+=1
  plt.title ('Random Forrest Survival Prob.- 4 Test Cases',  fontsize = 14)
  plt.ylabel('probability of survival S(t) ->',  fontsize = 14)
  plt.xlabel('time (days) -> ∞', fontsize = 14)
  plt.legend()
  plt.grid(True)
  plt.legend(loc='best')
```
See Figure 9-6: Area Under The Curve for 4 Subjects. AUC is discussed in section 9.5

Tutorial 9.4.2.4 Feature Ranking

```
Setup Data / Libraries as in   - Tutorial 9.4.1 Data Setup.
Build and Test RFS Model as in Tutorial 9.4.2 Build and Test RFS Model.
```

```
Install eli5:
```
We will use the module 'Permutation Importance' for computing feature importance by measuring the effect on the model score, with the entry or exit of a feature. Note that this method is similar to the step-wise regression method.

Fig. 9.6 Area under the curve for 4 subjects {0, 1, 2, 3}

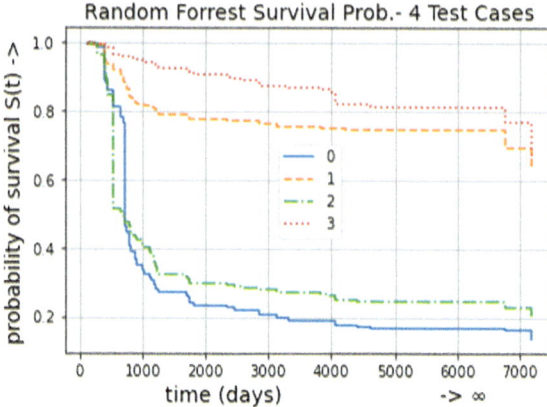

Fig. 9.7 ROC curve—TPR versus FPR

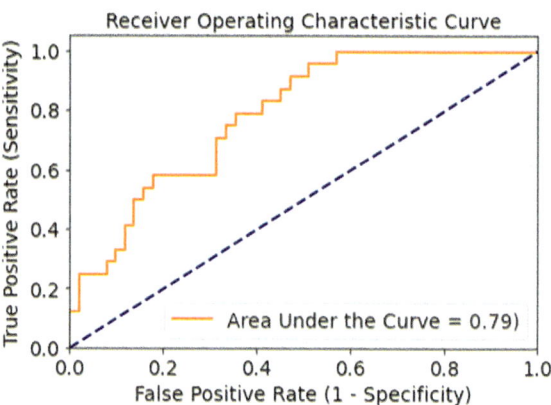

```
import eli5
from eli5.sklearn import PermutationImportance

features_list = list(X.columns)
```

Score the features: -
```
perm = PermutationImportance(RFS, n_iter=15, random_state=0)
perm.fit(X, Y)
eli5.show_weights(perm, feature_names=features_list)
```

Feature Ranking List: -
```
0.0254 ± 0.0499  X202240_at
0.0233 ± 0.0386  X204014_at
0.0183 ± 0.0181  X203306_s_at
0.0178 ± 0.0408  X219724_s_at
0.0058 ± 0.0275  X218883_s_at
0.0054 ± 0.0032  X203391_at
0.0036 ± 0.0036  X207118_s_at
```

```
0.0030 ± 0.0079 X201663_s_at
0.0030 ± 0.0106 X208180_s_at
0.0018 ± 0.0049 X204888_s_at
0.0018 ± 0.0034 X212014_x_at
0.0015 ± 0.0024 X220886_at
0.0015 ± 0.0122 X221882_s_at
0.0015 ± 0.0030 X221344_at
0.0012 ± 0.0040 X210028_s_at
0.0010 ± 0.0030 X212567_s_at
0.0010 ± 0.0044 X221816_s_at
0.0008 ± 0.0090 X221916_at
0.0008 ± 0.0082 X211040_x_at
0.0008 ± 0.0035 X201664_at
… 62 more …
```

9.5 Performance Metrics

Some of the performance metrics used in survival analysis are listed below. We will discuss them in the subsequent sections.

- Common Metrics: Sensitivity, Specificity, PPV, NPV, Accuracy
- Brier Score
- Receiver Operating Characteristic (ROC) curves, Area under the ROC Curve (AUC)
- Concordance Index (C-Index).

9.5.1 Score Based on Outcome Counts

The performance measures, Sensitivity, Specificity, PPV, NPV, and Accuracy are shown in Table 9.4. We have discussed the basic concepts underlying these measures in the Chapter on classification. AUC and concordance index provide discriminatory performance indices to judge the goodness of fit of a model. They can be extended to survival analysis models by incorporating the constraints of right-censored data. We will discuss this in the subsequent sections.

9.5.2 Brier Score

The **Brier score** is used to evaluate the accuracy of a survival prediction at a given time t. It represents the *mean value of the squared distances between the observed survival status and the estimated survival probability*. It is a value between 0 (best) and 1 (worst). The **Brier score can be expressed as,**

$$B(t) = \frac{1}{N} \sum_{t=1}^{N} (f_e - f_o)^2$$

where,

f_o is the observed survival status
f_e is the survival probability estimated by one of the models
N is the total number of observations

9.5.3   Area Under the Curve (ROC, AUC)

The area under the ROC (AUC) and concordance index provide discriminatory performance indices to judge the goodness of fit of a model. They can be extended to survival analysis models, by incorporating the constraints of right-censored data (James et al. 2021).

ROC (receiver operating characteristic curve) is a probability distribution obtained by plotting the True Positive Rate (y-axis) against False Positive Rate (x-axis) depicting the performance of a classifier at all classification thresholds (See Sect. 11.6.6). AUC is the area under the ROC curve. The AUC value is expected to be between 0.5 (random prediction) and 1 (perfect prediction). In Fig. 9.7, AUC=0.79. This implies that the accuracy of the model is 79%.

From Table 9.4, it may be noted that,

$$\textbf{Sensitivity} = \text{True Positive Rate } (\textbf{TPR}) = \text{Recall} = \text{TP} / (\text{TP} + \text{FN})$$

$$\text{Specificity} = \text{True Negative Rate} = \text{TN} / (\text{TN} + \text{FP})$$

$$\begin{aligned}
1-\textbf{Specificity} &= 1 - [\text{TN} / (\text{TN} + \text{FP})] \\
&= [(\text{TN} + \text{FP}) - \text{TN}] / (\text{TN} + \text{FP}) \\
&= \text{FP} / (\text{TN} + \text{FP}) \\
&= \text{False Positive Rate } (\textbf{FPR})
\end{aligned}$$

Table 9.4   Performance measurement of classification models

		Predicted labels		
		1	0	
Observed labels	1	True Positives (TP)	False Negatives (FN)	Sensitivity, Recall, Hit **Rate, True Positive Rate (TPR) = TP/(TP + FN)**
	0	False Positives (FP)	True Negatives (TN)	**False Positive Rate (FPR) =FP/ (TN+FP)**Specificity, Selectivity, True Negative Rate (TNR) =TN/(TN+FP)
		Precision, Positive Predictive Value (PPV) = TP/(TP+FP)	False Positive Rate (FPR) = FP/(TN+FP)	Accuracy = (TP+TN)/ (TP+TN+FP+FN)

The ROC curve and AUC can be extended to survival data by defining sensitivity (true positive rate) and specificity (true negative rate) as time-dependent measures and considering the data as right-censored. Assume cumulative cases as the number of all the individuals who experienced an event by time t (ti ≤ t), and dynamic cases are those with ti > t. The corresponding 'cumulative by dynamic' AUC statistic shows the goodness of fit of a model in discriminating the subjects who encounter the event at a given time (ti ≤ t) from the subjects who encounter the event after (ti > t).

The scikit-learn software module, sksurv.metrics.cumulative_dynamic_auc provides an estimate of cumulative/dynamic AUC for right-censored time-to-event data. Uno has developed a censored-pairs estimator of the concordance index based on inverse probability weighting. Given an estimator $\hat{f}(X_i)$ of the ith subject's risk score, the cumulative/dynamic AUC at time t can be defined using 'the inverse probability of censoring weights' (IPCW) of the features. For details, refer to (Brentnall and Cuzick 2018). Our tutorial will use IPCW's computed using the Kaplan–Meier estimator. Censoring is assumed to be at random. Time alone is considered, and the feature-set is not considered. To estimate IPCW, access to survival times from the training data is required. Therefore, check and ensure that the survival times of test data are within the range of survival times of training data. Otherwise, the system will throw an error indicating the above.

9.5.4   Concordance Index

Like AUC, the concordance index or C-Index also provides discriminatory performance indices to judge the goodness of fit of a model. They can be extended to survival analysis models by incorporating the constraints of right-censored data.

An example for manual identification of a concordant pair is shown below. Table 9.5 shows the survival data of six cases down the rows. The probability of survival after 289 days (X_{i1}) and after 404 days (X_{i2}) are shown in columns two and three respectively.

Let us compare Case 1 (X_1) with the other cases $(X_i, i = 1, 2, 3, 4, 5, 6)$. If the probability of survival of two cases increases or decreases over time, the two cases are called a concordant pair. If the probability of the cases match at time point one as well as time point two, the two cases are called a tied pair. Otherwise, the cases form a discordant pair. The computations are shown in Table 9.6. From Table 9.6, we observe that $\{X_1, X_2\}$ forms a concordant pair, as their probability of survival increases over time. The pair $\{X_1, X_3\}$ is also a concordant pair, as their probability of survival decreases over time. The pair $\{X_1, X_6\}$ is tied as their probabilities are all exactly the same. The pairs $\{X_1, X_4\}, \{X_1, X_5\}$ are discordant pairs.

The concordance index is defined as the proportion of observations that the model can order correctly in terms of survival time. This can be expressed as,

$$\text{concordance index} = \text{concordant pairs} / \text{total number of possible evaluation pairs}$$

$$= \text{concordant pairs} / (\text{concordant pairs} + \text{discordant pairs})$$

Table 9.5 CI—comparison of survival probabilities of six cases over two time periods

Case X_i	Probability of survival	
	After 289 days (X_{i1})	After 404 days (X_{i2})
1	0.97	0.94
2	0.99	0.96
3	0.96	0.92
4	0.95	0.95
5	0.98	0.93
6	0.97	0.94

Table 9.6 CI—investigation of concordance between 5 pairs of cases

Compare case 1 (X1) with other cases (Xi, i = 2, 3, 4, 5, 6)

	X11–Xi1		X12–Xi2		Are X11–Xi1 and X12–Xi2 of the same sign?
X1 versus X2	0.97–0.99 =	−0.02	0.94–0.96 =	−0.02	Yes. Concordant pair
X1 versus X3	0.97–0.96 =	0.01	0.94–0.92 =	0.02	Yes. Concordant pair
X1 versus X4	0.97–0.95 =	0.02	0.94–0.95 =	−0.01	No. Dis-Concordant pair
X1 versus X5	0.97–0.98 =	−0.01	0.94–0.93 =	0.01	No. Dis-Concordant pair
X1 versus X6	0.97–0.97 =	0	0.94–0.94 =	0	Tied pair

The concordance index calculation can be extended to survival analysis to accommodate censored data, by defining sensitivity and specificity as time-dependent measures. The index provides a reliable ranking of the survival time based on the individual risk scores. The index is also called Harrell's C-index, or simply C-index. The concordance index is a generalization of the area under the ROC curve (AUC) that can account for censored data. However, we are not providing any mathematical proof here.

Tutorial 9.5.1   BC Case - Compare ROC Curve and Concordance Index

!pip install scikit-survival
Setup Data / Libraries

```
import numpy as np
import matplotlib.pyplot as plt
import pandas as pd
from sksurv.datasets        import load_breast_cancer
from sklearn.model_selection import train_test_split
from sksurv.preprocessing    import OneHotEncoder, encode_categorical
```

```
from sksurv.metrics          import (
                                  concordance_index_censored,
                                  concordance_index_ipcw,
                                  cumulative_dynamic_auc,
                                  integrated_brier_score
                                  )

x, y = load_breast_cancer()
x.columns # ['X200726_at', ... , 'age', 'er', 'grade', 'size'])

x = OneHotEncoder().fit_transform(x)
x.info()      # Total 82 columns

numeric_columns = list(x.columns)
```

There are 82 features. If we plot them all, the plot will be cluttered. So we are selecting the top 9 features identified in Tutorial 9.4.

```
numeric_columns = list(('X202240_at', 'X204014_at', 'X203306_s_at',
                 'X219724_s_at', 'X218883_s_at', 'X203391_at',
                 'X207118_s_at', 'X201663_s_at', 'X208180_s_at'))
```

Train (80%) Test (20%) Split

```
x_train, x_test, y_train, y_test = train_test_split(x, y, test_size=0.2,
        random_state=0)
```

Convert x_train/x_test from pandas DataFrame to NumPy Array

```
x_train = np.array(x_train)
x_test  = np.array(x_test)
y.dtype # ('e.tdm', '?'), ('t.tdm', '<f8')

# Divide Survival Time to 10 slices of 10 percentiles
times = np.percentile(y["t.tdm"], np.linspace(10, 91, 10))
print(times)
# [943.6 1725.3 2680.0 3602.17 ... 5015.68 5502.66 5718.86 6180.17]
```

Compute Concordance Index, and Plot ROC

```
# Function to plot AUC
def plot_cumulative_dynamic_auc(risk_score, label):
    auc, mean_auc = cumulative_dynamic_auc(y_train, y_test, risk_score,
times)
    plt.plot(times, auc, marker="o", label=label)
    plt.xlabel("days from enrolment")
    plt.ylabel("time-dependent AUC")
    plt.axhline(mean_auc, linestyle="--", color='black',linewidth=0.5)
    plt.title('Area Under ROC Curve')
    plt.legend()
```

See Figure 9-8: ROC Curve of the Top 9 Features

```
ci_ipcw_matrix = pd.DataFrame()
ci_ipcw_matrix['Co-Variate'] = list(numeric_columns)
ci_ipcw_matrix['ci_ipcw'] = 0.0
plt.rcParams['figure.figsize'] = [7.2, 4.8]  #plot area
```

For Each Feature, weighted compute Concordance index (ipcw) and invoke plot
function

```
for i, col in enumerate(numeric_columns):
    plot_cumulative_dynamic_auc(x_test[:, i], col)
    ci_ipcw = concordance_index_ipcw(y_train, y_test,
                    x_test[:, i], tau=times[-1])
    ci_ipcw_matrix['ci_ipcw'][i] = ci_ipcw[0]
```

See Figure 9-8: ROC Curve of the Top 9 Features. The plot shows the estimated
area under the time-dependent ROC curve. The average across all time points
is shown as a dashed line.

Print the computed concordance_index for each feature

```
ci_ipcw_matrix.sort_values(by='ci_ipcw',ascending=False )
     Co-Variate    ci_ipcw
1.   3  X219724_s_at  0.566479
2.   6  X207118_s_at  0.562888
3.   7  X201663_s_at  0.542749
4.   0     X202240_at  0.487923
5.   2  X203306_s_at  0.485421
6.   5     X203391_at  0.405913
7.   1     X204014_at  0.349195
8.   4  X218883_s_at  0.344145
9.   8  X208180_s_at  0.323700
```

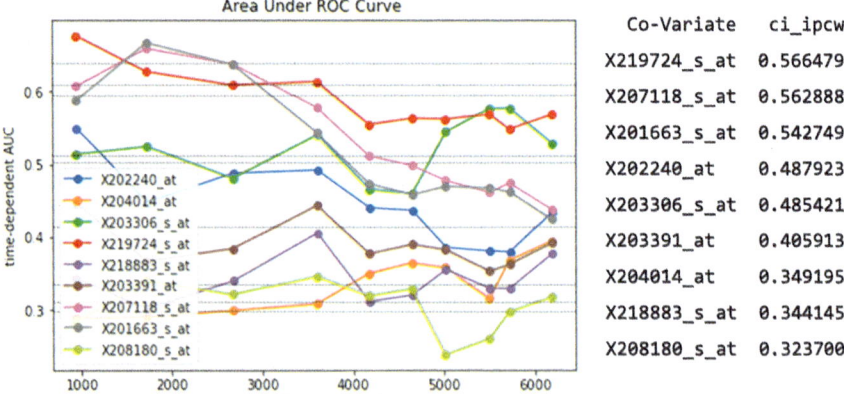

Fig. 9.8 ROC curve of the Top 9 features and their concordance index

Interpret ROC curve and Concordance Index

See Figure 9-8 for the ROC curve and Concordance Index (ci_ipcw_matrix).
1. From the figure, it is observed that the following are the most discriminating features in order (they cover a larger area) – the most discriminating features or the best predictors
'X219724_s_at'
'X207118_s_at', 'X201663_s_at'
2. The following are the least discriminating features, as they cover a smaller area
'X218883_s_at'
'X208180_s_at'
3. The concordance index of the top 9 features is shown by the side of Figure 9-9, in rank order. The ranking corresponds to the area under the ROC curve

Tutorial 9.5.2 BC Case - Compare the Performance of Estimators CPH, RSF

Data Setup

```
import numpy as np
import matplotlib.pyplot as plt
import pandas as pd
from sksurv.datasets        import load_breast_cancer
from sklearn.model_selection import train_test_split
from sksurv.preprocessing   import OneHotEncoder, encode_categorical
from sksurv.metrics         import (
                               concordance_index_censored,
                               concordance_index_ipcw,
                               cumulative_dynamic_auc,
                               integrated_brier_score
                               )
from sksurv.linear_model    import CoxPHSurvivalAnalysis
from sksurv.ensemble        import RandomSurvivalForest

x, y = load_breast_cancer()
```
There are 82 features. We are selecting the top 9 features identified in Tutorial 9.4 and used in Tutorial 9.5
```
numeric_columns = list(('X202240_at', 'X204014_at', 'X203306_s_at',
                  'X219724_s_at', 'X218883_s_at', 'X203391_at',
                  'X207118_s_at', 'X201663_s_at', 'X208180_s_at'))
```

Split 80% to training data (pandas data frames); 20% to training
```
x_train, x_test, y_train, y_test = train_test_split(x, y, test_size=0.2,
          random_state=0)
x_train = np.array(x_train[numeric_columns])
x_test  = np.array(x_test[numeric_columns])
y.dtype # ('e.tdm', '?'), ('t.tdm', '<f8')
times = np.percentile(y["t.tdm"], np.linspace(10, 91, 10))
```

```
print(times)
# [943.6 1725.3 2680.0 3602.17 … 5015.68 5502.66 5718.86 6180.17]
```

Build CPH Model

```
CPH = CoxPHSurvivalAnalysis()
CPH.fit(x_train, y_train)

CPH_Risk = CPH.predict(x_test)

CPH_AUC, CPH_AUC_Mean = cumulative_dynamic_auc(
    y_train, y_test, CPH_Risk, times)
```

Build RSF Model

```
RSF = RandomSurvivalForest(n_estimators=100, min_samples_leaf=7,
random_state=0)

RSF.fit(x_train, y_train)

RSF_CH_Functions = RSF.predict_cumulative_hazard_function(
                     x_test, return_array=False)
RSF_Risk = np.row_stack([chf(times) for chf in RSF_CH_Functions])
RSF_AUC, RSF_AUC_Mean= cumulative_dynamic_auc(
                     y_train, y_test, RSF_Risk, times)
```

Plot CoxPH Versus RSF

```
CoxPHleg ="CoxPH mean AUC = " +str(np.round(CPH_AUC_Mean,2))
RSFleg =  "RSF mean AUC = "   +str(np.round(RSF_AUC_Mean,2))
Leg = [CoxPHleg, RSFleg]
plt.plot(times, CPH_AUC, "o--")
plt.plot(times, RSF_AUC, "s-.")
plt.xlabel("days from enrolment", fontsize=14)
plt.ylabel("time-dependent AUC", fontsize=14)
plt.title("CoxPH estimate Versus RSF estimate", fontsize=14)
plt.legend(Leg)
plt.grid(True)
```

See Figure 9-9: AUC from Cox PH versus RSF Model

Interpret Model Performance (CoxPH versus RSF)

There are 82 features and covariates. We selected the top 9 features identified in Tutorial 9.4 and were used in Tutorial 9.5.
Figure 9-9 shows that both the estimators CoxPH (score 0.75) and Random Survival Forrest (score 0.81) perform in comparable terms. RSF appears to be
more accurate, with a higher score of 0.81

Fig. 9.9 AUC from cox PH versus RSF model

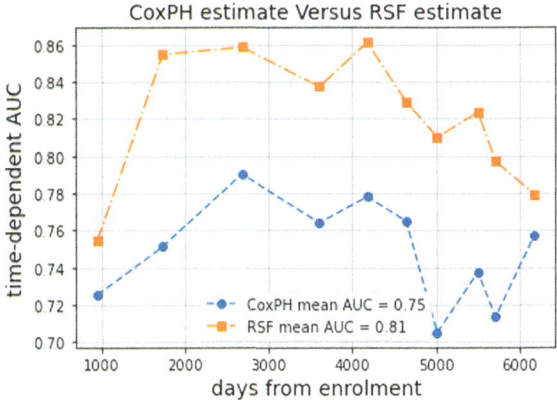

The concordance index finds wide application in areas such as medicine and finance to measure a model's ability to discriminate against subjects based on their risk (Ahuja and Van der Schaar 2019a, b). A Simple concordance index addresses one event (e.g., discharged alive). However, it is not suitable for assessing the prognostic ability of a model when multiple event types are involved, such as being on ventilation, experiencing organ failure, and being discharged alive simultaneously. Early warning systems incorporating feeds from multiple event types may require an improved approach, such as a joint concordance index or considering one covariate versus all other covariates together (Ahuja and Van der Schaar 2019a, b).

Data Analytics in Action

Survival Analysis of Patients with COVID-19 in India by Demographic Factors: Quantitative Study (Kundu et al. 2021).
The study period was January 30, 2020, up to June 30, 2020. To contain the pandemic, the government of India had declared a lockdown on March 25, 2020. Despite that, COVID-19 cases increased and surpassed 450,000, over the period of five months, reportedly. Data with missing values and spurious/inconsistent data were removed from the data collected. A sample of 26,815 subjects was considered for analysis.

The **Kaplan–Meier survival function** indicated that the probability of survival of patients with COVID-19 declined during the study period of 5 months. The **log-rank test (P<0.001)** and Wilcoxon test (P<0.001) confirmed this conjecture. Significant variability was observed in the age groups. Age increased the risk of COVID-19 casualties. The Cox proportional hazard model indicated that male patients with COVID-19 had a higher risk of fatality compared to female patients (hazard ratio 1.14; standard error 0.11; for a confidence interval of 95%). The survival rates in the south, northeast, and east were observed to be better than in the north, central, and western regions, in order.

Summary

Survival analysis is a statistical study of time and events. The event of interest may be death, recovery, or any specific experience of interest. Time is the duration from the beginning of the follow-up of an individual until an event occurs. Numerous features influence the prognosis (the likely course of a medical condition)—e.g., Age, Gender, Genomics (a comprehensive set of genetic information), months since occurrence/diagnosis, prior therapy, current treatment, etc. An individual may not experience the event during follow-up, or we may lose contact with him/her, or he/she might withdraw from the study. Despite follow-up, we may not know the exact survival time. This leads to the incompleteness of data, called censoring. Data is called right censored when we are not aware of the exact end time. Similarly, data is left censored when we do not know the exact start time. Data is interval-sensitive, or interval-censored when we have a time interval, rather than the exact time of an event.

The survivor function, also known as the survival or reliability function, provides the likelihood that an individual will survive beyond a specified time. The Kaplan–Meier survival estimate is a straightforward calculation of the probability of survival, derived from data on "survival time". The log-rank test compares the distribution of survival probabilities among different subject groups, such as those with or without co-morbidities. Both Kaplan–Meier estimation and the log-rank test are non-parametric methods for survival estimation. To examine the impacts of multiple numeric predictors and covariates, survival regression models like the Cox proportional hazards model can be employed. Another approach is the random survival forest, which comprises an ensemble of decision trees capable of handling numeric and categorical predictors.

A parsimonious model is characterized by having the least number of parameters while still delivering robust explanatory power. Constructing such a model involves pinpointing the essential features, a challenging endeavor when faced with many features. Several techniques, such as the Cox Proportional Hazards Model, Grid Search, and Random Survival Forest, among others, exist for feature selection. While the Cox proportional hazards model aids in ranking features, it does not specify the optimal number of variables for achieving parsimony. The scikit-learn software library addresses this concern with its 'grid search' module, providing a solution for determining the right balance between simplicity and explanatory effectiveness in model construction.

A random survival forest (RSF), an ensemble of decision trees, provides a single aggregated result. RSF could be built on multiple numeric and categorical predictors and use classification and regression models. The common approach to evaluate the prediction accuracy of an RSF model is Harrell's concordance index.

AUC and the concordance index are commonly employed performance metrics in survival analysis. The ROC curve, depicting the True Positive Rate against the False Positive Rate, forms the basis for AUC, representing the area under the curve. AUC values range from 0.5 (random prediction) to 1 (perfect prediction). The concordance index gauges a model's ability to correctly order survival times by considering the proportion of concordant pairs among all possible pairs. It serves as a generalized metric to accommodate censored data and applies to binary classification, offering a dependable ranking of survival times based on individual risk scores. However, the concordance index is limited to assessing a single event and cannot evaluate the prognostic ability of models for multiple event types. An enhanced approach like a joint concordance index is necessary for the effective evaluation of models in early warning systems predicting event times across diverse types.

Artificial neural network (ANN) based models are extensively applied to biomedical fields such as imaging analysis and diagnosis. Various performance metrics are used in survival analysis, which includes metrics derived from classification problems such as sensitivity, specificity, accuracy, brier score, AUC, concordance index, etc.

Questions

Comprehension

1. What are the primary goals of survival analysis, and how do they contribute to our understanding of time-related events?
2. Explain the different types of censored data (right-censored, left-censored, interval-censored) with examples. How does each type affect the analysis? Discuss the challenges and considerations when dealing with left-censored data in survival analysis.
3. Describe the survivor function and hazard function.
4. Explain stratification in sample selection for estimating survival probability.
5. How does the Kaplan–Meier survival estimate calculate the survival probability at a given time point? Explain the significance of the conditional probability term in the Kaplan–Meier survival estimate formula.
6. Elaborate on the grid search technique for building a parsimonious survival model. What advantages does this method offer in terms of model simplicity and performance?

7. Describe the role of the concordance index in survival analysis. State and explain the formulae.
8. How does the concordance index differ from traditional metrics like the area under the ROC curve?
9. What is the Brier score, and how does it assess the accuracy of survival predictions? Provide an example interpretation.

Analysis

10. How does survival analysis differ from a traditional binary classification problem?
11. Compare and contrast the Kaplan–Meier survival estimation method with the Cox proportional hazards survival model. In what scenarios would you prefer one over the other?
12. How does the log-rank test contribute to the comparison of survival curves? Please provide an example scenario where it would be beneficial.
13. Describe the key components of the Cox Proportional Hazards Model formula. How does the Cox model handle the influence of categorical covariates on survival? What does a hazard ratio (HR) greater than 1, equal to 1, and less than 1 signify in the context of the Cox model?
14. Explain the Random Survival Forest (RSF) concept and its application in survival analysis. How does the RSF method address the issue of overfitting? What is permutation importance, and how is it used in the context of feature ranking with RSF?
15. Discuss whether we can use the concordance index for predicting risk based on a set of covariates in the case of multiple failures.
16. How do the Receiver Operating Characteristic (ROC) curve and Area Under the Curve (AUC) extend to survival analysis, considering right-censored data?

Application

17. In a clinical trial setting, how might survival analysis be applied to assess the effectiveness of a new treatment compared to a standard treatment? Discuss the key variables and metrics involved.
18. Imagine you are working in a financial institution and want to predict the time until a customer defaults on a loan. How can survival analysis be utilized to model this scenario? What variables would you consider?
19. In a public health campaign aimed at reducing the incidence of a particular disease, how could survival analysis be applied to evaluate the impact of the intervention over time? What challenges might arise in such an analysis?
20. Suppose you are an HR analyst in a company and want to understand the average tenure of employees in different departments. How could survival analysis help you in this scenario? What factors might influence employee retention?

21. A manufacturing company wants to assess the reliability of a new product in the market. How could survival analysis be employed to estimate the probability of the product functioning without failure for a certain duration? What considerations are important in this analysis?

22. An insurance company wants to assess the risk of policyholders experiencing a certain event, such as a health-related incident, over time. How could survival analysis contribute to risk assessment in the insurance industry? What factors would be crucial to consider?

23. Imagine you are conducting a study on the survival of cancer patients. How would you utilize the Cox Proportional Hazards Model to assess the impact of various factors on patient survival?

24. Consider a dataset with a large number of features, including both categorical and numerical variables. How would you address the challenge of feature selection when building a survival analysis model? Discuss a scenario where feature selection using the Cox Proportional Hazards Model might be particularly beneficial.

25. Provide an example of how Random Survival Forest (RSF) could be employed in biomedical research to predict the risk of a specific health event.

26. Imagine you have developed a survival analysis model for predicting patient outcomes. How would you interpret the Brier Score, AUC, and Concordance Index values to assess the model's performance?

Exercises

We used a dataset with 198 samples and 80 features for modeling breast cancer survival estimates, based on 76 gene signatures. In this exercise, we have a dataset from a German breast cancer study group, having 686 samples and 8 features (gbsg2 dataset from sksurv.datasets). The endpoint is recurrence-free survival, which occurred for 299 patients (43.6%). This dataset is part of a study by the German Breast Cancer Study Group (GBSG) in 1984. This randomized clinical trial compared the effectiveness of different drug dosages on recurrence-free and overall survival.

Refer: https://www.ncbi.nlm.nih.gov/geo/query/acc.cgi?acc=GSE7390

Programming Hints

Programming Hint - breast cancer, gbsg2 dataset

```
from sksurv.datasets import load_gbsg2
X, Y = load_gbsg2()
X.info()
```

```
0    age        686 non-null    float64
1    estrec     686 non-null    float64
2    horTh      686 non-null    categories -> ['no', 'yes']
3    menostat   686 non-null    category -> ['Pre', 'Post']
4    pnodes     686 non-null    float64
5    progrec    686 non-null    float64
6    tgrade     686 non-null    category -> ['I', 'II', 'III']
7    tsize      686 non-null    float64

     Y.dtype        # 'cens', Boolean; 'time', float (survival in days)
```

Exercise 9.1 KM Survival Estimate—German Breast Cancer Study Group 2
Use the numeric features ['age', 'estrec', 'pnodes', 'progrec', 'tsize']. Compute the overall Kaplan–Meier survival estimate and plot it. Compute whether the estimates differ across the category variables.

Exercise 9.2 Cox Proportional Hazard Model and HR
Model the survival time estimate, considering the influence of the features/covariates. Compute the log hazard ratios (HR) and interpret them. Compute the bi-variate influence of features/covariates with survival and interpret them.

Exercise 9.3 Grid Search
Through a Grid Search of hyperparameters, identify a parsimonious survival model. Rank the features/covariates by their weights.

Exercise 9.4 Random Forrest
Build a random forest model. Plot the survival probability of select subjects.

Exercise 9.5 Random Forrest—Feature Selection
Build a random forest model. Identify the feature/covariates' importance.

Exercise 9.6 ROC / AUC and Concordance Index
Plot the ROC curve and Concordance Index for the top variables, considering data censoring - for a specific period within the study. Interpret the results.

Exercise 9.7 Comparison of CPH and RFS Models
Do one-hot-encoding to transform categorical features. Impute missing values, if any. Build and Compare the estimators—Cox Proportional Hazards Model (CPH) and Random Forest Survival Analysis (RFS).

References

Ahuja K, Van der Schaar M (2019a) Joint concordance index. In: 53rd Asilomar conference on signals, systems, and computers, pp 2206–2213

Ahuja K, van der Schaar M (2019b) Joint concordance index. In: 53rd Asilomar conference on signals, systems, and computers, pp 2206–2213

Brentnall AR, Cuzick J (2018) Use of the concordance index for predictors of censored survival data. Stat Methods Med Res 27(8):2359–2373

Ching T, Zhu X, Garmire LX (2018) Cox-nnet: an artificial neural network method for prognosis prediction of high-throughput omics data. PLoS Comput Biol 14(4). https://doi.org/10.1371/journal.pcbi.1006076

David GK, Mitchel K (2012) Survival analysis: a self-learning text, 3rd edn. In Public Health, Springer

James G, Witten D, Hastie T, Tibshirani R (2021) An introduction to statistical learning with applications in R. Curr Med Chem (10). Springer

Kundu S, Chauhan K, Mandal D (2021) Survival analysis of patients with covid-19 in india by demographic factors: quantitative study. JMIR Format Res 5(5). https://doi.org/10.2196/23251

Sundararajan S (2023) MVA-ML. https://github.com/sun-sri/MVA-ML

Chapter 10
Computational Techniques

Learning Objectives

- Explore frequent itemset mining, association rule mining, and market basket analysis, including the Apriori algorithm and other techniques for handling big data.
- Understand graph theory fundamentals, social network analysis, and community detection algorithms with practical applications.
- Explain the principles of recommendation systems and apply them in practice.

Overview

This chapter introduces computational techniques for data analytics in two sections—market basket analysis and social network analysis. The first section explores frequent itemsets and market basket analysis (MBA). We discuss the Apriori algorithm and its use for identifying frequent itemsets and mining association rules. We then look at typical applications of frequent itemsets mining, other algorithms, and improvements needed for handling big data. The Sect. 10.3 introduces social network analysis. We discuss social network analysis algorithms for community detection and demonstrate using the Girvan-Newman and correlation clustering algorithms using two popular methods. In Sect. 10.4, we discuss the principles of recommendation systems.

Supplementary Information The online version contains supplementary material available at https://doi.org/10.1007/978-981-99-0353-5_10.

Definitions

Apriori algorithm: a popular tool used for market basket analysis in retail analytics.
Apriori pruning principle: An association rule implies that if an itemset A occurs,
then itemset B also occurs with a certain probability. Any subset of a frequent
itemset must be frequent. Therefore, if an itemset is infrequent, its superset should
not be generated—this is known as the Apriori pruning principle.
Computational Techniques: Data may be modeled based on the parameters of the
underlying probability distribution. Where such an approximation (to an underlying probability distribution) is not possible, we use computational algorithms.
Some of these algorithms may use certain statistical summarizations while remaining predominantly nonstatistical.
Collaborative filtering: Collaborative filtering recommends an item based on (a) a
user's rankings of similar items or (b) the ratings of users with similar profiles who
have rated the item in question. The former is item-item collaborative filtering, and
the latter is user-user collaborative filtering.
Confidence: Confidence is a measure of the probability that the presence of
Itemset A implies the presence of itemset B.
Content-based filtering: Content-based filtering recommends items with 'characteristic features' similar to the ones a user has rated.
Frequent pattern growth algorithm: A popular tool used for market basket analysis
in retail analytics.
Frequent patterns: Some patterns in data may occur repeatedly. Those patterns that
have a high frequency of occurrence are called frequent patterns. The patterns can
be substructures, sequences, or items.
Homophily: Homophily is the principle that individuals with similar traits are like-
lier to form ties.
Market basket analysis: Frequent itemset mining leads to discovering associations
and correlations among items in large transactional datasets. Market basket analysis (MBA) is one such application. In this process, we analyze customer buying
habits by finding associations and correlations between the items customers place
in their 'shopping baskets'.
Social network analysis: Graphs can model structures and capture the underlying
relations/processing dynamics. One common application is social network analysis, which is extensively used by marketers.
Social network graph: A social network structure can be represented as a graph.
Here, the vertices are individuals or organizations, and the links are interdependencies between the vertices, representing friendship, shared interests, or collaborative activities. If there is a weightage associated with the relationship, that can be
represented by labeling the edges with an appropriate number. These graphs can
be directed or undirected.
Support: Support indicates the itemset frequency in a transaction dataset.

UV decomposition: One way to predict the missing ratings in a utility matrix is to find two matrices U and V, whose product approximates the given utility matrix. By doing so, we summarize and reduce the features to d dimensions that allow us to characterize users and items closely. Since the matrix product UV gives values for all user-item pairs, that value can be used to predict the value of a blank in the utility matrix.

10.1 Introduction to Computational Techniques

Data models may be classified into three categories using the techniques employed—statistical, computational, or machine learning. In statistical modeling, the data is modeled and analyzed based on the parameters of the probability distribution(s) underlying the data drawn. Where such an approximation (to an underlying probability distribution) is not possible, we use other computational algorithms. Some of these algorithms may use specific statistical summarizations while remaining predominantly nonstatistical. Machine learning is about learning from data, and it makes use of statistical and computational techniques.

Let us look at some of the common computational techniques. Google PageRank is a form of web mining where the complex structure of the web is summarized by assigning a single number for each constituent page (Leskovec et al., 2020). Frequent itemsets make use of the feature extraction approach. The popular market basket analysis (MBA) is a frequent itemsets application. Similar itemsets are another feature extraction approach used in recommendation systems such as Amazon or Netflix for product recommendations. Clustering in high dimensions demands special algorithms such as BFR, CURE, GRGPF, and BDMO. The graphs model numerous natural and human-made structures that capture the relations and processing dynamics. Graphs find applications in computer science, physical, biological, and social systems. A popular application is social network analysis. This chapter discusses market basket analysis, social network analysis, and recommendation systems.

10.2 Frequent Itemsets and Market Basket Analysis (MBA)

Data mining is about analyzing patterns and trends in data. Some patterns in data may occur repeatedly. Those patterns that have a high frequency of occurrence are called frequent patterns. The patterns can be substructures, sequences, or items. Here, we will look at one of the repeated patterns—frequent items.

10.2.1   Frequent Itemsets

The weekly purchase of a family may include a set of items such as {bread, milk, yogurt, eggs, oats, cornflakes}, {onion, potato, spinach, cabbage, cauliflower, cucumber}, or {apple, banana, orange, grapes, pomegranate, pineapple}. A monthly purchase may include {wheat flour, rice, lentils, oil, pickles}. These are all examples of itemsets. If we check the customer baskets, we may find that certain sets of items are more frequently purchased than others. The set of items that appear most frequently is called a *frequent itemset*. Consider that we analyze the weekly purchases of all the customers in a retail store. We may find the following set of items to be frequent {bread, eggs, milk}, {onion, potato}, or {apple, banana}. See Fig. 10.1. A retailer may stock them on adjacent shelves for improved sales.

Frequent itemset mining leads to the discovery of associations and correlations among items in large transactional datasets and helps in business decision-making processes such as customer behavior analysis (Han & Micheline Kamber, 2014). A typical example of frequent itemset mining is *market basket analysis (MBA)*. In this process, we analyze customer buying habits by finding associations and correlations between the different items that customers place in their shopping baskets click or tap here to enter text. Frequent itemsets find applications in many domains other than retail. In telecom, each customer is associated with a set of call records between chosen destinations. Each debit/credit card account is associated with a set of payments in financial services. In medical practice, each patient is characterized by a set of diseases and treatments.

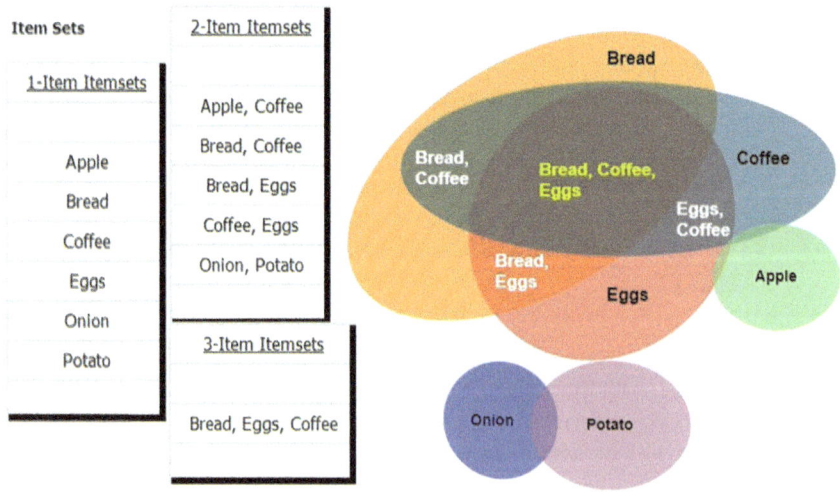

Fig. 10.1   Market basket analysis

10.2.2 *Association Rule Mining*

An association rule implies that if an itemset A occurs, then itemset B also occurs with a certain probability. The **association rule mining procedure** (Han & Micheline Kamber, 2014) may be summarized as follows:

1. Find all frequent itemsets: By definition, each of these itemsets will occur at least as frequently as the specified minimum support count.
2. Generate strong association rules from the frequent itemsets: These rules must satisfy minimum support and minimum confidence.

Let us examine some terms commonly used by business personnel in association rule mining, such as support, confidence, lift, the downward closure property, and the Apriori pruning principle.

Support

Support indicates the frequency of the itemset in a transaction dataset.

Let 'I' = {i1, i2, ..., in} be the set of all the items in a retail store. Let 't' be a customer transaction that consists of a set of items such that $t \subseteq I$. Assume that the transaction database *TDB* consists of a set of all customer transactions 't' in a given period.

Let A be a set of items, called itemset, which we want to investigate. For example, A = {bread, eggs} may be a frequent 2-item itemset, or A = {bread, eggs, coffee} may be a frequent 3-item itemset.

- The frequency of the itemset A in transaction database *TDB*, termed as the support count of itemset A, can be expressed as follows:

  The support count of itemset-A $= SC_A = n(A)$

- The probability that a randomly selected customer transaction 't' may contain the set of items A can be expressed as follows:

  The support of itemset-A $S_A = n(A)/n(D)$
  Note: An itemset A is frequent if A's support count is greater than or equal to the minimum support threshold set by the business.

Confidence

Confidence is the probability that the presence of Itemset A implies the presence of itemset B.

$$\text{Confidence } (A -> B)$$
$$= P(B/A)$$
$$= \text{Support } (A \cup B)/\text{Support } (A)$$
$$= \text{Support} - \text{Count } (A \cup B)/\text{Support} - \text{Count } (A)$$
$$= n\,(A \cup B)/n(A)$$

Lift

The lift of a rule A $\Rightarrow$ B is the ratio of observed support to the expected support if A and B were independent. A lift >1 implies a positive correlation between A and B; a lift <1 implies a negative correlation. This enhances the interestingness of frequent itemsets, over and above the confidence measure, as it computes the posterior probability of confidence (A $\Rightarrow$ B), given the presence of B. For example, consider retail sales. The seasonal hike in the sales of A and B will not influence Lift (A $\Rightarrow$ B).

$$\text{Lift } (A \Rightarrow B) = \text{Support } (A \cup B)/\big(\text{Support } (A)^*\text{Support } (B)\big)$$

In other words, support (A) indicates the probability of itemset A occurring in a transaction, support (B) indicates the probability of itemset B occurring in a transaction, and support (A $\cup$ B) represents the probability of both itemsets A and B occurring together in a transaction. Lift (A $\Rightarrow$ B) is Confidence (A $\Rightarrow$ B) / Support (B).

The downward closure property and Apriori pruning principle

- The downward closure property of frequent patterns states that any subset of a frequent itemset must be frequent (Han & Micheline Kamber, 2014).
- The Apriori pruning principle states that if an itemset is infrequent, its superset should not be generated.

10.2.3   The Apriori Algorithm

The Apriori algorithm identifies the frequent itemsets from a transaction database. The list of frequent itemsets can be used to mine the association rules (association of items). Apriori algorithm is shown below (Han & Micheline Kamber, 2014):

- Assume that we have a transaction database *TDB*.
- Let C_k represent the candidate itemset of size k.
- Let L_k represent the frequent itemset of size k.
- Initially, scan *TDB* and prepare *L1* the list of frequent single items.
- The algorithm iterates through the following three steps until L_k is empty.
  1. Generate C_{k+1} from L_k through join operation of L_k with L_k.
  2. Inspect all the transactions t in *TDB*. If t contains one or more itemsets of C_{k+1}, increment the count of those itemsets in C_{k+1}.
  3. Obtain L_{k+1} by selecting itemsets with minimum support count in from C_{k+1}.
- Generate L, the list of all frequent itemsets of size $1 \ldots k$, by the union over all L_k.

The steps of the Apriori algorithm are described below, followed by an example.

Initially, we scan all the transactions in *TDB* and list all the items whose counts exceed the minimum support count. This gives a list of frequent 1-item itemset *L1*. Then we generate *L1* join *L1*, the candidate itemset of two items each –> *C2*. For example, if *L1* contains {A}, {B}, {C} then *C2* will contain {A, B}, {A, C}, and {B, C}. We go through TDB, count the number of 2-item sets, and discard infrequent 2-item itemsets in *C2*. This will give a shorter list of 2-item itemsets that are frequent –> *L2*.

Let us continue the process for one more iteration. We generate *L2* join *L2*, the candidate itemset of three items each –> *C3*. For example, if *L2* contains {A, B} and {C, D}, then *C3* will contain {A, B, C} and {A, B, D}, and so on. We then go through the TDB, count the number of occurrences of 3-item itemsets in TDB, and discard infrequent *3-item* itemsets in *C3*. This gives a list of frequent *3-item* itemsets –> *L3*.

The above process is repeated till the emerging list L_{k+1} is empty or we do not have any more frequent *(k+1)-item* itemsets that satisfy the support count requirement. Then all the frequent itemsets *L1* … L_k are printed out.

10.2.4   Generation of Frequent Itemsets—An Example

Table 10.1 shows a database of transactions in a retail store. Generate frequent itemsets with a minimum support count of 2.

Initiation

Initially, scan the database of transactions *TDB* to get a list of items and their frequency. Identify items with support of 2 and above. Table 10.2 shows that {D} occurs only once in *TDB*. However, the items {Apple, Bread, Coffee, Eggs} have support >=2. They are selected for the next iteration.

- *L1* = {Apple, Bread, Coffee, Eggs}

Iteration-1

We get *L1* = {Apple, Bread, Coffee, Eggs} from the previous step. Join *L1* with itself and enumerate all 2-item itemsets. The candidate itemset $C2 = L1 \bowtie L1$

Transaction	Items
1	Apple, Coffee, Sugar
2	Bread, Coffee, Eggs
3	Apple, Bread, Coffee, Eggs
4	Bread, Eggs

Table 10.1 List of transactions

Table 10.2 Identifying frequent 1-item itemsets

Itemset	Support	Itemset	Support
{Apple}	2	{Apple}	2
{Bread}	3	[Bread}	3
{Coffee}	3	{Coffee}	3
{Sugar}	1		
{Eggs}	3	{Eggs}	3
C1 (Candidate Itemset of Size 1)		L1 (Frequent Itemsets of Size 1)	

Table 10.3 Identifying frequent 2-item itemsets

Itemset	Support	Itemset	Support
{Apple, Bread}	1		
{Apple, Coffee}	2	{Apple, Coffee}	2
{Apple, Eggs}	1		
{Bread, Coffee}	2	{Bread, Coffee}	2
{Bread, Eggs}	3	{Bread, Eggs}	3
{Coffee, Eggs}	2	{Coffee, Eggs}	2
C2 (Candidate Itemset of Size 2)		L2 (Frequent Itemset of Size 2)	

= [{Apple, Bread}, {Apple, Coffee}, {Apple, Eggs}, {Bread, Coffee}, {Bread, Eggs}, {Coffee, Eggs}].

Table 10.3 shows the frequency of the two-item itemsets (obtained by scanning all the transactions in *TDB*). We can observe that the pair of items {Apple, Coffee}, {Bread, Coffee}, {Bread, Eggs}, and {Coffee, Eggs} occur two or more times in *TDB*. They have support >=2. They are selected for the next iteration.

Iteration 2: From the previous step, we get $L2$ = [{Apple, Coffee}, {Bread, Coffee}, {Bread, Eggs}, {Coffee, Eggs}].

Self Join

- Drop {Apple, Coffee}, as no other 2-item itemset start with the item Apple.
- {Bread, Coffee} and {Bread, Eggs} start with a common key Bread. Join them. $L2 \bowtie L2$
  -> {Bread, Coffee} $\bowtie$ {Bread, Eggs}
  -> {Bread, Coffee, Eggs}
- Drop {Coffee, Eggs}, as no other itemsets start with the item Coffee.

Prune

- Immediate subsets of {Bread, Coffee, Eggs} are {Bread, Coffee}, {Bread, Eggs}, {Coffee, Eggs}.
- All these 3-item itemsets are in $L2$ (frequent itemsets of size 2).
- Therefore {Bread, Coffee, Eggs} is a candidate 3-item itemset of size 3. $C3 =$ {Bread, Coffee, Eggs}.

Table 10.4   Identifying frequent 3-item itemsets

Itemset	Support
{Bread, Coffee, Eggs}	2

Scan TDB and Count

See Table 10.4. We find that {Bread, Coffee, Eggs} occur two times in TDB. Therefore, {Bread, Coffee, Eggs} is selected as a frequent 3-item itemset. $L3 =$ {Bread, Coffee, Eggs}.

Termination of the algorithm

The algorithm terminates when there are no more frequent itemsets to generate. We are left with only one item in $L3$. No more joins are possible. Therefore, we terminate further iterations. The output (list of all frequent itemsets with minimum support of 2) is shown below:

1. Frequent 1-item itemsets: $L1 =$ {Apple}, {Bread}, {Coffee}, {Eggs}
2. Frequent 2-item itemsets: $L2 =$ {Apple, Coffee}, {Bread, Coffee}, {Bread, Eggs}, {Coffee, Eggs}
3. Frequent 3-item itemsets: $L3 =$ {Bread, Coffee, Eggs}

10.2.5   Pruning—An Example with 3-item Itemset

Let $L3 =$ {abc, abd, acd, ace, bcd} be a set of frequent itemsets of size 3. Identify candidate itemsets of size 4 ($C4$).

1. Self-join: $L3 \bowtie L3$

   Pairs to be considered for joining:

   {abc}, {abd}: can be joined since they have a common prefix {ab}
   {abc}, {acd}: discard, since they do not have a 3-item common prefix
   {abc}, {ace}: discard, since they do not have a 3-item common prefix
   {abc}, {bcd}: discard, since they do not have a 3-item common prefix
   {abd}, {acd}: discard, since they do not have a 3-item common prefix
   {abd}, {ace}: discard, since they do not have a 3-item common prefix
   {abd}, {bcd}: discard, since they do not have a 3-item common prefix
   {acd}, {ace}: can be joined since they have a common prefix {ac}
   {acd}, {bcd}: discard, since they do not have a 3-item common prefix
   {ace}, {bcd}: discard, since they do not have a 3-item common prefix

So we narrow down the following pairs:

{abc} and {abd} can be joined since the prefix **ab** is common to both sets
abc, abd -> abcd
{acd} and {ace} can be joined since the prefix **ac** is common to both sets
acd , ace -> acde

2. Pruning

Immediate subsets of {abcd} are: {abc}, {abd}, {acd}, {bcd}.
All the above subsets are in *L3* (frequent itemsets of size 3). Therefore,
{abcd} is a candidate for further processing.
Immediate subsets of {acde} are: {acd}, {ace}, {ade}, {cde}
{ade} is not in *L3*. Therefore, discard {acde}

3. Result: Candidate itemsets of size 4 (*C4*): {acde}

10.2.6   Association Rule Mining—An Example

Table 10.5 shows the frequency of itemsets mined from a transaction database
TDB.
Consider the frequent itemset {I1, I2, I5}. Generate the association rules.

1. Enumerate the subsets: I1, I2, I5, {I1, I2}, {I1, I5}, {I2, I5}, {I1, I2, I5}.
2. Generate the frequent itemsets lists of sizes 1, 2, 3.
3. Generate the association rules, s => (1−s)
   I1 => {I2, I5}, I2 => {I1, I5}, I5 => {I1, I2}
   {I1, I2} => I5, {I1, I5} => I2, {I2, I5} => I1
   (see Table 10.5).
4. Enumerate the association rules. Get the support count from Table 10.5 and
   calculate the respective confidence.
   Confidence A ⇒ B= Support-Count (A U B)/Support-Count (A)
   Get the support counts from Table 10.5. The results are shown in Table 10.6.

Table 10.5 Identifying frequent itemsets

L1 (frequent itemsets of size 1)		*L2* (frequent itemsets of size 2)		*L3* (frequent itemsets of size 3)	
Itemset	Support	Itemset	Support	Itemset	Support
I1	6	I1, I2	4	I1, I2, I3	2
I2	7	I1 , I3	4	I1, I2, I5	2
I3	6	I1 , I5	2		
I4	2	I2, I3	4		

Table 10.6 Enumerating association rules

AR for {I1, I2, I5}	Confidence = SC (A U B)/SC (A)	Confidence
{I1, I2} => I5	SC (I1,I2,I5) / SC (I1,I2)	$2/4 = 50\%$
{I1, I5} => I2	SC (I1,I2,I5) / SC (I1,I5)	$2/2 = 100\%$
{I2, I5} => I1	SC (I1,I2,I5) / SC (I2,I5)	$2/2 = 100\%$
I1 => {I2, I5}	SC (I1,I2,I5) / SC (I1)	$2/6 = 33\%$
I2 => {I1, I5}	SC (I1,I2,I5) / SC (I2)	$2/7 = 29\%$
I5 => {I1, I2}	SC (I1,I2,I5) / SC (I5)	$2/2 = 100\%$

Tutorial 10.2   Identifying Frequent Itemsets

Customer purchases in a fruit store are listed below. Identify frequent item-
sets satisfying minimum support of 0.3
['Litchi', 'Banana', 'Plum'],
['Litchi', 'Grapes'],
['Grapes', 'Orange'],
['Litchi', 'Banana', 'Plum'],
['Pears', 'Plum'],
['Pears'],
['Pears', 'Grapes'],
['Litchi', 'Banana', 'Mango', 'Plum'],
['Mango', 'Orange'],
['Litchi', 'Banana']

Tutorial 10.2.1 Apriori Algorithm

Install apyori package using the following python command
!pip install apyori. Refer: https://pypi.org/project/apyori/

```
import numpy as np
from apyori import apriori

TxnDB = [
['Litchi', 'Banana', 'Plum'],
['Litchi', 'Grapes'],
['Grapes', 'Orange'],
['Litchi', 'Banana', 'Plum'],
['Pears', 'Plum'],
['Pears'],
['Pears', 'Grapes'],
['Litchi', 'Banana', 'Mango', 'Plum'],
['Mango', 'Orange'],
['Litchi', 'Banana']
]
```

Invoke Apriori algorithm, after setting minimum support threshold.   Let min_support be 0.3 (3 out of 10 transactions in TxnDB).

```
mba_results = list(apriori(TxnDB, min_support = 0.3))
```

Explore the result mba_results, for getting familiar with it

```
mba_results[0]
mba_results[0].support
mba_results[0].ordered_statistics
set(mba_results[0].ordered_statistics[0][1])
```

Print the results of the Apriori algorithm

```
nTxn = len(TxnDB)
for i in mba_results:
    MarketBasket = set(i.ordered_statistics[0][1])
    if (len(MarketBasket) > 0):
        print(
            MarketBasket,
            'support=', round(i.support,2),
            'support count=', int(i.support*nTxn)
            )
```

Final Result:-

```
{'Banana'} support= 0.4 support count= 4
{'Grapes'} support= 0.3 support count= 3
{'Litchi'} support= 0.5 support count= 5
{'Pears'}  support= 0.3 support count= 3
{'Plum'}   support= 0.4 support count= 4

{'Litchi', 'Banana'} support= 0.4 support count= 4
{'Banana', 'Plum'}   support= 0.3 support count= 3
{'Litchi', 'Plum'}   support= 0.3 support count= 3

{'Litchi', 'Banana', 'Plum'} support= 0.3 support count= 3
```

Explanation:-

Round-1: 1-item itemsets

```
{'Banana'} sup-count= 4 (>= minimum support count of 3) #select
{'Grapes'} sup-count= 3 (>= minimum support count of 3) #select
{'Litchi'} sup-count= 5 (>= minimum support count of 3) #select
{'Mango'}  sup-count= 2 (less than min.supp.count of 3) #discard
{'Orange'} sup-count= 2 (less than min.supp.count of 3) #discard
{'Pears'}  sup-count= 3 (>= minimum support count of 3) #select
{'Plum'}   sup-count= 4 (>= minimum support count of 3) #select
```

Round-2: 3-item itemsets

```
{'Banana', 'Litchi'} sup-count= 4 (>= minimum support count of 3) #select
{'Banana', 'Mango'}  sup-count= (less than min.supp.count of 3)   #discard
{'Plum', 'Banana'}   sup-count= 3 (>= minimum support count of 3) #select
{'Grapes', 'Litchi'} sup-count= (less than min.supp.count of 3)   #discard
```

```
{'Orange', 'Grapes'} sup-count= (less than min.supp.count of 3)   #discard
{'Pears', 'Grapes'}  sup-count= (less than min.supp.count of 3)   #discard
{'Litchi', 'Mango'}  sup-count= (less than min.supp.count of 3)   #discard
{'Plum', 'Litchi'}   sup-count= 3 (>= minimum support count of 3) #select
{'Orange', 'Mango'}  sup-count= 1 (less than min.supp.count of 3) #discard
{'Plum', 'Mango'}    sup-count= 1 (less than min.supp.count of 3) #discard
{'Pears', 'Plum'}    sup-count= 1 (less than min.supp.count of 3) #discard
```

Round 3: 3-items itemsets

```
{'Litchi','Banana','Mango'} sup-count=1(<min.supp.countof3)       #discard
{'Plum','Banana','Litchi'}  sup-count=3(>=min.supp.countof3)      #select
{'Plum','Banana','Mango'}   sup-count=1(<min.supp.countof3)       #discard
{'Plum','Litchi','Mango'}   sup-count=1(<min.supp.countof3)       #discard
{'Mango','Plum','Banana','Litchi'} sup-count=1(<min.supp.count3) #discard
```

Tutorial 10.2.2 Association Rule Mining

Perform association rule mining satisfying minimum support of 0.3

Importing the libraries
```
import numpy as np
from apyori import apriori

TxnDB = [
['Litchi', 'Banana', 'Plum'],
['Litchi', 'Grapes'],
['Grapes', 'Orange'],
['Litchi', 'Banana', 'Plum'],
['Pears', 'Plum'],
['Pears'],
['Pears', 'Grapes'],
['Litchi', 'Banana', 'Mango', 'Plum'],
['Mango', 'Orange'],
['Litchi', 'Banana']
]
```

Set the minimum support threshold: minimum support = 0.3 (3 out of 10 transactions in TxnDB). Invoke the Apriori algorithm, to get all frequent itemsets

```
mba = apriori(TxnDB,min_support =0.3)
mba_results = list(mba)
```

Explore the result mba_results, for getting familiar with it
```
mba_results[0]
mba_results[0].support
mba_results[0].ordered_statistics
set(mba_results[0].ordered_statistics[0][1])
```

```
Print the results of Association Rule Mining
    for i in mba_results:
        MarketBasket = set(i.ordered_statistics[0][1])
For all frequent itemsets, print the support
        if (len(MarketBasket) > 0):
            print('\nItemset ', MarketBasket, 'support=', i.support)
```

if the number of items in the frequent itemset > 1, list all the association rules underlying the selected itemset. Print the corresponding confidence, and lift

```
        if (len(MarketBasket) > 1):
            print('Association Rules for the Itemset:-')
            rule_set=list(i.ordered_statistics)
            for r in rule_set:
                if r[3] > 1:
                    print(set(r[0]), '->', set(r[1]),
                            ' confidence:', round(r[2],2),
                            ' lift:', round(r[3],2))
```

```
Final Result:-
Itemset  {'Banana'} support= 0.4
Itemset  {'Grapes'} support= 0.3
Itemset  {'Litchi'} support= 0.5
Itemset  {'Pears'} support= 0.3
Itemset  {'Plum'} support= 0.4

Itemset  {'Litchi', 'Banana'} support= 0.4
Association Rules for the Itemset:-
{'Banana'} -> {'Litchi'}   confidence: 1.0  lift: 2.0
{'Litchi'} -> {'Banana'}   confidence: 0.8  lift: 2.0

Itemset  {'Banana', 'Plum'} support= 0.3
Association Rules for the Itemset:-
{'Banana'} -> {'Plum'}   confidence: 0.75  lift: 1.87
{'Plum'} -> {'Banana'}   confidence: 0.75  lift: 1.87

Itemset  {'Litchi', 'Plum'} support= 0.3
Association Rules for the Itemset:-
{'Litchi'} -> {'Plum'}   confidence: 0.6  lift: 1.5
{'Plum'} -> {'Litchi'}   confidence: 0.75  lift: 1.5

Itemset  {'Litchi', 'Banana', 'Plum'} support= 0.3
Association Rules for the Itemset:-
{'Banana'} -> {'Litchi', 'Plum'}   confidence: 0.75  lift: 2.5
{'Litchi'} -> {'Banana', 'Plum'}   confidence: 0.6  lift: 2.0
{'Plum'} -> {'Litchi', 'Banana'}   confidence: 0.75  lift: 1.87
{'Litchi', 'Banana'} -> {'Plum'}   confidence: 0.75  lift: 1.87
{'Banana', 'Plum'} -> {'Litchi'}   confidence: 1.0  lift: 2.0
{'Litchi', 'Plum'} -> {'Banana'}   confidence: 1.0  lift: 2.5
```

10.2.7 Algorithms Used in Market Bask Analysis

The Apriori algorithm is a popular tool used for market basket analysis in retail analytics. MBA may be used for (a) identifying products that may be purchased together and placing them nearby, in a physical store, website, or catalog to encourage the customer to buy related items; (b) cross-sell, up-sell, and bundling opportunities, so that customers may buy more items when certain products are bundled together; and (c) customer retention through promotions based on buying behavior, among other things.

Like the Apriori algorithm, the frequent pattern growth algorithm also offers an efficient approach to extracting association rules from the large datasets of transaction logs. These algorithms are now commonly applied in clickstream analysis, cross-selling recommendation engines, information security, bioinformatics, etc.

When it comes to processing large datasets, Apriori algorithms need large memory and processing time (Leskovec et al., 2020). The PCY (Park-Chen-Yu) algorithm improves on Apriori by creating hash tables on the first pass, hashing pairs of items, and identifying the frequent item pairs. The multistage algorithm improves it further by including additional passes between the first and second pass of the PCY algorithm to hash pairs. The multi-hash algorithm modifies the first pass of the PCY algorithm by creating several hash tables. Toivonen's algorithm is another improvement over PCY. The primary focus of Toivonen's algorithm is to reduce memory requirements by working with random data samples rather than the entire dataset. The algorithm works in two phases. In the first phase, a random sample of the transactions is taken, and frequent itemsets are discovered. Based on the frequent itemsets found in the sample, the algorithm generates candidate itemsets for the entire dataset. In the second phase, the algorithm verifies the generated candidates against the entire dataset to ensure no frequent itemsets are missed during the sampling phase.

Randomized algorithms choose a random sample of the baskets with a sample size that fits the main memory. The SON algorithm improves this by allocating multiple segments for the baskets so that all frequent itemsets for each segment can be found in the main memory. The decaying window technique is used for counting Frequent Itemsets in Streams (like the one IMDB uses for movie rating).

10.3 Social Network Analysis

Social media is a major channel used extensively by marketers and other researchers. This section provides an overview of social network analysis, which is an integral part of social media analysis, and illustrates some algorithms for community detection.

10.3.1 Social Networks—Introduction and Overview

A collection of entities, typically people or organizations, participate in a social network. There are various categories of networks, such as friendship networks (Facebook), follower networks (Twitter, LinkedIn), preference similarity networks (Instagram), interaction networks (emails, WhatsApp), and co-authorship networks (Wikibooks).

A network structure can be represented as a graph. Here, the vertices are individuals or organizations, and the links are interdependencies between the vertices, representing friendship, common interests, or collaborative activities. If there is a weightage associated with the relationship, that can be represented by labeling the edges with an appropriate number. These graphs can be directed or undirected. For example, friendship network graphs of Facebook users are undirected. On Twitter, a person 'A' may follow another person 'B', but 'B' may not follow 'A'. This exemplifies a unidirectional (or 'directed') relationship.

Common categories of measurements in network analysis include the following (Al-Taie, 2017):

- **Network connection** includes transitivity, multiplexity, homophily, mutuality, and reciprocity.
- **Network distribution** includes the distance between nodes, degree, centrality, geodesic distance, and density.
- **Network segmentation**, which includes cohesive subgroups, cliques, clustering coefficients, k-cores, and block models.

A network graph can be very sophisticated, involving features such as edge weights/probabilities, and directed/undirected graphs. A graph can have multiple vertices, which results in high dimensionality. A large graph can be sparse. All these scenarios may necessitate non-exhaustive and non-exact solutions. The distance measures used in common cluster analysis include Euclidean distance, Jaccard distance, cosine distance, etc. However, in social networks, we use different measurements such as the degree of a node, the betweenness of an edge, the closeness of a node, network connection, network distribution, and network segmentation. It follows that the common clustering techniques that we learned earlier cannot be applied as such for clustering social network graphs.

10.3.2 Basics of Graph Theory

A graph is a data structure that helps specify objects and their relationship. A graph consists of a finite set of nodes (or vertices) and edges. The edges are sets of pairs of nodes. Graphs may be broadly classified as follows. See Fig. 10.2 for a pictorial representation.

- Complete graphs contain edges between all pairs of nodes.
- Signed graphs contain edges with either a positive or a negative sign.

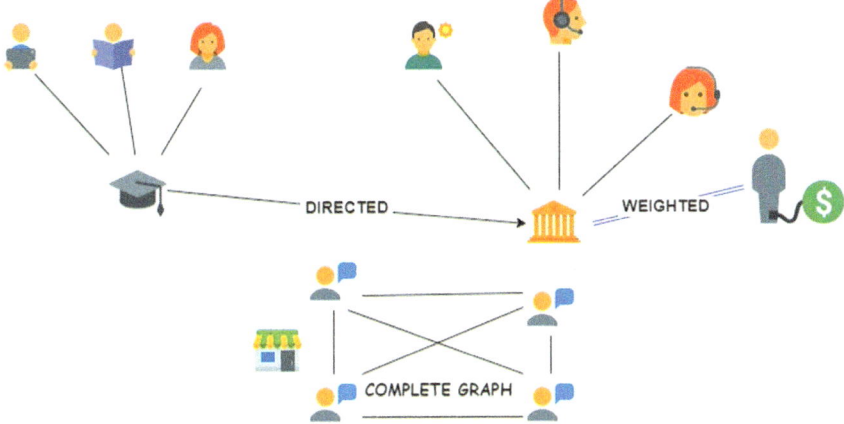

Fig. 10.2 Types of graphs

	Edges									A	B	C	D	E	F	G
	A	A	B	C	D	D	F									
	B	C	C	D	E	F	G		A	0	1	1	0	0	0	0
A	1	1							B	1	0	1	0	0	0	0
B	1		1						C	1	1	0	1	0	0	0
C		1	1	1					D	0	0	1	0	1	1	0
D				1	1	1			E	0	0	0	1	0	0	0
E					1				F	0	0	0	1	0	0	1
F						1	1		G	0	0	0	0	0	1	0
G							1									

Fig. 10.3 Graph, incidence matrix, adjacency matrix

- Directed graphs contain edges that have a specified direction. Edge e (u, v) indicates that the direction is from u to v. Edge e (u, v) is not the same as e (v, u).
- Weighted graphs contain weights on the edges.

In this chapter, we will do computations with complete signed graphs that are undirected and unweighted.

How do we represent a graph for computations? There are many ways. Let us look at some possibilities here—adjacency list, adjacency matrix (see Fig. 10.3), and incidence matrix (see Fig. 10.3). An adjacency list is a collection of unordered lists representing a finite graph. Each unordered list within an adjacency list describes the set of neighbors of a particular vertex in the graph. Though the space required is less, the access time for an element is high. An adjacency matrix is a two-dimensional matrix in which the rows represent source vertices, and the columns represent destination vertices. The incidence matrix is a two-dimensional matrix in which the rows represent the vertices, and the columns represent the edges. The entries indicate the incidence relation between the vertex at a row and the edge at a column.

10.3.3 *Understanding NetworkX Package*

NetworkX is a Python library designed to build, manipulate, and analyze a complex networks' structure, dynamics, and functions. In NetworkX, nodes can be any hashable object, such as a text string, an image, an XML object, another graph, a customized node object, and more. Let us delve into NetworkX through various tutorials.

Tutorial 10.3.1   NetworkX Graph Examples

Import NetworkX Python package. Refer: https://networkx.org/
```
   import networkx as nx
```

Create an empty graph with no nodes and no edges.
```
  G = nx.Graph()
```

Build a std Graph from the library - petersen_graph
```
  G = nx.petersen_graph()
```

Draw Graph - Figure 10-4
```
   nx.draw(G, with_labels=True, font_weight='bold')
   nx.draw_shell(G, nlist=[range(5, 10), range(5)],
                 with_labels=True, font_weight='bold')
```

Getting Familiar with NetworkX Commands:
```
   import matplotlib.pyplot as plt
   import networkx as nx
```

Build a std Graph from the library - karate_club_graph
```
  G = nx.karate_club_graph()
  nx.density(G)     # 0.139
  nx.info(G)        # 34 nodes and 78 edges
  G.nodes()
(0, 1, 2, 3, 4, 5, 6, 7, 8, 9, 10, 11, 12, 13, 14, 15, 16, 17, 18, 19, 20,
21, 22, 23, 24, 25, 26, 27, 28, 29, 30, 31, 32, 33))
```

Find the node with the maximum degree
```
  n = len(G.nodes())
  for i in range(n):
      deg = nx.degree(G,i)
      if (maxDeg < deg):
          maxDeg = deg
          nn = i
      print(i, nx.degree(G,i))
  print('node number', nn, 'has the maximum degree', maxDeg)
  The node number 33 has a maximum degree 17
```

```
Plot the histogram of the degree of nodes
    plt.plot(nx.degree_histogram(G))
    plt.xticks(range(0,19,2))
    plt.ylabel('frequency',fontsize=14)
    plt.xlabel('degree of node',fontsize=14)
See Figure 10-5: Histogram of Degree of Nodes
    G._adj            # Adjacency Matrix
```

Commonly used NetworkX functions are listed below (networkx, 2023). The tutorials help to better understand some of these NetworkX functions (Fig. 10.4).

- degree(G[, nbunch, weight]): A view of the degree for a single node or a set of nodes in the graph.
- degree_histogram(G): Generates a list detailing the frequency of each degree value in the graph.
- density(G): Computes and returns the density of the graph.
- info(G[, n]): Furnishes a summary of information for either the entire graph G or a specific node $\(n\)$.
- is_empty(G): Returns a Boolean value indicating whether the graph G has no edges.
- number_of_nodes(G): Determines and returns the count of nodes present in the graph.
- neighbors(G, n): Supplies a list of nodes connected to the specified node $\(n\)$ in the graph.
- common_neighbors(G, u, v): Identifies and returns the common neighbors of two nodes u and v in the graph.
- number_of_edges(G): Provides the count of edges in the graph G.
- density(G): Computes and returns the density of the graph.
- is_weighted(G[, edge, weight]): Returns a Boolean value indicating whether the graph G has weighted edges.
- is_negatively_weighted(G[, edge, weight]): Determines whether the graph G has negatively weighted edges and returns a Boolean value accordingly.

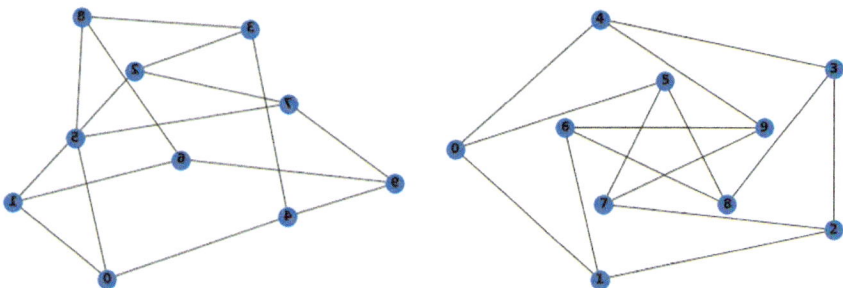

Fig. 10.4 NetworkX graph examples

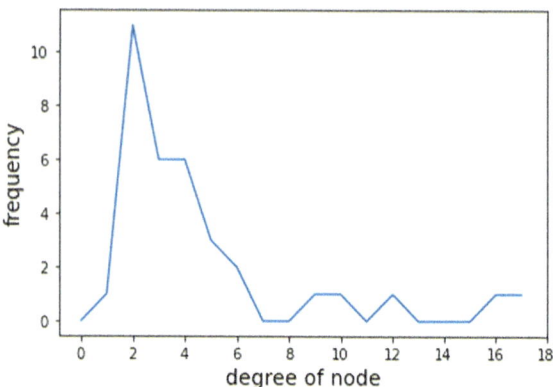

Fig. 10.5 Histogram of degree of nodes

- is_path(G, path): Checks and returns a Boolean value indicating the existence of the specified path in the graph.
- path_weight(G, path, weight): Calculates and returns the total cost associated with the specified path, considering the specified weight (Fig. 10.5).

10.3.4 *Analysis of Social Networks*

Social media provides opportunities for billions of individuals around the globe to interact, share, post, and participate in numerous activities. Social media enables connectivity and interaction anytime and anywhere. This allows a marketing researcher to observe human sentiments and behavior closely and unobtrusively, at a scale and magnitude unheard of in traditional marketing (Zafarani et al., 2014). Social network analysis helps to understand the entities, their relationships, and their behavior in social networks. Homophily is the tendency of similar individuals to get connected. Influence is the process by which an individual affects another individual. Confounding is the environment's effect on making individuals similar. The similarity in social networks is based on many factors—sociology, economics, culture, geography, interests, activities, behavior, etc.

A major task in social network analysis is community detection. This section aims to familiarize oneself with community detection algorithms and discuss/demonstrate two algorithms for the same.

In social networks, individuals (nodes) typically participate in multiple communities, posing a challenge for conventional clustering algorithms in community detection. A practical approach to segregate nodes into communities involves assessing the betweenness of edges, a concept integral to the Girvan-Newman algorithm. Complete bipartite graphs consist of two sets of nodes with all potential edges connecting nodes from different sets. Detecting complete bipartite graphs utilizes techniques like those used for identifying frequent itemsets.

Another method for community identification involves iteratively partitioning a graph into segments of approximately equal sizes. Spectral methods leverage the eigenvalues of a matrix linked to the graph to achieve an optimal partition. In social networks, friendship between two individuals may be inferred by their shared membership in multiple communities. Given knowledge of community memberships, maximum likelihood estimates (MLE) can be derived using techniques like gradient descent. The strength of membership in communities offers enhanced insights into individuals' friendships.

Outlined below are three prevalent methods for community detection in social networks.

- Girvan-Newman algorithm: The Girvan-Newman algorithm detects communities by progressively removing edges that have the highest betweenness from the original network.
- Correlation clustering algorithm: Correlation clustering aims to cluster items based on item similarity. The objective could be to minimize the number of disagreements (popular method) or to maximize the number of agreements.
- Building bipartite graphs: A complete bipartite graph consists of two distinct groups of nodes, with edges connecting every possible pair of nodes, each chosen from a different group. No edges exist between nodes within the same group. Identifying complete bipartite graphs from a node set involves employing the methodology for discovering frequent itemsets. In this context, a node's corresponding basket is considered the set of its adjacent nodes, akin to items. Viewing a complete bipartite graph with node groups of sizes t and s can be conceptualized as identifying frequent itemsets of size t with specified support levels.

10.3.5 The Betweenness of Edges

One way to separate nodes into communities is to measure the betweenness of edges. Let us assume a simple case—an undirected unweighted graph G. Assume a total of P shortest paths, considering all possible pairs of nodes from {a1 ... an} of graph G. Assume that 'p' is the number of shortest paths that go through a given pair of nodes {ai, aj}. The betweenness of the edge {ai, aj} can be indicated as the fraction p/P.

Figure 10.6a illustrates the connectivity of the members of a community. Table 10.7a shows the shortest path between each pair of nodes. Note that there

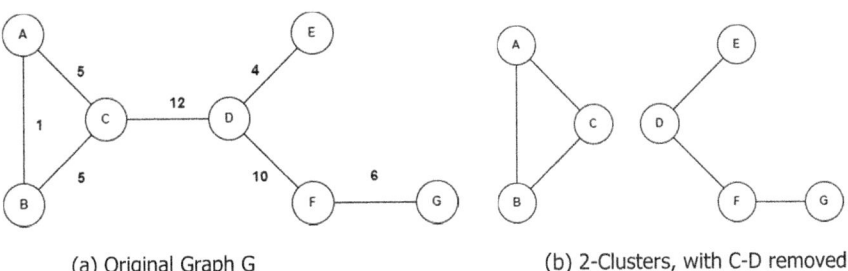

(a) Original Graph G   (b) 2-Clusters, with C-D removed

Fig. 10.6 Betweenness of edges—an example

Table 10.7   Betweenness of edges—an example

	(a) Shortest path between each pair			(b) Betweenness of each edge			
	From node	To node	Shortest path	Edge	p	P	Betweenness = p/P
1	A	B	AB	AB	1	21	0.05
2	A	C	AC	AC	5	21	0.24
3	A	D	ACD	BC	5	21	0.24
4	A	E	ACDE	CD	12	21	0.57
5	A	F	ACDF	DE	4	21	0.19
6	A	G	ACDFG	DF	10	21	0.48
7	B	C	BC	FG	6	21	0.28
8	B	D	BCD				
9	B	E	BCDE				
10	B	F	BCDF				
11	B	G	BCDFG				
12	C	D	CD				
13	C	E	CDE				
14	C	F	CDF				
15	C	G	CDFG				
16	D	E	DE				
17	D	F	DF				
18	D	G	DFG				
19	E	F	EDF				
20	E	G	EDFG				
21	F	G	FG				

are 21 shortest paths (P = 21). Table 10.7b shows the betweenness of each edge, computed according to the formula mentioned above (p/P). For example, the edge AC is between five shortest paths, as shown in Table 10.7a. They are AC, ACD, ACDE, ACDF, and ACDFG. Therefore (p = 5).

So, the betweenness of the edge AC is = p/P = 5/21 = 0.24

The edge CD has the highest betweenness of 0.57. Therefore, it is removed, generating two clusters, as shown in Fig. 10.6b.

The above process is repeated with individual clusters until we arrive at the optimal number of clusters.

10.3.6   Community Detection: Girvan-Newman Algorithm

The Girvan-Newman algorithm (Girvan & Newman, 2002) provides an efficient technique for computing the betweenness of edges using a breadth-first search. Then the edges with the highest betweenness are removed, one by one. The network breaks down into smaller pieces or communities (Fig. 10.7).

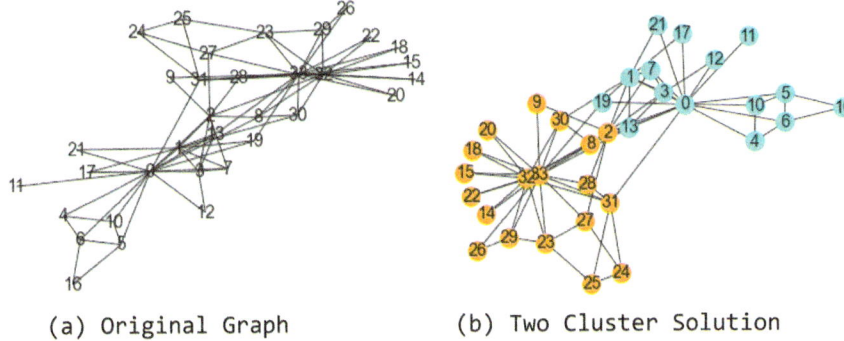

| (a) Original Graph | (b) Two Cluster Solution |

Fig. 10.7   Girvan-Newman algorithm

Tutorial 10.3.2 Community Detection Tutorial – Girvan Newman Algorithm

```
import networkx as nx
from networkx.algorithms.community.centrality import girvan_newman
import matplotlib.pyplot as plt
```

Build a std Graph from the library - karate_club_graph
```
kClubGraph = nx.karate_club_graph()
```

Girvan_newman algorithm for identifying clusters
```
community_structure = girvan_newman(kClubGraph)
clusters = []
for cluster in next(community_structure):
    clusters.append(list(cluster))
    color_map = []
```

Draw the Original Graph - Figure 10-7 (a)
```
nx.draw(kClubGraph, node_color=color_map, with_labels=True, font_size=16)
plt.show()
```

Mark the Community Clusters
```
color_map = []
for node in kClubGraph:
    if node in clusters[0]:
        color_map.append('cyan')
    else:
        color_map.append('orange')
```
Show the Clusters that Emerged - Figure 10-7 (b)
```
nx.draw(kClubGraph, node_color=color_map, with_labels=True, font_size=16)
plt.show()
```

10.3.7 Community Detection: Correlation Clustering

Correlation clustering aims to cluster items based on the similarity between items. The objective could be to minimize the number of disagreements (popular method) or to maximize the number of agreements.

Correlation clustering (CC) uses network-graph-based solutions. Given a graph with n vertices, each edge u-v is labeled '+' (nodes u, v are similar) or '−' (nodes u, v are dissimilar). A partition of the nodes is produced to agree as much as possible with the edge labels. The intention is to maximize the number of '+' edges within clusters and the number of '-' edges across clusters. Unlike other clustering algorithms such as k-means and k-median, the number of clusters in correlation clustering need not be specified beforehand. This could also be extended to real-life scenarios where we deploy a similarity metric to identify similar and dissimilar entities in a dataset and group them accordingly. We will discuss the CC-Pivot correlation clustering algorithm proposed by Ailon et al. (2008) and demonstrate it through a tutorial.

CC-Pivot Algorithm

1. Pick a random node in graph G.
2. Form a cluster with all the nodes with which it has '+' edges.
3. Remove the above cluster from graph G to form G'.
4. Repeat steps 1 to 3 for G' until no more positive edges remain.

CC-Pivot—Example

In the following tutorial, we analyze the closeness of a student fraternity in West Virginia. The cell values indicate the number of times they interacted over a week. We intend to form different groupings of the students for various New Year's parties. We expect the parties to be organized such that the students are most known to each other (Figs. 10.8 and 10.9).

Fig. 10.8 West Virginia
student fraternity community

Fig. 10.9 West Virginia student fraternity-clustering

Tutorial 10.3.3   Community Detection - West Virginia Student Fraternity

Data Setup
```python
import pandas as pd
```

Download the file from GitHub; Refer {(Sundararajan, 2023)} for Data File
```python
d=pd.read_csv('WV-fraternity.csv')
d.shape          # (116, 58)
dataset = d.iloc[0:58]
dataset.shape    # (58, 58)
```

We take the mean of all values in the matrix.
```python
mean = dataset.mean().mean()    # 1.892
```

Tutorial 10.3.3.1 Building a Graph Showing Student Connectivity

Build a graph with 58 nodes representing students
```python
import networkx as nx
G=nx.Graph()              # Create an empty graph
for i in range(58):       # Create 58 nodes, each node indicates one
student
    G.add_node(i)
```

Check each cell of the dataset with 58x58 cells. Values greater than the mean indicate friendship between students. if the cell value > mean, add an edge connecting the nodes represented
```python
friendship_threshold = 2
for i in range(len(dataset)):
    for j in range(len(dataset)):
        if dataset.iloc[i][j] >= friendship_threshold:
            G.add_edge(i,j)
G.number_of_edges() # 596 edges are there in the graph.
```

Draw the graph
```python
nx.draw(G, with_labels=True, font_weight='bold', font_size=10,
        font_color='w',
        node_size=1024)
```
See Figure 10-8: West Virginia Student Fraternity Community
There can be 58x58 = 3364 possible edges. Off these, we form 596 edges, which have a strength of connection >=mean

Tutorial 10.3.3.2 CC-Pivot Clustering Algorithm

Now, we use a simple clustering algorithm on this dataset (algorithm by Ailon
and Charikar). We pick random nodes and build clusters around their neigh-
bours in sequential order

```
import random
class charikar:
    def __init__(self,G):
        self.clusters = []
        self.G = G
    def ccpivot(self):
        " Treats existing edge as '+' and non-existing edge as '-' "
        while len(self.G.nodes())>0:
            # Randomly picks a node as the pivot
            vs = random.sample(self.G.nodes(),1)
            v  = vs[0]
```

Obtain the cluster from its set of positive neighbors

```
            Av = [v]
```

Note: neighbors() returns a list of nodes connected to node n.

```
            for u in self.G.neighbors(v):
                Av.append(u)
            self.clusters.append(Av)
            # Remove this cluster from the graph
            self.G.remove_nodes_from(Av)
            # the loop now iterates on v'
        return self.clusters
```

To return the cluster number of each node

```
    def clusterise(cluster):
        clusternumber={}
        c=0
        for i in cluster:
            c=c+1
            for v in i:
                clusternumber[v]=c
        return clusternumber
```

Tutorial 10.3.3.3 Clustering Nodes using the CC-Pivot Algorithm

The clusters that emerge indicate the party group to which a particular student could be assigned to, based on his/her links

```
    T=G.copy()
```

Random seed helps in the reproducibility of results on repeated trials

```
    random.seed(0)
    test=charikar(T)
    c=test.ccpivot()
```

Check the number of clusters that emerge

```
    print(len(test.clusters))   # ~ 7
```

Clustering All the Nodes with 1 Or 2 Edges Only. Remove nodes having one /
two edges only and form a cluster of them

```
clusters = test.clusters.copy()
to_be_added=[]
for i in clusters:
    if len(i)>2:
        pass
    else :
        print(i)
        for j in i:
            to_be_added.append(j)
        test.clusters.remove(i)
test.clusters.append(to_be_added)
```

Tutorial 10.3.3.4 Performance Metrics

```
TP=0
FP=0
TN=0
FN=0
```

clusternumber is a list which stores the cluster number of each node

```
clusternumber=clusterise(test.clusters)
for n in G.nodes():
    for v in G.nodes():
        if v!=n:
```

neighbors(): Returns a list of nodes connected to node n.

```
            if v in G.neighbors(n):
                if clusternumber[n]==clusternumber[v]:
                    TP=TP+1
                else :
                    FN=FN+1
            else:
                if clusternumber[n]==clusternumber[v]:
                    FP=FP+1
                else:
                    TN=TN+1
```

Print the accuracy of the model.

```
Accuracy=(TP+TN)/(TP+TN+FP+FN)
print('Accuracy of the model= ',Accuracy) #  0.70
```

Plot the Clusters for Visual Understanding

```
import numpy as np
cp =['b','g','r','c','m','y','b']
ClusterLabel = list(np.arange(0,58))
nNum    = list(np.arange(0,58))
```

```
for i in nNum: ClusterLabel[i] = clusternumber[i]
nx.draw(G, with_labels=True,
        font_weight='bold', font_color='w', font_size=10,
        node_color = ClusterLabel, node_size=1024)
See Figure 10-9: West Virginia Student Fraternity - Clustering

ClusterLabel
Cluster Labels for Nodes 0..57, in order
[7, 3, 1, 1, 4, 1, 1, 4, 1, 6, 1, 5, 3, 1, 1, 1, 1, 1, 5, 1, 1, 3, 2, 7,
3, 7, 1, 7, 4, 7, 5, 7, 6, 1, 1, 1, 2, 2, 1, 6, 5, 7, 7, 4, 4, 7, 2, 3, 1,
5, 2, 7, 2, 1, 1, 1, 1, 5]
```

10.4   Recommendation Systems

The limitation of shelf space prohibits a physical retailer from storing all types of products in a shop. While the fast-moving ones can be displayed, the others go to oblivion. Online shopping websites are free of this limitation as they can maintain an online catalog of a vast number of items and their features.

Websites offer recommendations for items such as movies, books, products, services, or content based on personal preferences. These websites build a personal profile based on our earlier ratings, purchases, website behavior, etc.; or based on the interests shown by people of similar profiles. They use recommendation systems to help users discover new, old, popular, or esoteric content, products, services, etc. There are two primary types of recommendation systems—collaborative filtering and content-based filtering.

10.4.1   *Content-Based Recommendation*

Content-based filtering recommends items with 'characteristic *features*' similar to the ones a user has rated. See Table 10.8. You can make recommendations even if no other users have rated the items. However, there is a tendency to recommend

Table 10.8 Content-based recommendation

Feature	Samsung M13	Realme 50A	Nokia G21	Redmi Note 11
Screen size	6.6	6.6	6.5	6.43
Screen type	LCD	LCD	TFT	Amoled
RAM in GB	6	4	6	6
Storage in GB	128	64	128	128
Weight (g)	207	193	191	179
Price (₹)	13,000	11,500	13,300	15,000
User rating	7	6	8	9

only items similar to those the user under consideration has already rated. This results in a lack of diversity and novelty in recommendations. Moreover, content-based recommendation systems rely on the capture of relevant features of items, which is challenging.

Table 10.8 shows that the scales differ for each feature. The features must be scaled to a single scale before computation using min-max standardization $(x-\mu)/$ range, z-score standardization $(x-\mu)/\sigma$, etc. This is to avoid bias by the features having a larger range of values and to truly represent the features' variations.

10.4.2   *Collaborative Filtering*

Collaborative filtering recommends an item based on (a) a user's *ratings* (actually, rankings) of *similar items* that the user has rated or (b) the ratings of *users with similar profiles* who have rated the item in question (Leskovec et al., 2020). The former is known as item-item collaborative filtering and later as user-user collaborative filtering.

Collaborative filtering algorithms use a user-item matrix representing the user's ratings of items (e.g., movies, news, products, etc.). See Table 10.9. This *utility matrix* can be used to identify similarities between users and items. Based on this information, an algorithm can then make personalized recommendations for specific users. These matrices are typically sparse because users might not be aware of the vast majority of items, and even if they are aware, they might have rated only a few. If a user rated similar items, item-item collaborative filtering is more effective and easier to implement compared to user-user collaborative filtering algorithms. It is also computationally efficient as the similarity matrix can be pre-computed and stored for quick lookup.

Collaborative filtering can be formulated as a machine learning problem relying on iterative optimization. The predicted rating by User X, denoted as Rx, is a function of the utility (User × Item):

Predicted Rating by User $X = R_x = f\,[\text{Utility (User} \times \text{Item)]}$

Assume,

G-Mean = mean global rating of the item,

Table 10.9 Utility matrix y (user-item)

Legend: Kung Fu Panda: KP; The Exorcist: Ex; Item-Item and User-User Rating										
	KP-1	KP-2	KP-3	KP-4	EX-1	EX-2	Omen-1	Omen-2	Schindler's List	Forrest Gump
User1	7.6				8.1	3.8			?	
User2	6.0	7.2	7.1	?					8.0	
User3	8.0						7.5	6.2	9.2	
User4	7.5								9.0	8.8

U-Mean-Diff = difference of user X's mean rating from the global mean concerning items of similar category.

Weighted Average of K Nearest Neighbor's Ratings =

$$\Sigma_{i=1..k} \text{ Similarity } (U_x, U_i) * \text{ Rating } (U_i)$$

The general formula for collaborative filtering can be expressed as follows:

Predicted Rating by $X = R_x =$

$G - \text{Mean} + \text{U-Mean-Diff} + \text{Similarity } (U, U_i) * \text{ Rating } (U_i).$

10.4.3   Collaborative filtering (User-User)—Example

Let us do an exercise on user-user collaborative filtering to predict User1's rating of Schindler's List, based on the utility matrix shown in Table 10.10, based on a scale of 0 to 10. The steps are enumerated below.

Step-I Center and scale

The psychology of users in rating items differs. Some users liberally rate in the higher end, e.g., 5–10, whereas some may be stingy, awarding ratings in a lower

Table 10.10 User-user collaborative filtering—data preprocessing

(a) Find row mean (user-wise average rating)

	KP-1	KP-2	KP-3	KP-4	EX-1	EX-2	Omen-1	Omen-2	Schindler's List	Forrest Gump	Mean
User1	7.6				8.1	3.8					6.5
User2	6	7.2	7.1		9				8		7.46
User3	8				8		7.5	6.2	9.2		7.64
User4	7.5				8				9	8.8	8.325

(b) Center (X-μ)

	KP-1	KP-2	KP-3	KP-4	EX-1	EX-2	Omen-1	Omen-2	Schindler's List	Forrest Gump	Row Sum
User1	1.1				1.6	-2.7					0
User2	-1.46	-0.26	-0.36		1.54				0.54		0
User3	0.36				0.36		-0.14	-1.44	0.86		0
User4	-0.825				-0.325				0.675	0.475	0

(c) Impute missing values with new row mean (zero)

	KP-1	KP-2	KP-3	KP-4	EX-1	EX-2	Omen-1	Omen-2	Schindler's List	Forrest Gump
User1	1.1	0	0	0	1.6	-2.7	0	0	0	0
User2	-1.46	-0.26	-0.36	0	1.54	0	0	0	0.54	0
User3	0.36	0	0	0	0.36	0	-0.14	-1.44	0.86	0
User4	-0.825	0	0	0	-0.325	0	0	0	0.675	0.475

range, e.g., 3–7. For someone consistently awarding ratings in the range of 5–10, a rating of '6' may suggest a negative perception. However, for an individual accustomed to ratings in the range of 3–7, a '6' rating might be considered positive. To avoid such bias, we will first center the ratings by subtracting the cell values from his/her mean rating $(X-\mu)$. Table 10.10a shows the user-wise (row-wise) mean. Table 10.10b shows each nonnull cell value of X centered around the mean. Table 10.10c shows missing values imputed with zero. (Note that zero is the row-wise mean of the centered data.) We must standardize the data to one scale if different scales are used.

Step-II Compute the cosine similarity of User-1 with the other users who have rated the item under consideration

The next step is to compute the similarity of user-1 with the other users who have rated the movie Schindler's List. As shown in an example in Chap. 8, cosine similarity is expressed as follows:

Cosine (User1, User2) = User1.User2 / ||User1||.||User2||

Cosine similarity values range from -1 to 1, with higher values indicating greater similarity. Table 10.11a shows the computation of the *L2* norm of user-wise ratings—||User-X||. This is computed as the square root of the row-wise squared sum of squares of the user ratings. Table 10.11b shows the computation of the dot products—User1.User2, User1.User3, and User1.User4. Table 10.11c shows the computation of the cosine similarity of User1 with User2, User3, and User4.

Table 10.11   User-user collaborative filtering—cosine similarity

(a) ||UserX||

	KP-1	KP-2	KP-3	KP-4	EX-1	EX-2	Omen-1	Omen-2	Schindler's List	Forrest Gump	√(Sum of Squares)
User1	1.210	0.000	0.000	0.000	2.560	7.290	0.000	0.000	0.000	0.000	3.326
User2	2.132	0.068	0.130	0.000	2.372	0.000	0.000	0.000	0.292	0.000	2.234
User3	0.130	0.000	0.000	0.000	0.130	0.000	0.020	2.074	0.740	0.000	1.758
User4	0.681	0.000	0.000	0.000	0.106	0.000	0.000	0.000	0.456	0.226	1.211

(b) Dot product of U1 with U2, U3, and U4   Sum

U1.U2	-1.606	0.000	0.000	0.000	2.464	0.000	0.000	0.000	0.000	0.000	0.858
U1.U3	0.396	0.000	0.000	0.000	0.576	0.000	0.000	0.000	0.000	0.000	0.972
U1.U4	-0.907	0.000	0.000	0.000	-0.520	0.000	0.000	0.000	0.000	0.000	-1.428

(c) Cosine similarity of U1 with U2, U3, and U4

Legend: Kung Fu Panda: KP; The Exorcist: Ex; User-User Rating; Cosine Similarity = U1.U2/||U1||.||U2||										
	KP-1	KP-2	KP-3	KP-4	EX-1	EX-2	Omen-1	Omen-2	Schindler's List	Forrest Gump	Cosine Similarity
User1	7.6				8.1	3.8			8.6		
User2✔	6	7.2	7.1	?	9				8		0.115
User3✔	7				8		7.5	6.2	9.2		0.166
User4	7.5				8				9	8.8	-0.354

For example,

```
||User1||     = Square root of the Sum of squares of all the ratings of User1
              = √ (1.1**2 + 1.6**2 + (-2.7)**2)
              = √ (1.21 + 2.56 + 7.29) = 3.326
||User2||     = Square root of the Sum of squares of all the ratings of User2
              = √ ((-1.46)**2 + (-0.26)**2 + (-0.36)**2 + 1.54**2 +0.54**2)
              = √ (2.132 + 0.068 + 0.13 + 2.732 + 0.292) = 2.234
User1.User2 = Dot Product {Multiply Corresponding Cells and Sum Up)
              = 1.1 * (-1.46) + 1.6 * 1.54 = 0.858
Cosine (User1, User2) = User1.User2/ ||User1||.||User2||
                      = 0.858/ (3.326 x 2.234) = 0.115
Similarly,
Cosine (User1, User3) = 0.972/ (3.326 x 1.758) = 0.166
Cosine (User1, User4) = - 1.428/ (3.326 x 1.211) = - 0.354
```

Step-III Predict User-1's rating, based on K nearest neighbor's rating

Consider 'k' nearest neighbor's similarity rating. Let 'k' = 2. From Table 10.11c, we observe that the users User2 and User3 appear to be closest to User-1, with similarity measures of 0.115 and 0.166, respectively. Taking the weighted average, we can predict User-1's rating of Schindler's List as follows:

```
User-1's Rating of Schindler's list =
[User2's rating of Schindler's List * Cosine (User1, User2) +
User3's rating of Schindler's List * Cosine (User1, User3)] / [Cosine (User1, User2) +
Cosine (User1, User3)]= [8 * 0.115 + 9.2 * 0.166] / [0.115 + 0.166] = 8.6
So User-1's Rating of Schindler's list is predicted to be 8.6!
```

Interpretation: The predicted rating for Schindler's List is 8.6, suggesting that, based on the collaborative filtering model, User-1 will likely have a positive opinion of the movie. This prediction is a statistical estimation based on users' ratings with similar preferences. Individual tastes and subjective factors may influence user experiences, and the predicted rating serves as a recommendation rather than a definitive outcome.

10.4.4 Collaborative Filtering (Item-Item)

In the above exercise on user-user collaborative filtering, we took the similarity of the user under consideration with k-users, by computing the cosine similarity between 'k' row-wise ratings available. Similarly, we can compute item-item collaborative filtering if we compute the similarity between 'k' column-wise ratings. This approach helps predict a user's rating based on k similar items that the user has rated earlier. For example, we can predict User2's rating for Kung Fu Panda-4, based on his/her rating of Kung Fu Panda-1, Kung Fu Panda-2, and Kung Fu Panda-3.

10.4.5   UV Decomposition

Note that movies are rated based on numerous features such as genre, language, director, actors, cinematography, music, and story. These ratings may be obtained explicitly or from user review texts. In any case, a comprehensive rating of multiple product or service features is never available for analysis. The UV decomposition method approaches this from the angle of underlying feature summarization.

Assume a utility matrix M with m rows and n columns. Our task is to find two matrices U with n rows and d cols, and V with d rows and m cols, such that U.V closely approximates M.

$$[M]_{n \times m} \approx [U]_{n \times d} [V]_{d \times m}$$

By doing so, we summarize and reduce the features to d dimensions that allow us to characterize users and items closely. The dimension d emerges from iterative trials. Machine learning techniques can be used to implement this effectively. Once the U and V matrices are determined we can compute the missing values of M from them.

In our example, M is 4×10, U is 10×2, V is 2×4; and 2 is the reduced dimension of underlying features we expect to compute. See Table 10.12. Initially, we populate the U and V matrices with random values. We will compute U.V and find the deviation from the actual values of matrix M. The previous step will be treated many times to minimize deviation or error. Once the error converges, we have U and V with nonnull values in every cell. From this, we can compute the predicted value of missing values (ratings) in the matrix M.

Table 10.12 UV decomposition

7.6			8.1	3.8			?			U01	U02		V11	V12	V13	V14
6	7.2	7.1	?	9			8		=	U11	U12	x	V21	V22	V23	V24
7			8		7.5	6.2	9.2			U21	U22					
7.5			8				9	8.8		U31	U32					
										U41	U42					
										U51	U52					
										U61	U62					
										U71	U72					
										U81	U82					
										U91	U92					

Summary

Data may be modeled based on the parameters of the underlying probability distribution. Where such an approximation (to an underlying probability distribution) is not possible, we use general computational algorithms. Some of these algorithms may use certain statistical summarizations while remaining predominantly nonstatistical. Common examples of computational techniques include Google PageRank, market basket analysis, recommendation systems, social network analysis, etc.

Market Basket Analysis

Some patterns in data may occur repeatedly. Those patterns that have a high frequency of occurrence are called frequent patterns. The patterns can be substructures, sequences, or items. Frequent itemset mining leads to discovering associations and correlations among items in large transactional datasets. Market basket analysis (MBA) is one such application. In this process, we analyze customer buying habits by finding associations and correlations between the items customers place in their 'shopping baskets'.

Support indicates the frequency of the itemset set in a transaction dataset. Confidence is the probability that the presence of Itemset A implies the presence of itemset B. An association rule implies that if an itemset A occurs, then itemset B also occurs with a certain probability. Any subset of a frequent itemset must be frequent. Therefore, if an itemset is infrequent, its superset should not be generated—this rule is called the Apriori pruning principle.

The Apriori algorithm is a popular tool used for market basket analysis in retail analytics. Like the Apriori algorithm, the frequent pattern growth algorithm also offers an efficient approach to extracting association rules from large datasets of transaction logs. Both are now commonly applied in clickstream analysis, cross-selling recommendation engines, information security, bioinformatics, etc.

Apriori algorithms need large memory and processing time when processing large datasets. Therefore, enhanced algorithms such as PCY, Multistage, Multi-hash, Randomized, and SON were developed to handle big data, especially data streams.

Social Network Analysis

Social network analysis helps us to understand entities and their relationship in a community. A social network structure can be represented as a graph. Here, the vertices are individuals or organizations, and the links are interdependencies between the vertices, representing friendship, common interests, or collaborative activities. If there is a weightage associated with the relationship, that can be represented by labeling the edges with an appropriate number. These graphs can be directed or undirected.

Common categories of measurements in social network analysis include—network connection, network distribution, and network segmentation. Social network relationships tend to cluster. However, numerous challenges in clustering network graphs make them computationally expensive. Moreover, the usual statistical methods used in clustering may not be applicable here.

There are various methods for community detection in social networks. The Girvan-Newman algorithm detects communities by progressively removing edges from the original network. Correlation clustering aims to cluster items based on the similarity between items. The objective could be to minimize the number of disagreements (popular method) or to maximize the number of agreements. A complete bipartite graph has two groups of nodes. We can find complete bipartite graphs from a set of nodes using the technique used to find frequent itemsets.

Recommendation Systems

The limitation of shelf space prohibits a physical retailer from storing all types of products in a shop. While the fast-moving ones can be displayed, the others go to oblivion. Online shopping websites are free of this limitation as they can maintain an online catalog of a huge number of items and their features. Websites offer recommendations for items such as movies, books, products, services, or content based on personal preferences. These websites build a personal profile based on our earlier ratings, purchases, website behavior, etc.; or based on the interests shown by people of similar profiles. They employ recommendation systems to help users discover new, old, popular, or esoteric content, products, services, etc.

There are two primary types of recommendation systems—collaborative filtering and content-based filtering. Content-based filtering recommends items with 'characteristic features' similar to the ones a user has rated. However, they tend to recommend only the items similar to those the user under consideration has already rated. This results in a lack of diversity and novelty in recommendations. Moreover, content-based recommendation systems rely on the capture of relevant features of items, which is challenging.

Collaborative filtering recommends an item based on (a) a user's ratings (actually, rankings) of similar items that the user has rated or (b) the ratings of users with similar profiles who have rated the item in question. The former is known as item-item collaborative filtering and later as user-user collaborative filtering. If a user rated similar items, item-item collaborative filtering is more effective and easier to implement than user-user collaborative filtering algorithms. It is also computationally efficient as the similarity matrix can be pre-computed and stored for quick lookup.

Products or services are rated based on numerous features. A comprehensive rating of multiple features is never available for analysis. The UV decomposition method approaches this from the angle of underlying feature summarization. Once the U and V matrices are determined, we can compute the missing values of X from them.

Questions

Comprehension

1. Provide examples of computational techniques in data mining.
2. Describe frequent itemsets and their applications in various domains.
3. Summarize the key concepts and applications of market basket analysis in a few sentences.
4. Define the Apriori principle.
5. Describe the Apriori algorithm.
6. List the basic principles in association rule mining, using Apriori algorithm.
7. Discuss association rule mining in high dimensional data space.
8. Explain the concepts of support, confidence, and lift in association rule mining. How do these metrics help in understanding patterns in data?
9. Name different types of social networks and mention how they differ.
10. Define weighted graphs.
11. How to find the neighbors of a node in the graph?
12. Describe the betweenness of edges with an example.
13. Write a brief note on the incidence matrix and adjacency matrix
14. Write a brief note on bipartite graphs.
15. Describe the Girvan-Newman algorithm for community detection. What role does the betweenness of edges play in this algorithm?
16. Write a brief note on correlation clustering.
17. Write a brief note on the CC pivot algorithm for correlation clustering.
18. Define the distance measures Jaccard, Cosine, and TF.IDF, with suitable examples and applications.
19. Describe UV decomposition in recommendation systems.

Analysis

20. How does the Apriori pruning principle contribute to the efficiency of the Apriori algorithm?
21. Compare and contrast the Apriori algorithm, PCY algorithm, Multistage Algorithm, Multi-hash Algorithm, and Toivonen's Algorithm in terms of their memory requirements and efficiency.
22. Explain the challenges in social network clustering compared to traditional clustering.
23. How does the choice of distance measures in cluster analysis differ between traditional datasets and social network graphs?
24. Explain the concept of community detection in social networks and why traditional clustering techniques may not be directly applicable.
25. Compare and contrast the three ways of representing a graph for computations: adjacency list, adjacency matrix, and incidence matrix.
26. Discuss how long-tail phenomena necessitate recommendation systems.

27. Describe content-based and collaborative filtering for recommendation systems. Compare and contrast their strengths and weaknesses.

Application

28. Imagine you are a retail manager. How would you apply market basket analysis to improve the layout of your store and boost sales? Based on market basket analysis, can you propose specific strategies for cross-selling, up-selling, and bundling products?
29. Consider a scenario in financial services where each debit/credit card account is associated with a set of payments. How could frequent itemsets be valuable in this context? Provide examples.
30. Suppose you are a marketing strategist. How could the insights gained from community detection algorithms like the Girvan-Newman algorithm help to prepare a strategy for targeted advertising or product promotion on social media?
31. Discuss how community detection algorithms could enhance content recommendation systems on social media platforms. What benefits might this bring to users and content creators?
32. How could a content-based recommendation system be implemented for an e-commerce platform selling diverse products?
33. How would you implement a collaborative filtering system for a new online bookstore with a limited user base?
34. In what scenarios might user-user collaborative filtering outperform item-item collaborative filtering and vice versa?
35. Can you provide examples of real-world applications where collaborative filtering has been successfully employed?
36. Describe the steps in building a collaborative filtering algorithm for a movie recommendation website.

Exercises

Exercise 10.1 Apriori Algorithm to Identify Frequent Itemsets

Customer purchases in a retail store are listed below. Do association rule mining, satisfying minimum support of 0.5 and minimum confidence of 0.5. (Note: strong association rules satisfy both minimum support and minimum confidence).

```
['A', 'C', 'D'],
['B', 'C', 'E'],
['A', 'B', 'C', 'E'],
['B', 'E']
```

Exercise 10.2 Association Rule Mining—Snacks Store

Consider the customer purchases in the snacks section of a retail store given below. Perform association rule mining satisfying minimum support of 0.15, and minimum confidence of 0.6.

```
['cashews', 'cookies', 'chocolate'],
['cake', 'mixture', 'cookies'],
['dates', 'peanuts'],
['biscuits', 'peanuts'],
['crackers', 'peanuts'],
['cookies', 'dates'],
['mixture', 'apricots', 'peanuts', 'chocolate'],
['peanuts', 'cashews', 'cake', 'crackers'],
['apricots', 'dates', 'crackers', 'chocolate', 'cashews', 'peanuts']
```

Exercise 10.3 West Virginia Fraternity—Most Influential Individual(s)

Refer to Tutorial 10.6, where we analyzed the closeness of a student fraternity in West Virginia. Here, we have a similar dataset, where the cell values indicate the closeness of students ranked on a scale of 1 ... 5 (No interaction ... Very High Interaction). Find the most influential individual(s) in the network.

    # Download the dataset 'WV-fraternity.csv' from GitHub

    # Data File – (Sundararajan, 2023)

Exercise 10.4 Social Network Analysis

Compute the betweenness of edges and partition the network graph in Fig. 10.10 into two clusters.

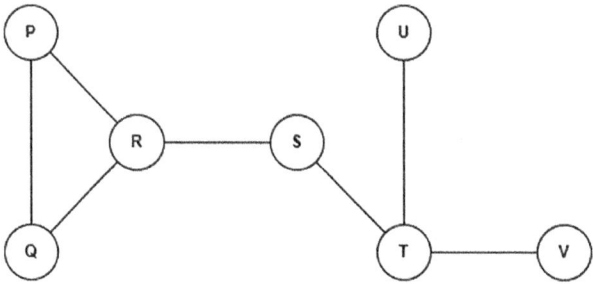

Fig. 10.10 Clustering problem

Exercise 10.5 Recommendation Systems

Predict the Kungfu Panda-4 rating by User2, using the utility matrix below.

Legend: Kung Fu Panda: KP; The Exorcist: Ex; Item-Item Rating									
	KP-1	KP-2	KP-3	KP-4	EX-1	EX-2	Omen-1	Omen-2	Schindler's List	Forrest Gump
User1	7.6				8.1	3.8				
User2	6	7.2	7.1	?	9				8	
User3	8	8.5	8.3	7.8	8		7.5	6.2	9.2	
User4	7.5	8	7.8	7.3	8				9	8.8

References

Ailon, N., Charikar, M., & Newman, A. (2008). Aggregating inconsistent information: Ranking and clustering. *Journal of the ACM, 55*(5). https://doi.org/10.1145/1411509.1411513

Girvan, M., & Newman, M. E. J. (2002). Community structure in social and biological networks. *Proceedings of the National Academy of Sciences of the United States of America, 99*(12). https://doi.org/10.1073/pnas.122653799

Han, J., & Micheline Kamber, J. P. (2014). Data mining. Concepts and techniques, 3rd Edition (The Morgan Kaufmann Series in Data Management Systems). In *Proceedings—2013 International Conference on Machine Intelligence Research and Advancement, ICMIRA 2013.*

Leskovec, J., Rajaraman, A., & Ullman, J. D. (2020). Mining of massive datasets. In *Biometrics* (Issue 4). Cambridge University Press. https://doi.org/10.1111/biom.12982

networkx. (2023). *networkx.* https://networkx.org/documentation/stable/tutorial.html

Sundararajan, S. (2023). *MVA-ML.* https://github.com/sun-sri/MVA-ML

Zafarani R, Abbasi MA, Liu H (2014) Social media mining: An introduction. Cambridge University Press

Chapter 11
Machine Learning

Learning Objectives

- Understand the fundamentals of supervised machine learning techniques.
- Describe optimization techniques.
- Explain the gradient descent optimization technique in regression and classification and demonstrate its application.
- Explain regularization techniques in regression and classification and demonstrate their application.
- Demonstrate cross-validation techniques in regression and classification.
- Explore bias-variance trade-off, test-train convergence, and hyperparameter tuning.
- Understand ensemble methods for model performance improvement.

Overview

This chapter will discuss supervised machine learning techniques and their application in regression and classification. The topics include introduction to supervised machine learning, optimization, gradient descent, regularization, resampling, bias-variance trade-off, test-train convergence, hyperparameter tuning, and ensemble methods. We will apply our learning to solve regression and classification problems using Python.

Supplementary Information The online version contains supplementary material available at https://doi.org/10.1007/978-981-99-0353-5_11.

349
S. Sundararajan, *Multivariate Analysis and Machine Learning Techniques*,
Transactions on Computer Systems and Networks,
https://doi.org/10.1007/978-981-99-0353-5_11

Definitions

Bias-variance trade-off: Increasing a model's complexity will typically increase its variance and reduce its bias. Reducing the model complexity increases its bias and reduces its variance. So, there is a trade-off between the two.

Bootstrap: Bootstrap resamples at random, with replacement. Also, see 'resampling'.

Data pipeline: A machine learning job may consist of a sequence of tasks. Pipelines are functions available in the scikit-learn software package to define and automate such workflows.

Elastic net regularization: A combination of L1 and L2 regularization is used in elastic net regularization. Elastic net is useful when multiple features are correlated with one another. Also, see 'regularization'.

Gradient descent: Gradient descent is an iterative optimization technique for finding the local minimum of a differentiable function. This technique can find the model parameters (coefficients) that optimize a specified loss function.

Hyperparameters: A machine learning model is a mathematical model with several parameters learned from the data. Another set of parameters is manually set during the model development phase to facilitate effective model development. These are known as hyperparameters. Examples include the regularization parameters, the learning rate α, the k in k-nearest neighbors, the hyperparameters for support vector machines—C and σ, etc.

K-fold cross-validation: K-fold cross-validation is a resampling procedure. In this, we divide the entire dataset into k subsets at random. We do k training sessions. In each session, we use one subset as the test sample, while the other $k - 1$ subsets are combined to form the training sample. See also 'resampling'.

Lasso regression is a regularization method that solves the minimization of the sum of the magnitude (L1 norm) of the coefficients with a penalty term added. This method is helpful for feature elimination or feature selection. Also, see 'regularization'.

Machine learning: While humans learn from experience, machines learn from data and improve their accuracy over time, without being programmed to do so. Machine learning involves identifying the correct model parameters that optimize the predictive capability by deriving information from available data.

Mini-batch gradient descent: This is a technique for optimization. In the mini-batch gradient descent method, a subset of 'm' $(1 < m > n)$ data points are randomly selected in each iteration. This method is a compromise between the speed of SGD and the goodness of fit of batch gradient descent. The gradient descent method is repeated numerous times to reach a specified convergence criterion. See also 'Gradient descent'.

Regularization: Regularization is a class of techniques that reduce model complexity and prevent overfitting. Ridge regression and lasso regression are examples of regularization methods.

Resampling: Resampling procedures are based on repeating the training and testing on different randomly chosen subsets or splits of the original dataset.

Ridge regression is a regularization method that solves the minimization of the least squares (L2 norm) with a penalty term added. Also, see 'regularization'.

Stochastic gradient descent: This is a technique for optimization. In Stochastic Gradient Descent (SGD), we randomly select one data point from among the n data points available in each iteration. The gradient descent method is repeated numerous times to reach a specified convergence criterion. Also, see gradient descent.

Supervised learning: In machine learning, the sample data consists of input features and the desired outputs (targets). Based on the data samples, an algorithm learns to predict the output (target), given a set of features.

11.1 Exploring Supervised Machine Learning

As we know, a computer system is programmed to execute a set of actions based on the instructions stored in it. However, a machine 'learning' (ML) system learns from the data to which it is exposed.

Artificial intelligence (AI) agents and systems are built to match or exceed human capabilities in information analysis and complex decision-making. Machine learning is a specific approach within AI that focuses on developing algorithms and models capable of learning from data and refining their accuracy over time, all without explicit programming. Machine learning involves identifying the correct model parameters that optimize the predictive capability and deriving information from available data. The learning process can be broadly categorized as follows:

- Supervised learning, where the training data includes desired outputs.
- Unsupervised learning, where the training data does not include desired outputs.
- Semi-supervised learning, where the training data includes a few desired outputs.
- Reinforcement Learning (RL), where an agent learns to make decisions by interacting with an environment to maximize cumulative rewards.

In this chapter, our primary focus centers on one prominent machine learning paradigm—supervised learning. The next chapter delves into the domain of artificial intelligence and explores a specific machine learning subset known as deep learning.

Supervised Learning

In supervised learning, the sample data consists of input features and the desired outputs (targets). The sample is divided into two sets—the training dataset and the

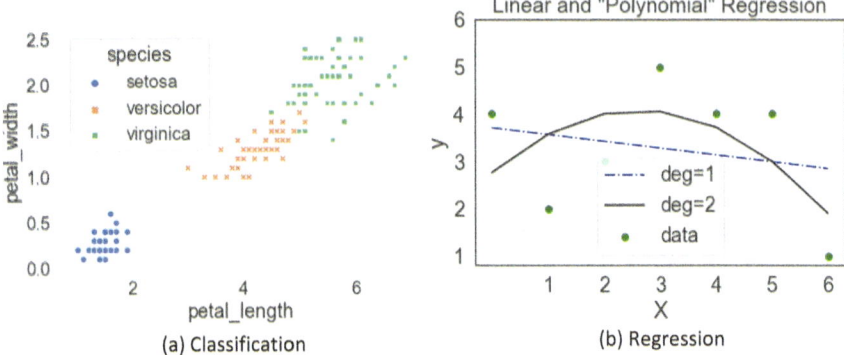

Fig. 11.1 Supervised learning—classification and regression

test dataset. A training dataset is used to train algorithms such as Bayes belief networks, linear discriminant analysis, decision trees, support vector machines, and many others. The model is then tested over the test dataset. The train-test process is repeated till an accurate model emerges.

Algorithms for classification are trained to predict categorical class labels (nominal or ordinal). See Fig. 11.1a—classification of Iris flowers. Typical applications include medical diagnosis (whether a patient tests positive or negative for a disease, based on biochemical or histopathological test results), the recognition of a handwritten digit (in this case, there are 10 classes labeled 0 to 9), credit/loan approval, fraud detection (whether a financial transaction is fraudulent or not), webpage categorization, and email spam. Regression algorithms are continuous-valued functions trained to predict numeric values (real numbers). Typical applications include annual sales turnover, revenue, or yield of crops.

11.2 Learning by Gradient Descent Optimization

The fundamental components of machine learning include (a) the model, such as a classifier or regressor and (b) the learning method, exemplified by gradient descent or its variations.

Gradient descent is a popular optimization strategy in machine learning and deep learning. It is an iterative technique used in regression and classification problems. Variations to gradient descent include stochastic gradient descent, mini-batch gradient descent, batch gradient descent, Adagrad, and Adam. Let us first look at the limitations of pure algebraic solutions in optimization and why iterative techniques are necessary.

Fig. 11.2 Linear regression

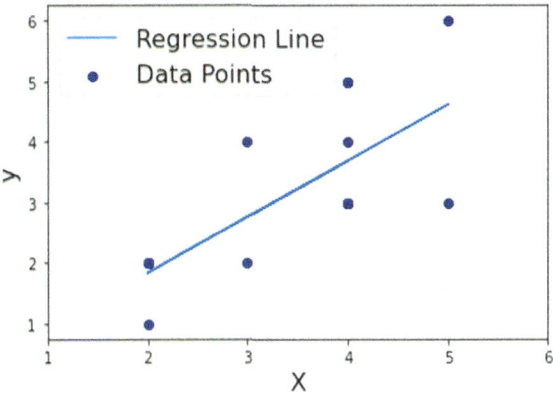

11.2.1   *Regression Analysis: From Exact Solutions to Gradient Descent*

Regression analysis examines associative relationships between a target varia-ble (y) and feature variables X $\{X_1 \dots X_k\}$, where k is the number of features. The following equation shows one instance of y:

$$y^{(i)} = W_0 + W_1 X_1^{(i)} + W_2 X_2^{(i)} + \cdots + W_k X_k^{(i)} + \text{error}$$

where w's are the weights, or coefficients, associated with the feature variables.

The above equation may be rewritten as

$$y^{(i)} = W^T X^{(i)} + \text{error}$$

We can optimize the above equation by altering (W). To solve for k variables, we need at least k distinct equations. In practice, we may have a large number of observations $n \gg k$. As discussed in Chap. 5, we may use an exact solution like the matrix method. However, the solution does not guarantee exact results. For example, no single line y = f(X) can predict all the data points in the scatter plot in Fig. 11.2.

Instead of an exact solution, we look for one that minimizes the difference between the observed value of y and the value estimated from the equation $(\hat{y})$. The difference is called error or residual. We assume that the features $(X_1 \dots X_k)$ are uncorrelated.

$$\text{Residual Sum of Squares (RSS)} = \sum_{i=1}^{n} (y^{(i)} - \hat{y}^{(i)})^2 \qquad (11.1)$$

$$\text{Mean Squared Error (MSE)} = \sum_{i=1}^{n} \left(y^{(i)} - \hat{y}^{(i)}\right)^2 / n \qquad (11.2)$$

$$\text{Mean Absolute Error (MAE)} = \sum_{i=1}^{n} |y^{(i)} - \widehat{y}^{(i)}|/n \tag{11.3}$$

Huber loss combines the benefits of MSE and MAE, and is more robust in handling outlier (see Sect. 2.8.2) data than MSE.

$$\text{Huber Loss} = \frac{1}{n} \sum_{i=1}^{n} \left(y^{(i)} - \widehat{y}^{(i)} \right)^2, \text{if} \left| y^{(i)} - \widehat{y}^{(i)} \right| \leq \delta(\sim\text{MSE})$$

$$= \frac{1}{n} \sum_{i=1}^{n} \delta(\left| y^{(i)} - \widehat{y}^{(i)} \right| - \frac{1}{2}\delta), \text{if} \left| y^{(i)} - \widehat{y}^{(i)} \right| > \delta(\sim\text{MAE})$$

where
'δ' determines the error (or loss) threshold for the transition from MSE to MAE.

The ordinary least squares method, or linear least squares method, estimates the parameters of a regression model by minimizing the sum of the squared errors. We have assumed k features and sample size n. Given that X is a matrix of size (n, k), OLS has a computation cost of O (nk^2) if we look for an exhaustive algebraic solution.

Optimization can generally be stated as minimizing or maximizing some function f(W, X) by altering W. The function that we want to optimize is called the objective function or criterion (or cost function, loss function, or error function). In other words, a loss function gives a measure of the error, or the difference between the predicted output of a model and the actual output. The loss functions RSS, MSE, and MAE discussed above are used in developing regression-based machine learning models. During training, our objective is to minimize prediction errors.

When the data is large, an exhaustive mathematical solution may be prohibitive. Moreover, it may not automatically guarantee a reliable model fit, as the presence of noise, outliers, and other factors may prevent insight into the actual parameters of the underlying distribution. An alternative is to use an iterative solution using methods such as gradient descent, giving due consideration to the model's reliability.

11.2.2   Gradient Descent Optimization Techniques

Several cost functions (or loss functions) are used in machine learning. Examples include the loss functions used in regression such as absolute error, squared error, and Huber loss; and the functions used in classification such as cross entropy, log loss, exponential loss, and hinge loss. In this section, we explore an iterative method for optimizing a cost function (alternatively, minimizing the loss function)—the gradient descent.

Gradient descent is an **iterative optimization** technique to find the **local minimum** of a **differentiable function** (Chollet 2019). Gradient descent is used to find the values of a function's parameters (coefficients) that optimize a cost function. See Fig. 11.3a. The **gradient descent** method picks training samples up at a time. The average prediction error is computed. Based on that, each coefficient value (W_i's) is adjusted. The adjustment is based on the prediction error and parameter α. This process is repeated many times for convergence.

Assume we have 1 data point (sample size 1) and 1 feature (X). The observed outcome is y, whereas the estimated or computed outcome is $\widehat{y}$. The residual is $(y - \widehat{y})$. We will take a derivative of $(y - \widehat{y})^2$ upon W and do one weight adjustment (from W to W'). Assume that we have 10 features (X_i, $i = 1 \ldots 10$) and the corresponding coefficients (W_i, $i = 1 \ldots 10$). We have to take 10 partial derivates $d(y - \widehat{y})^2/dW_i$, for each W_i, and adjust the respective coefficients W_i. Therefore, we have 10 adjustments to be done.

Now let us consider **batch gradient descent**. Assume we have 100,000 data points (sample size = 100,000) and 10 features (X_i, $i = 1 \ldots 10$). We will have 100,000 residuals. We need to compute 100,000 * 10 partial derivates. Then do 1 million adjustments in an iteration. In the Stochastic Gradient Descent (**SGD**) method, in an iteration, we pick up just one data point **at random** from among the 100,000 data points available. Therefore, it will be fast.

In the **mini-batch gradient descent** method, a subset of 'm' data points is picked up (sample size = m; $1 < m > 100,000$), in each iteration, at random. This is a compromise between picking up, just one data point at random in SGD and all the 100,000 data points in batch gradient descent. This results in a compromise between the speed of SGD and the goodness of fit of batch gradient descent.

Mini-Batch Gradient Descent Algorithm

The general steps involved in the gradient descent algorithm are listed below:

Consider a regression model in k variables,

$$\widehat{y}^{(i)} = W_0 + W_1 X_1^{(i)} + W_2 X_2^{(i)} + \cdots + W_k X_k^{(i)}$$

1. Initial step: Initialize coefficients ($W_0 \ldots W_k$) with random values. (There are guidelines for initialization, which are discussed in the next chapter.) Shuffle the data. Split the data into a fixed number of 'm' mini-batches of size m.
2. Repeat over each mini-batch 1 … m.

   a. Cost evaluation: Compute the sum of the cost over all the data instances in the mini-batch. For example, assume the cost function is sum of squared errors, $\Sigma (y - \widehat{y})^2$.
   b. Derivative calculation: Compute the sum of partial derivatives $\Sigma [\delta(\text{cost-function})/\delta W_i]$, for each variables W_i of the cost function, over all

the data instances in the mini-batch. (The sum Σ [δ(cost-function)/δW_j] is also called the gradient of W_j.)

c. Update the coefficients. Move the weights in the opposite direction of the gradient to reduce the loss by a small step.

$$W_i = W_i - r * [\Sigma \delta(\text{cost-function})/\delta W_i],$$

where r is a fraction, aka 'learning rate'.

3. Iterate: Repeat step 2 until a termination criterion is reached. The criterion may be set as follows—the loss or mismatch is within an acceptable limit, there is no significant difference in the loss over a set of iterations, or the number of iterations has reached a set limit.

The SGD, mini-batch, or batch gradient descent is repeated several times to reach the specified convergence criteria. The convergence process will be prolonged if the learning rate (r) is very low. If r is very high, oscillations may occur, so the model may not converge. See Fig. 11.3b. When the entire training dataset passes through the machine learning algorithm, we say one **epoch** is completed. The entire training dataset is used in each epoch in full-batch gradient descent. In mini-batch gradient descent, the training dataset is divided into mini-batches, and each epoch consists of multiple iterations, with each iteration updating the model based on a mini-batch. A pass over all the mini-batches completes one epoch.

SGD needs low memory and converges very fast. Therefore, this method is preferred for large data. SGD gives a satisfactory solution, but it may not be the optimal. However, it can come out of the local minimum fast. See Fig. 11.4. Samples are selected randomly; therefore, data must be shuffled before the start of every epoch. The batch gradient descent method needs large memory and takes

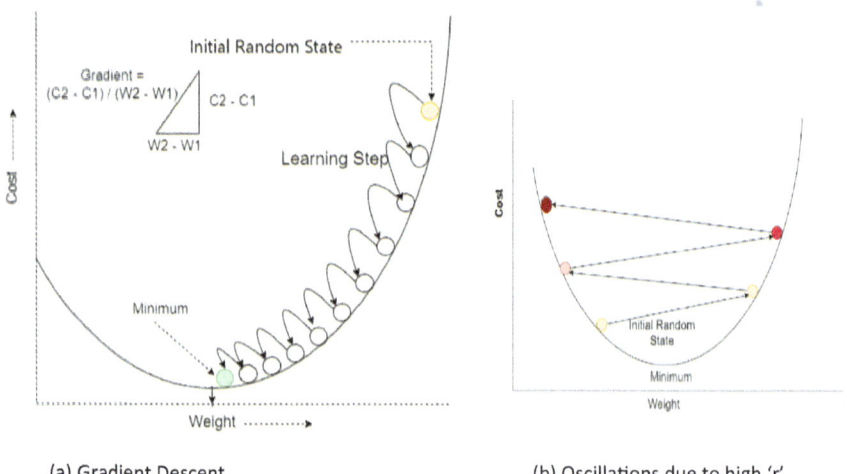

(a) Gradient Descent   (b) Oscillations due to high 'r'

Fig. 11.3   Gradient descent and oscillations

Fig. 11.4 Local minimum, global minimum

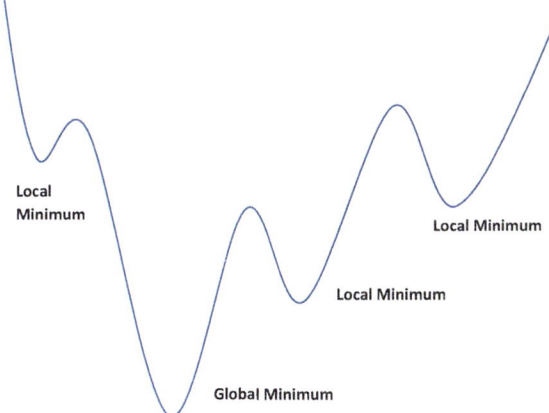

high computation time. Therefore, this method is not preferred for large data. The convergence is slow and cannot escape the local minimum quickly. However, the method converges to the optimal solution ultimately. As mentioned earlier, the mini-batch gradient descent method is a compromise between the speed of SGD and the goodness of fit of batch gradient descent.

There are several other gradient descent algorithm variants. Adaptive learning rate methods adjust the learning rate based on the history to improve convergence. Examples include **Adagrad, RMSProp, and Adam**. These methods are very popular in deep learning and are discussed in the next chapter.

Data Standardization and Pipeline

A machine learning job may consist of a sequence of tasks. In Python scikit-learn, **pipelines** help to define and automate such workflows. See Fig. 11.5. Data **standardization is a task that is recommended** before the use of learning algorithms. Standardization or normalization transforms the features to some chosen scale such as [0.0: 1.0], or [−1.0: +1.0]. z-score normalization is a popular technique used in machine learning. It rescales features to have a mean of 0 and a standard

Fig. 11.5 Processing pipeline

deviation of 1. Some of the standardization methods are discussed in Chap. 2. Many algorithms, especially SGD, are sensitive to scaling. Therefore, standardized data is usually provided as input.

11.2.3 Stochastic Gradient Descent Regressor

In the previous section, we saw that Stochastic Gradient Descent (SGD) is a subcategory of the gradient descent optimization method, where training samples are picked up one at a time, at random (sample size = 1). SGD finds successful applications in large-scale and sparse machine learning problems, especially deep learning. Some of the applications include text classification and natural language processing.

SGD requires hyperparameters such as the regularization method ($L2$, elastic net); α; the number of epochs. To get good results, shuffle the training data before starting any epoch. SGD is sensitive to feature scaling. In general, features are standardized before use. With standardization, SGD converges in much fewer iterations. Please note that the parameters are different for SGD regression and SGD classification.

scikit-learn Parameters for Stochastic Gradient Descent Regressor

Some of the parameters applicable for SGD regression are mentioned below. They are for information only. You may skip this section altogether. Those who want to learn about SGD parameters may refer to scikit-learn documentation.

The loss function ('loss' parameter) takes the default value of 'squared_loss' in regression. The loss functions allowed include 'squared_error' for ordinary least squares; 'huber', a modification of 'squared_error', to reduce the impact of outliers; 'epsilon_insensitive' for linear SVM regression; and 'epsilon_sensitive' which ignores errors less than epsilon.

In machine learning, we use regularization for smoothening (discussed in the next section). The parameter 'alpha' indicates the regularization rate or the multiplication factor in the regularization term. The learning rate could be independent or related to the alpha parameter. The parameter alpha defaults to 0.0001. The parameter 'penalty' refers to the regularization method (see next section). The possible values are {'l2' for ridge regression, 'l1' for lasso regression, and 'elasticnet' for a combination of the two}. 'l1_ratio' is used only if the penalty is 'elasticnet'; the default value is 0.15, which implies 15% lasso and 84% ridge.

The parameter 'max_iter' is the number of epochs and defaults to 1000. The 'shuffle' parameter indicates whether to shuffle training data after each epoch. It defaults to True. 'epsilon' is the tolerance in the prediction. It defaults to 0.1.

Multiple parameters affect early stopping. Convergence is checked against the training or validation loss depending on several parameters. If there is no change in performance, training will stop. The parameters include early_stopping, validation_fraction, n_iter_no_change, 'tol' (tolerance).

Tutorial 11.2   Regression: Stochastic Gradient Descent (SGD)

We demonstrate regression model building using Stochastic gradient descent. We are using diamonds dataset. The objective is to predict diamond price (y) based on its features (X): 'carat', 'cut', 'color', 'clarity', 'depth', 'table', 'x', 'y', 'z'. The dataset diamonds is described in Chapter 1. We will use scikit library extensively in this Chapter. You may refer to https://scikit-learn.org/stable/

Tutorial 11.2.1 Regression Data Setup / Standardisation

```
import seaborn as sb
from sklearn.preprocessing    import StandardScaler
import numpy as np

d = sb.load_dataset('diamonds')
d = d.dropna()  # drop null valued rows
```

Convert category labels to integers for use in the regression model
```
d['cuti'] = d.cut.astype("category").cat.codes
d['colori'] = d.color.astype("category").cat.codes
d['clarityi'] = d.clarity.astype("category").cat.codes

X = d[['carat', 'cuti', 'colori', 'clarityi',
        'depth','table', 'x', 'y', 'z']]
y = d.price
```

Standardize features using z score transformation
```
Xz = StandardScaler().fit(X).transform(X)
```

Tutorial 11.2.2 Build Regression Model Using SGD

```
import numpy as np
```

Import stochastic gradient descent regressor (SGDregressor model)
```
from sklearn.linear_model import  SGDRegressor
```

Setup SGD regressor
```
regressor = SGDRegressor(max_iter=50)
```

Run regression
```
model = regressor.fit(Xz, y)
```

Check the regression score and regression results
```
model.score(Xz,y) # 0.907
np.round(model.intercept_,1) # 3926.1
np.round(model.coef_,1)

[5090.6, -138.4, -525.5, -794.4, -121.5, -54.3, -943.5, 38.3, -16.3]
```

```
model.get_params()

'alpha': 0.0001,
 'epsilon': 0.1,
 'eta0': 0.01,
 'l1_ratio': 0.15,
 'learning_rate': 'invscaling',
 'loss': 'squared_loss',
 'max_iter': 50,
 'n_iter_no_change': 5,
 'penalty': 'l2',
 'power_t': 0.25,
 'shuffle': True,
 'tol': 0.001,
 'validation_fraction': 0.1,
 ..
model.intercept_   # 3947.012
model.score(Xz,y) # 0.907
model.n_iter_      # 30
```

```
Inference:
SGD method parameters and model coefficients are listed above. SGD converged
in 30 iterations. Score is 0.907
```

11.3   Regularization for Smoothening

Regularization is a class of techniques used to reduce model complexity and prevent overfitting. For example, see Fig. 11.6, which shows the $y = f(X)$ regression model. The blue line represents linear regression (with degree = 1). This is an underfit. We get an exact fit with a polynomial of degree 6, as shown by the red line that touches all the data points. However, we call this model an overfit model. This model will 'memorize' and 'recall' the seven given sets of data points, without any error. However, it may fail to predict the outcome of an unknown data. The black line (with degree = 2) is considered an optimal fit.

Let us consider regularization in the context of a linear model (degree = 1). Ridge and Lasso regression are simple techniques to reduce model complexity and prevent overfitting, which may result from linear regression. The ridge regression shrinks the weights by applying a penalty term called L2 norm (the sum of the squared weights) iteratively. In this process, the features with minor contribution to the target get their weights reduced and may become closer to zero. In lasso regression, the features with smaller weights get dropped altogether. Lasso is good for feature selection when some features alone have relatively high weights.

Fig. 11.6 Regularization in polynomial regression

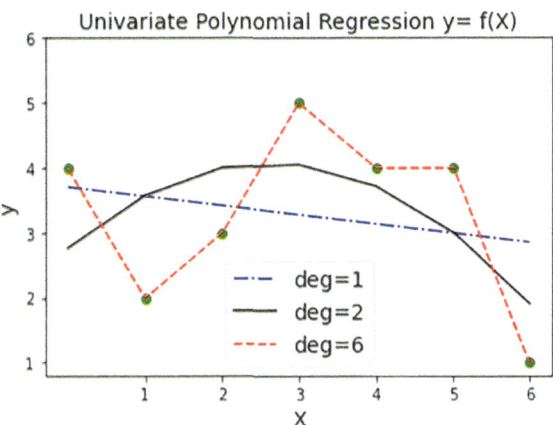

11.3.1 Ridge Regression

Ridge regression is a regularization technique used in linear models. It is also called 'L2 regularization'. Ridge regression uses a loss function with an added term for regularization—the 'penalty'.

$$\text{L2 Penalty} = \sum_{i=1}^{k} W_i^2 = W_1^2 + W_2^2 + \cdots + W_k^2$$

Ridge regression solves the minimization of the least squares with L2 penalty added. The objective of Ridge regression may be stated as

$$\text{Minimize (Loss Function} + \alpha * \text{L2 Penalty).}$$

$$\text{Minimize (Mean Squared Error} + \alpha * \text{sum of the square of coefficients)}$$

$$\text{Minimize } [\text{MSE}(y, f(W, X)) + \alpha \sum_{i=1}^{k} W_i^2]$$

where

MSE	is the mean squared error in prediction (see Eq. 11.2),
$X \{X_i \ldots X_k\}$	is a set of k features,
W_i's	are the coefficients of Xi's, and
α,	the multiplication factor which determines the learning rate.

It may be noted that 'α' is one of the model *hyperparameters* determined manually during the learning process ($0 < \alpha < \infty$).

11.3.2   Lasso Regression

The Lasso regression is another regularization technique used in linear models. Lasso stands for *Least Absolute Shrinkage and Selection Operator.* It is also called 'L1 regularization'. Like ridge regression, the lasso regression uses a loss function with an added term for regularization—the 'L1 penalty'.

$$\text{L1 Penalty} = \sum_{i=1}^{k} |W_i| = |W_1| + |W_2| + \cdots + |W_k|$$

The lasso regression can result in some of the coefficients becoming zero leading to feature reduction. In other words, lasso regression helps in feature selection, as well as the reduction of overfitting. For large datasets with numerous features, it is challenging to use stepwise regression or other methods for feature reduction. In this context, lasso's feature reduction ability is considered particularly useful in certain applications. The objective of lasso regression can be stated as

$$\text{Minimize (Loss Function} + \alpha * \text{L1 Penalty)}.$$

Minimize (Mean Squared Error $+ \alpha *$ sum of the absolute value of the coefficients)

$$\text{Minimize [MSE}(y, f(w, X)) + \alpha \sum_{i=1}^{k} |W_i|]$$

where

MSE	is the mean squared error in prediction (see Eq. 11.2),
$X\{X_i \dots X_k\}$	is a set of k features,
W_i's	are the coefficients of Xi's, and
α,	the multiplication factor which determines the learning rate.

11.3.3   Elastic Net Regression

Suppose a combination of *L1* regularization (Lasso) and *L2* regularization (Ridge) is introduced as penalty in the minimization function, the combo is called elastic net regularization. Elastic net is useful when multiple features are correlated with one another. The lasso penalty helps to drop some of those highly correlated features. Ridge regression gives stability under rotation. The objective of elastic net regression can be stated as

$$\text{Minimize (Loss Function} + \alpha 1 * \text{L1 Penalty} + \alpha 2 * \text{L2 Penalty)}.$$

$$\text{Minimize [MSE}(y, f(W, X)) + \alpha_1 \sum_{i=1}^{k} |W_i| + \alpha_2 \sum_{i=1}^{k} w_i^2]$$

where

MSE   is the mean squared error in prediction (see Eq. 11.2),
$X \{X_i ... X_k\}$ is a set of k features,
W_i's   are the coefficients of Xi's,
$\alpha_1 = L1$ Ratio   penalty term}, and
$\alpha_2 = (1 - L1$ Ratio)   {penalty term}.

Tutorial 11.3   Regression: Regularization

Using the diamonds dataset, we will demonstrate regularization.
First do Data Setup as in Tutorial 11.2.1 Data Setup for Regressor

Tutorial 11.3.1 Ridge Regression: (L2 Regularization)

```
import numpy as np
Import Ridge (linear regression model with L2 penalty)
from sklearn.linear_model import Ridge

regressor = Ridge(alpha=0.5)
model = regressor.fit(Xz, y)
model.score(Xz,y) # 0.907
np.round(model.intercept_,1) # 3932.8
np.round(model.coef_,1)
[5091.5, -134.8, -548.9, -826.6, -114.2, -59.8, -983 , 49.8, -20.9]
```

Inference:
Ridge regression model coefficients are listed above. The score is 0.907 (same
as SGD).

Tutorial 11.3.2 Lasso Regression: (L1 Regularization)

First, do Data Setup as in Tutorial 11.2.1 Data Setup for Regressor

```
import numpy as np
Import Lasso (linear regression model with L1 penalty)
from sklearn.linear_model import Lasso

regressor = Lasso(alpha=0.5)
model = regressor.fit(Xz, y)
model.score(Xz,y) # 0.907
np.round(model.intercept_,1) # 3932.8
np.round(model.coef_,1)

[5070.7, -134.7, -548, -826.7, -112.9, -59.2, -946.5, 30.7, -18.3]
```

Inference:
Lasso regression model coefficients are listed above. The score is 0.907 (same
as SGD Regression and Ridge Regression).

11.4 Classification

As discussed earlier, regression and classification come under supervised learning methods. Whereas the outcome variable in the regression is continuous, the outcome in classification is a set of discrete values (corresponding to category labels). As we discussed in the regression case, the following also holds good in classification.

- The learning method—gradient descent or its variants.
- Regularization techniques (L1, L2, elastic net …) to reduce model complexity and prevent overfitting.

However, the major difference occurs in selecting classifiers and respective loss functions. Recall our discussion on classification methods in Chap. 6. We discussed linear classifiers such as logistic regression and linear discriminant analysis (LDA). We also discussed some nonlinear classifiers, such as support vector machine (SVM) and decision tree induction.

The loss function quantifies the disparity between a model's predicted and actual output. Throughout the training process, the goal is to reduce this loss, i.e., minimizing the prediction error. Common loss functions used in classification problems include binary cross entropy (log Loss) for binary classification problems, categorical cross entropy used in multi-class classification problems, and hinge loss used in Support Vector Machines (SVMs).

Binary Cross Entropy for Binary Classification

Binary cross-entropy loss is used for binary classification problems where the target variable has only two classes. It measures the difference between the predicted and the actual probability of positive classes. Binary cross entropy is also called log loss or logistic regression loss function) for binary classification (Han et al. 2014).

Assume we have 'n' objects to be classified into two classes $\{1, 0\}$.

Binary cross entropy or log loss for binary classification $=$

$$-\frac{1}{n} \sum_{i=1}^{n} [y_i \log(p_i) + (1 - y_i)\log(1 - p_i)]$$

where

y_i	is the label $\{1 \text{ or } 0\}$ of the object to be classified,
p_i	is the predicted probability of y_i belonging to class 1,
$\therefore (1 - p_i)$	is the probability of y_i belonging to class 0,
n	is the sample size.

Categorical Cross Entropy for Multi-class Classification

Assume we have 'n' objects to be classified into c classes $\{1 \dots m\}$. We can extend the formula for binary classification to include all the 'm' classes as shown below:

Categorical cross-entropy loss for multi-class classification =
Sum of (binary cross entropy or log loss) over all the class labels $c = \{1 \dots m\}$

$$= -\frac{1}{n} \sum_{c=1}^{m} \sum_{i=1}^{n} [y_i \log(p_i) + (1 - y_i)\log(1 - p_i)]$$

Hinge Loss (SVM)

Hinge loss is used in Support Vector Machine (SVM) classification problems. It measures the difference between the predicted and true scores of the positive class. The formula for hinge loss is

$$\text{Hinge Loss} = \max(0, 1 - y * f(x)),$$

where

y	is the true label (either $+1$ or -1),
f(x)	is the predicted score for the input x,
max (0, …)	implies that the loss is 0, if the true label and predicted score have the same sign; otherwise, it is the difference between them.

11.4.1   Regularization in Classification

The primary objective of regularization is to strike a balance between fitting the training data well (minimizing training error) and ensuring the model's ability to generalize effectively to new, unseen data (minimizing test error). Without regularization, models tend to become overly complex, capturing the noise in the training data, making them less effective in identifying the underlying parameters or patterns.

Regularization techniques help to reduce overfitting. Section 11.3 discussed ridge, lasso, and elastic net regression. L1 regularization (Lasso) introduces a penalty to the loss function as a fraction of the sum of the absolute values of the coefficients. It promotes feature selection by driving certain coefficients to absolute zero. L1 regularization is valuable when we suspect only a subset of features is relevant. L2 regularization (ridge) imposes a penalty by adding a fraction of the sum of the squares of the coefficients to the loss function. It tends to shrink the coefficients to zero without forcing them to become precisely zero. L2 regularization is effective when we believe all the features contribute meaningfully to the model's performance. The Tutorials 11.4.1 and 11.4.2 demonstrate regularization techniques applied to classification problems.

Tutorial 11.4   Linear Classification using Ridge and Lasso Regressors

Tutorial 11.4.1 Classification Data Setup | Price -> 3 categories

```
import seaborn as sb
d = sb.load_dataset('diamonds')
d = d.dropna()
```

Creating a categorical variable pCat - three categories of price - 0 (low),
1 (medium), 2 (high)

```
import numpy as np
import pandas as pd
d['pCat'] = np.zeros(d.shape[0])
for i in  d.index:
    if    d.loc[i, 'price'] < 2000: d.loc[i, 'pCat'] = 0
    elif  d.loc[i, 'price'] < 8000: d.loc[i, 'pCat'] = 1
    else: d.loc[i, 'pCat'] = 2
```

Create DataFrame X that consists of feature variables
```
X = d[['carat', 'depth', 'table', 'x', 'y', 'z']]
```
Create DataFrame y for class labels (pCat =0,1,2)
```
y = d.pCat
```

Tutorial 11.4.2 Linear Classification using Ridge Regressor

Do Data Setup as in Tutorial 11.4.1 Classification Data Setup | Price -> 3
categories

Standardize features
```
from sklearn.preprocessing import StandardScaler
scaler = StandardScaler()
Xz = scaler.fit_transform(X)
```

Setup classifier with L2 regularization (Ridge regression). In L2 Regularization, we try to minimise RSS + α ΣWj^2
```
from sklearn.linear_model import RidgeCV
classifier = RidgeCV()
```

Fit the classifier
```
model = classifier.fit(Xz, y)
```

View the results
```
model.alpha_        # 0.1
model.score(Xz,y)   # 0.807
np.round(model.coef_,3)

0.087,    0.009,  -0.011, 0.484, 0.049, 0.018
'carat', 'depth', 'table', 'x',   'y',   'z'

Rank: x, carat, y, z, depth, table
```

Inference:

The Model score is 0.807. The model coefficients are listed. The features are ranked by their order of importance - x, carat, y, z, depth, table

Tutorial 11.4.3 Linear Classification using Lasso Regressor

Do Data Setup as in Tutorial 11.4.1 Classification Data Setup | Price -> 3 categories

Standardize features

```
from sklearn.preprocessing import StandardScaler
scaler = StandardScaler()
Xz = scaler.fit_transform(X)
```

Setup classifier with Lasso regression regulariser. In lasso regularization, we try to minimise RSS + α ∑Wj.

```
from sklearn.linear_model import Lasso
classifier = Lasso(alpha=0.5)
```

Fit the linear classifier

```
model = classifier.fit(Xz, y)
```

View coefficients

```
model.alpha     # 0.5
model.score(Xz,y)    # 0.2996
np.round(model.coef_,3)

[0.      , 0.      , 0.      , 0.131, 0.      ,      0.     ])
'carat', 'depth', 'table',   'x',    'y',      'z'
```

Inference:

The Model score is 0.299 (very low). The model coefficients are listed.
Only one variable emerges – x. All the features except 'x' stand eliminated!
It may be noted that Lasso Regularization is useful in variable reduction
when the number of variables is very large

11.4.2 Stochastic Gradient Descent Classifier

Section 11.2 discussed the gradient descent optimization technique and its sub-categories—batch, mini-batch, and stochastic gradient descent. We understand that gradient descent is an iterative optimization technique for finding the local minimum of a differentiable function. This is achieved by adjusting the weights (coefficients) to optimize a cost (loss) function. We also saw that stochastic gradient descent (SGD) is a subcategory of the gradient descent optimization method, where training samples are picked up one at a time, at random (sample size = 1).

SGD requires hyperparameters such as the regularization method (*L2*, elastic net); α; the number of epochs. To get good results, shuffle the training data before starting any epoch. SGD is sensitive to feature scaling. In general, features are standardized before use. With standardization, SGD converges in much fewer iterations.

scikit-learn Parameters for Stochastic Gradient Descent Classifier

Please note that the parameters for SGD regression and SGD classification differ. Some of the parameters applicable for classification are mentioned below. They are for information only. You may skip this section altogether. Those who want to learn about SGD parameters may refer to scikit-learn documentation.

The **loss** function ('loss' parameter) takes the default value of '**hinge**' in classification. The loss functions allowed are 'hinge' for linear SVM; 'log' for logistic regression; 'modified_huber' for smooth loss that brings tolerance to outliers and estimates; 'squared_hinge' which is hinge with squared loss; and 'perceptron' which is the linear loss used by the perceptron algorithm. Regression loss functions can also be used, where appropriate. This includes 'squared_error' for ordinary least squares; 'huber', a modification of 'squared_error', to reduce the impact of outliers; 'epsilon_insensitive' for linear SVM regression; and 'epsilon_insensitive' which ignores errors less than epsilon.

The parameter 'alpha' indicates the regularization rate or the multiplication factor in the regularization term. The learning rate could be independent or related to the alpha parameter. The parameter alpha defaults to 0.0001. The parameter 'penalty' refers to the regularization method (see next section). The possible values are {'l2' for ridge regression, 'l1' for lasso regression, and 'elasticnet' for a combination of the two}. 'l1_ratio' is used only if the penalty is 'elasticnet'; the default value is 0.15, which implies 15% lasso and 84% ridge.

The parameter 'max_iter' is the number of epochs and it defaults to 1000. The 'shuffle' parameter indicates whether to shuffle training data after each epoch. It defaults to True. 'epsilon' is the tolerance in the prediction. It defaults to 0.1.

Multiple parameters affect early stopping. Convergence is checked against the training or validation loss depending on several parameters. If there is no change in performance, training will stop. The parameters include early_stopping, validation_fraction, n_iter_no_change, 'tol' (tolerance).

Tutorial 11.4.4   Classification: Stochastic Gradient Descent

Using the diamonds dataset, build a classification model to predict diamond price category using Stochastic gradient descent.
Note that SGD Classifier is sensitive to scale. With standardization, SGD converges in much fewer iterations.

Do Data Setup as in Tutorial 11.4.1 Classification Data Setup | Price -> 3 categories

```
from sklearn.model_selection import train_test_split
from sklearn.preprocessing import StandardScaler
from sklearn.metrics import confusion_matrix
from sklearn.metrics import accuracy_score, f1_score
```

SGD Classifier is sensitive to scale. Without standardization, SGD needs too many iterations to converge. However, with standardization, SGD converges in fewer iterations
Standardize the features

```
Xz = StandardScaler().fit_transform(X)
```
Test - Train Split - 80% data for training, 20% for test
```
X_train, X_test, y_train, y_test = train_test_split(
          Xz, y, train_size= 0.8, random_state=1)
```

Import SGD Classifier
```
from sklearn.linear_model import SGDClassifier
```
Setup SGDclssifier
```
classifier = SGDClassifier(alpha=0.001,
               loss="hinge",
               penalty="l2",
               max_iter=20)
model = classifier.fit(X_train,y_train) # Train the model
```

Predict the class using the model
```
y_predicted = model.predict(X_test)
```
Generate Confusion report
```
report = confusion_matrix(y_test, y_predicted)
print(report)

[[4513  185    0]
 [ 235 4193  154]
 [   0  393 1115]]
ac = accuracy_score(y_test,y_predicted)
f1 = f1_score(y_test,y_predicted, average = 'weighted')
print('accuracy =', round(ac,4),'; wtd. f1 score =', round(f1,4))

accuracy = 0.91 ; wtd. f1 score = 0.91

model.n_iter_      # 11
model.get_params() # model parameters
{'alpha': 0.001,
 'average': False,
 'class_weight': None,
 'early_stopping': False,
 'epsilon': 0.1,
 'eta0': 0.0,
```

```
'fit_intercept': True,
'l1_ratio': 0.15,
'learning_rate': 'optimal',
'loss': 'hinge',
'max_iter': 20,
'n_iter_no_change': 5,
'n_jobs': None,
'penalty': 'l2',
'power_t': 0.5,
'random_state': None,
'shuffle': True,
'tol': 0.001,
'validation_fraction': 0.1,
'verbose': 0,
'warm_start': False}
```

```
Observation:
SGD method parameters and model coefficients are listed above. Hinge Loss with
elastic net regularization was used. (SGD converged in 11 iterations. F1
Score is 0.91. For more information on parameters, refer to https://scikit-
learn.org/stable/modules/sgd.html
```

11.5   Cross-Validation

Resampling is an iterative process **of assessing a machine learning model's performance and generalizability and identifying issues** like overfitting or underfitting. Resampling techniques involve drawing multiple samples, at random, from the available data, and exploring the model performance. Resampling techniques are computationally expensive and reuse the available sample to make statistical inferences. However, they are helpful when the training dataset is small. Two common resampling techniques are discussed in this section.

- Cross-validation.
- Bootstrapping.

Cross-Validation

The K-fold cross-validation generally divides the entire dataset into k subsets at random. We do k training sessions. In each session, we use one subset as the test sample, while the other k − 1 subsets are combined to form the training sample (see Table 11.1). The steps involved in cross-validation are shown below:

- Split the data into K roughly equal-sized parts.
- Designate the kth part as the test sample.
- Estimate the model using the other K − 1 part (training sample).

Table 11.1 Cross-validation splits

0	1	2	3	4		6	14	0	2	8	Split 1
5	6	7	8	9		15	13	22	18	11	Split 2
10	11	12	13	14		3	1	21	17	19	Split 3
15	16	17	18	19		10	24	23	12	9	Split 4
20	21	22	23	24		5	16	20	4	7	Split 5

- Calculate its prediction error on the test sample (which was left out).
- Do the above K times and combine (usually average) the K estimates of prediction error.

If K = n, the number of instances in the data, it is called **leave-one-out cross-validation**. Note that the test dataset will have one data instance, while the rest is used for training. **Stratified cross-validation** is another variation. In this method, each fold is created in such a way that it maintains the same class distribution as the original dataset. For example, in a binary classification problem in which 20% of samples belong to class A and 80% to class B, each fold in stratified cross-validation will also be constituted with a similar class distribution.

Some of the objectives of cross-validation are listed below:

- Cross-validation is used to estimate prediction error (or the model performance).
- Cross-validation results in k-models. The researcher may choose a model based on a criterion such as the median performance. The goal is to select a model that performs well on average, across all folds, and is less likely to overfit to a single data split.
- It sometimes limits the number of feature variables to a relevant few. We can analyze the importance of features by observing how the model's performance changes when different subsets of features are used during each fold.

Bootstrap

Bootstrap resamples at random, with replacement. See Table 11.2. Since we draw data with replacement, a bootstrapped dataset may contain multiple instances of

Table 11.2 Bootstrap splits

0	1	2	3	4		6	10	9	4	20	Split 1 (1 item duplicated)
5	6	7	8	9		7	0	24	5	1	Split 2 (no duplicates)
10	11	12	13	14		3	12	14	13	19	Split 3 (1 item duplicated)
15	16	17	18	19		17	18	22	9	11	Split 4 (1 item duplicated)
20	21	2	23	24		14	21	4	2	15	Split 5 (1 item duplicated)

the same original cases (e.g., 9 and 14) and completely omit other original cases (e.g., 19 and 23). However, from a statistical perspective, this method is considered superior.

Though they appear similar, bootstrapping is primarily used for statistical inference and assessing the uncertainty of population parameter estimates. In contrast, cross-validation is primarily used in machine learning to evaluate and compare the performance of predictive models.

Tutorial 11.5.1   Regression: k-fold Cross-validation

We demonstrate k-fold cross-validation of the diamond price (y) prediction model. Features (X), viz., 'carat', 'cut', 'color', 'clarity', 'depth', 'table', 'x', 'y', 'z'. The dataset diamonds is described in Chapter 1.
Note: - The following parameters are important for fine-tuning and configuring the cross-validation process for model evaluation.
1. The parameter n_splits determines how many subsets or "folds" the dataset will be split into during cross-validation.
2. If Shuffle is set to `True`, the dataset will be shuffled before creating the folds. This can help reduce inherent bias in data by way of order or structure.
3. Setting the random state to an integer value ensures that the random processes involved in shuffling or other operations are reproducible. Using the same random_state value will yield the same results in each run.

Do Data Setup as in Tutorial 11.2.1 Data Setup for Regressor

```
from sklearn import linear_model
from sklearn.model_selection import cross_validate, KFold

regressor = linear_model.LinearRegression()
```

K-fold Cross Validation

```
KF = KFold(n_splits=5, shuffle=True, random_state=1)
CV_results = cross_validate(regressor, Xz, y, cv=KF, return_estimator=True)

np.round(CV_results['test_score'],3)
[0.909, 0.904, 0.905, 0.909, 0.907]

np.median(CV_results['test_score'])   # 0.907
CV_results['test_score'].max()        # 0.909

for model in CV_results['estimator']:
    print(np.round(model.coef_,1))
```

```
carat       cuti  colori  clarity   depth   table      x        y       z
[5e+03 -1.3e+02 -5.4e+02 -8.2e+02 -1.0e+02 -5.5e+01 -9.3e+02 3.4e+01 -3.7]
[5e+03 -1.3e+02 -5.5e+02 -8.3e+02 -1.1e+02 -6.5e+01 -9.7e+02 3.3e+01 -2.2]
[5155.1  -136.9  -550.4  -821.9  -119.3  -57.4 -1192.4 198.8  -19.0]
[5101.4  -139.6  -554.8  -826.5  -105.0  -55.5  -915.7  57.0 -105.8]
[5074.2  -130.1  -540.8  -824.7  -114.6  -64.0  -948.9  45.7  -42.7]
```

Inference:

The coefficient estimates from each of the 5 splits are shown above. Here, the 5th model, score 0.907, has a median score. The coefficients are:-

[5074.2, -130.1, -540.8, -824.7, -114.6, -64.0, -948.9, 45.7, -42.7]

Cross-validation is used to compare the performance of different models and their ability to generalize (how accurately the model performs on unknown data)

Tutorial 11.5.2   Classification: cross_val_score method

We demonstrate k-fold cross-validation of the diamond price (y) prediction model. Features (X), viz., 'carat', 'cut', 'color', 'clarity', 'depth', 'table', 'x', 'y', 'z'. The dataset diamonds is described in Chapter 1.

Do Data Setup as in Tutorial 11.4.1 Classification Data Setup | Price -> 3 categories

```
from sklearn.model_selection import KFold, cross_val_score
from sklearn.pipeline import make_pipeline
from sklearn.preprocessing import StandardScaler
```

Import the classifier 'LogisticRegression'

```
from sklearn.linear_model import LogisticRegression
classifier = LogisticRegression()
```

Setup a pipeline with two steps (1) standardization and (2) classification

```
pipeline = make_pipeline(StandardScaler(), classifier)
```

Setup k-Fold cross-validation

```
KF = KFold(n_splits=10, shuffle=True, random_state=1)
```

Conduct k-fold cross-validation

```
CV_results = cross_val_score(
      pipeline,  # Pipeline
      X,         # DataFrame of feature variables
      y,         # DataFrame of Class Labels
      cv = KF,   # Cross Validation Technique
      scoring = "accuracy", # Loss function
      n_jobs=-1  # Use all CPU scores
      )

for acc in CV_results: print(np.round(acc,3))
# 0.91  0.912  0.907  0.912  0.902  0.906  0.911  0.91  0.907  0.911

# Calculate mean accuracy
print(CV_results.mean()) # 0.909
```

Inference:

Data Standardization helps in fast convergence. This is explained in a subsequent tutorial

The median score may be usually considered a classification model's expected accuracy. In this case, it is 0.909.

Cross-validation is used to compare the performance of different models and their ability to generalize (how accurately the model performs on unknown data)

Tutorial 11.5.3   Classification: cross_validate method

We are demonstrating k-fold cross-validation to classify diamond price (y) from its features (X), 'carat', 'depth', 'table', 'x', 'y', 'z' using the cross_validate method.

In this method, if the parameter return_estimator is set to 'True', the classifier's coefficients will be returned

Classification: without standardization

Do Data Setup as in Tutorial 11.4.1 Classification Data Setup | Price -> 3 categories

```
from sklearn.model_selection import cross_validate
import numpy as np
```

Import the classifier 'LogisticRegression'
```
from sklearn import datasets, linear_model
classifier = linear_model.LogisticRegression(
    tol=0.01, max_iter = 500)
```

Perform 5-fold cross-validation, with return_estimator=True
```
CV_results = cross_validate(classifier,
              X,
              y,
              cv=5,
              return_estimator=True)
```

Warning:

The program execution might have taken long time or ended prematurely. Data Standardization helps in fast convergence. This is illustrated in the following tutorial

```
sorted(CV_results.keys())
# ['estimator', 'fit_time', 'score_time', 'test_score']

np.round(CV_results['test_score'],3)
# [0.902, 0.937, 0.956, 0.841, 0.846]
```

Get the 5 estimators
```
for model in CV_results['estimator']:
    print(np.round(model.coef_,1))
```

The Model Parameters from the first Cross-Validation
There are three discriminant functions, as there are three classes
Six features ['carat', 'depth', 'table', 'x', 'y', 'z']]
Their coefficients are shown along with

```
[[-11.    0.2   0.1   0.3  -2.3  -0.1]
 [  3.2   0.    0.   -0.4   0.4  -0.1]
 [  7.8  -0.2  -0.2   0.1   1.9   0.1]]
```

The Model Parameters from the second Cross Validation
```
[[-11.2   0.1   0.1   0.5  -2.6  -0.1]
 [  3.3  -0.    0.   -0.1   0.5  -0.1]
 [  7.9  -0.1  -0.2  -0.4   2.2   0.2]]
```

The Model Parameters from the second Cross-Validation
```
[[-10.7   0.2   0.2  -0.3  -2.1  -0. ]
 [  3.6  -0.    0.   -0.5   0.7  -0. ]
 [  7.1  -0.2  -0.2   0.8   1.4   0. ]]
```

The Model Parameters from the third Cross Validation
```
[[-10.5   0.2   0.2   0.1  -2.8  -0.1]
 [  1.8  -0.    0.   -0.4   0.8  -0.1]
 [  8.6  -0.2  -0.2   0.2   2.    0.2]]
```

The Model Parameters from the fourth Cross Validation
```
[[-10.2   0.3   0.2   0.3  -4.5  -0.7]
 [  4.7  -0.   -0.    1.2  -0.4  -0.7]
 [  5.5  -0.3  -0.2  -1.5   4.9   1.4]]
```

The Model Parameters from the fifth Cross Validation
```
[[-7.073  0.316  0.372  1.104 -4.227 -0.721]
 [ 4.063 -0.181 -0.042 -0.175 -1.285  0.191]
 [ 3.01  -0.135 -0.33  -0.929  5.512  0.53 ]]
```

Warning:
The program execution might have ended prematurely. Data Standardization
helps in fast convergence. See the next tutorial

Inference:
The Model Parameters from the five cross-validations are listed. The scores
are [0.902, 0.937, 0.956, 0.841, 0.846]. The median score may be taken as the
estimate of model performance.
Each CV has three discriminant functions. There are six features ['carat',
'depth', 'table', 'x', 'y', 'z']].

Classification: with Standardization

Do Data Setup as in Tutorial 11.4.1 (Price->3 Categories)

```
from sklearn.model_selection import cross_validate
from sklearn.pipeline import make_pipeline
from sklearn.preprocessing import StandardScaler
```

Import the classifier 'LogisticRegression'

```
from sklearn import linear_model
classifier = linear_model.LogisticRegression(tol=0.01, max_iter = 500)
```

Setup a pipeline with two steps (a) standardize X (b) then run the classifier

```
pipeline = make_pipeline(StandardScaler(), classifier)
```

Perform 5-fold cross validation

```
CV_results = cross_validate(pipeline,
                X,
                y,
                cv=5,
                return_estimator=True)

sorted(CV_results.keys())
# ['estimator', 'fit_time', 'score_time', 'test_score']
np.round(CV_results['test_score'],3)
# [0.901, 0.937, 0.956, 0.842, 0.849]
```

Get the 5 estimators

```
for model in CV_results['estimator']:
    print(np.round(model[1].coef_,3))
```

Inference:

Data Standardization helps in fast convergence. Therefore, unlike the previous tutorial, this one completed normally.

The Model Parameters from the five cross-validations are listed. The scores are [0.901, 0.937, 0.956, 0.842, 0.849]. The median score may be taken as the estimate of model performance.

Each CV has three discriminant functions.

11.6   Building Robust Machine Learning Models

There are several factors that an ML engineer needs to consider in building robust machine learning models. The typical steps involved in the model-building process may be summarized as follows:

1. Data preprocessing: Address missing values, handle outliers, encode categorical variables, and scale/normalize features if needed (see Sect. 2.8). Note that data preprocessing takes about 70% of the total effort in a typical machine learning project.
2. Hyperparameter tuning: Find the best combination of hyperparameters for the model using techniques like grid search (see Sect. 9.4) to identify the combination that yields the best performance.

3. Model training: Trains the model using the selected hyperparameters on the training dataset.
4. Ensemble methods: Improve model performance by combining multiple models like bagging (e.g., Random Forest) or boosting (e.g., AdaBoost, Gradient Boosting). Bagging relies on averaging or other central tendencies. Boosting focuses on sequential learning.
5. Cross-validation: Provides a more robust estimate of the model's performance by assessing its generalization across different subsets of the data, to identify potential issues like overfitting (see Sect. 11.5).
6. Model evaluation: Evaluate the model's performance. Metrics include accuracy, precision, recall, and F1-score (see Sect. 6.3).
7. Test set evaluation: Get a final unbiased assessment of the model's generalization to new, unseen data, ensuring it performs well in real-world scenarios.
8. ROC/AUC analysis: Specifically applicable to classification models, ROC/ AUC analysis provides insights into the model's ability to discriminate between classes at different thresholds (see Sect. 9.5.3). This helps to study the trade-offs between sensitivity and specificity.
9. Bias-variance analysis: Aids in balancing model simplicity and complexity, guiding decisions to avoid underfitting (high bias) or overfitting (high variance).

Some of the topics are covered earlier. This section covers the topics such as overfitting and underfitting mitigation using bias-variance trade-off, test-train convergence, train-validate-split, hyperparameter tuning, ensemble methods, and ROC/ AUC Analysis.

11.6.1 Hyperparameter Tuning

In machine learning, a model is essentially a mathematical representation that learns patterns and relationships from data. This learning process involves adjusting parameters based on the input data to make accurate predictions or classifications. However, alongside these parameters, there is another set of parameters known as hyperparameters.

Hyperparameters are not learned from the data. They are set manually during the model development phase. They play a crucial role in facilitating effective model development and training. Here is a breakdown of the hyperparameters mentioned:

1. Train-test split ratio: Defines the proportion of the dataset used for training the model versus testing its performance.
2. Choice of cost or loss function: Determines how the model evaluates the difference between its predictions and the actual values. Examples include absolute error, square error, Huber loss, cross entropy, log loss, exponential loss, and hinge loss.

3. Choice of optimization algorithm: Dictates how the model adjusts its parameters during training to minimize the defined loss function. Options include mini-batch gradient descent, stochastic gradient descent, Adagrad, and Adam, among others.
4. Number of iterations (epochs) in training: Specifies the number of times the entire dataset is passed forward and backward through the machine learning model during training.
5. Learning rate in optimization algorithms: A crucial hyperparameter that controls the size of steps taken during the optimization process. It influences how quickly or slowly a model learns.
6. Multiplication factor for regularization term (α): Relates to regularization techniques, where a penalty term is added to the loss function to prevent overfitting. The multiplication factor (α) determines the strength of this regularization.
7. Penalty—e.g., L2 regularization: Specifies the type and strength of regularization applied. L2 regularization penalizes large weights in the model.
8. C and σ hyperparameters for support vector machines: In the context of Support Vector Machines (SVM), these parameters control the trade-off between achieving a smooth decision boundary and classifying the training points correctly.
9. k in k-nearest neighbors: In the k-nearest neighbors algorithm, 'k' denotes the number of nearest neighbors to consider when making predictions.

Understanding and appropriately tuning these hyperparameters are critical for achieving optimal model performance and avoiding issues such as overfitting or underfitting. It often involves a process known as hyperparameter tuning or optimization, where different combinations of hyperparameters are tested to find the most effective configuration for a given machine learning task.

11.6.2 Ensemble Methods

Ensemble methods involve using a set of multiple classifiers or **models** and combining their predictions to make decisions **with improved accuracy and robustness**. Ensemble deploys methods such as bagging, boosting, random forests, and using different classification models. Ensemble aims at improving the accuracy of prediction (Han et al. 2014).

Bagging

Bagging uses resampling technique. A data sample is selected at random with replacement from the given dataset. An appropriate classification model is built based on the sample. This process of sampling and model building is repeated to develop k classifiers. The data object to be predicted is passed the ensemble of k-models. The target label predicted by the k classifiers is inspected, and the

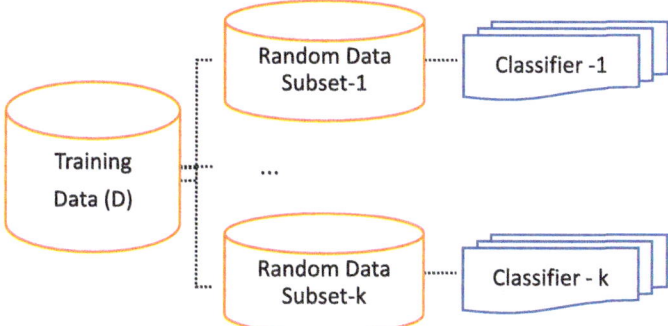

Fig. 11.7   Bagging

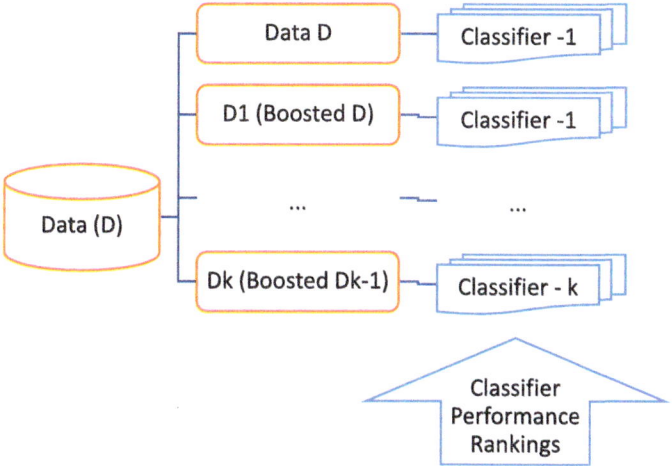

Fig. 11.8   Boosting

most frequent label is assigned to the new object. We can also use this method for regression by computing the average of the k predictions. Compared to a single classifier, bagging improves accuracy and is less affected by noise.

Boosting

Boosting is an improvisation of bagging. While building models $m_1 \ldots m_k$ in order from data samples $s_1 \ldots s_k$, the miss-classified data is given an added weightage after each model building/testing. The models with higher accuracy receive higher voting weightage from the ensemble of emerging classifiers. Boosting accuracy is expected to be higher but may tend to overfit miss-classified data (Figs. 11.7 and 11.8).

Random Forest

We discussed building a decision tree classifier by ranking and selecting the attributes by their contribution to classification. Instead of building one tree, we may build multiple trees using the following method:

- Random data sampling: Different subsets of the training data are randomly selected for each decision tree. This process is known as bootstrapping.
- Feature randomness: When building each decision tree, a random subset of features is considered at each split point. This helps in reducing the risk of overfitting and decorrelates the trees.

We will pass the data object to be predicted through all the individual trees and choose the target label that gets the maximum votes. Such an ensemble of decision trees is called a random forest. It is a popular method deployed in machine learning (Figs. 11.9 and 11.10).

Hard Voting Classifier

Yet another technique is to build a set of different classification models (e.g., logistic regression, linear discriminant analysis, random forest, support vector machines). We could pass the data to be predicted through all these classifiers and choose the target label that gets the maximum votes. This is called a hard voting classifier (Fig. 11.11).

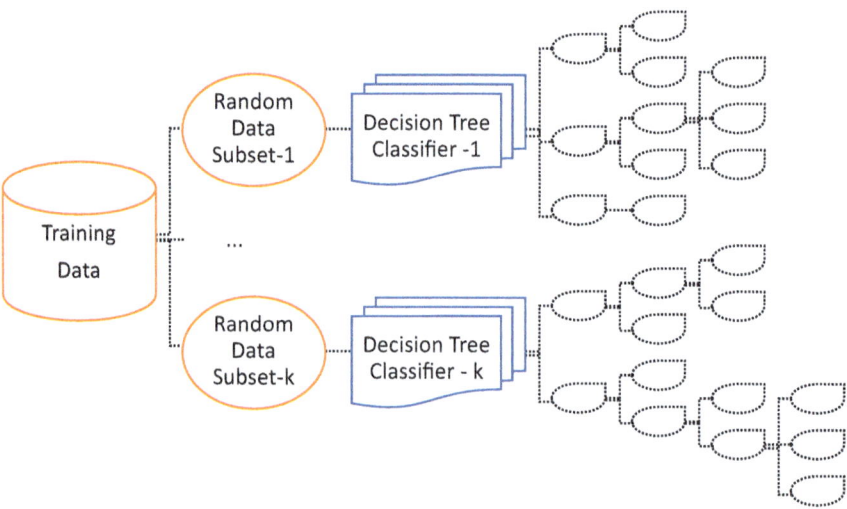

Fig. 11.9 Random forest by random data sampling

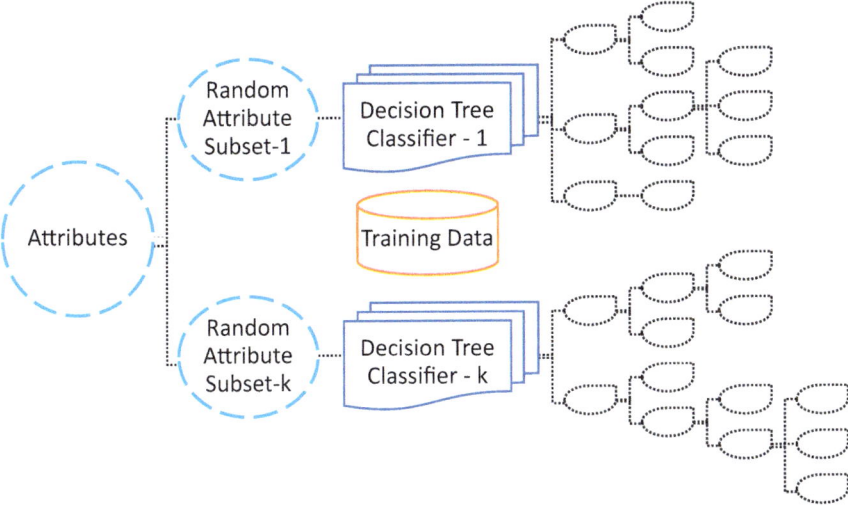

Fig. 11.10 Random forest by feature randomness

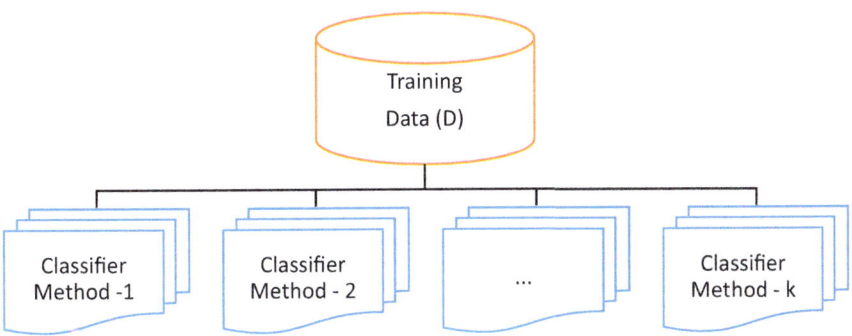

Fig. 11.11 Hard voting

11.6.3 Bias-Variance Trade-Off

The error in a model's prediction comprises three constituents—noise, bias, and variance. Noise is the unexplainable error resulting from data collection errors, outliers, and unknown features. Bias happens when simple assumptions are made by a model, whereas the actual mapping of features to target is complex. For example, proposing a linear model while the underlying distribution is quadratic or of a higher degree polynomial. A high bias may result in an underfit model. Examples of high-bias models include linear regression, logistic regression, and linear discriminant analysis. These models are fast to learn but they are less flexible. Examples of low-bias models include decision trees, k-nearest neighbors, and support vector machines.

Variance refers to the responsiveness to fluctuations in the data. Models with intricate structures, like higher degree polynomials, often exhibit elevated variance, which can arise from random noise in the training dataset. Generally, nonlinear models such as decision trees, k-nearest neighbors, and support vector machines possess heightened flexibility and variance, leading to overfitting. Conversely, models characterized by low variance, such as linear regression, logistic regression, and linear discriminant analysis, tend to be less susceptible to overfitting. To mitigate overfitting, one strategy involves regularization, which entails smoothing or constraining the model. For instance, reducing the degree of variables (see Fig. 11.6) is a regularization approach in polynomial models. In linear models, regularization can be accomplished by imposing weight constraints using ridge regression, lasso regression, and elastic net techniques.

Augmenting a model's complexity typically increases its variance while diminishing its bias. Conversely, decreasing complexity heightens bias and diminishes variance. Striking a balance between the two is crucial, and the subsequent section explores one of the solutions to achieve this trade-off. Figure 11.7 illustrates this idea.

11.6.4 Train-Test Convergence

This is an approach to ensure that the model can generalize well to new, unseen data, ensuring the model's reliability and effectiveness. We divide the dataset randomly into two—a large training dataset (consisting of 80% of the data) and a smaller test dataset consisting of the rest. The model built from the training dataset is tested on the test dataset. The training and test performances are compared, using a measure such as 'loss' or 'accuracy' (Fig. 11.12).

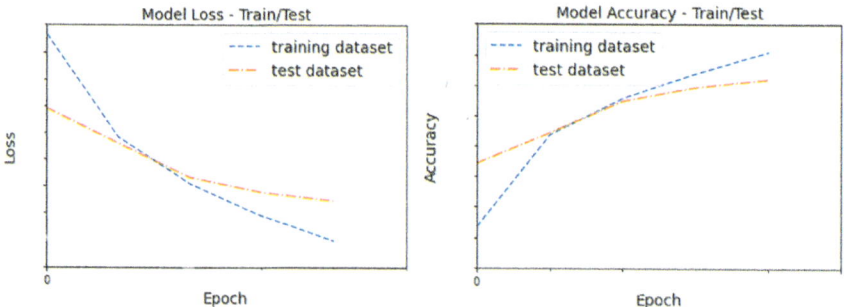

Fig. 11.12 Train-test convergence

If the model performs well on the training data but poorly on the testing data, it suggests overfitting, which means the model has learned to fit the training data too closely and may not generalize well to new, unseen data. To achieve test-train convergence, we use techniques like cross-validation, regularization, early stopping, and careful hyperparameter tuning to find the right balance between model complexity and generalization. The idea is to avoid overfitting and ensure that the model's performance remains consistent between the training and testing phases. We repeat the train-test iterations to achieve convergence in training and test performances. In other words, convergence happens after multiple epochs, as seen in Fig. 11.7.

11.6.5 Train, Validate, Test Split

We may divide the sample data during training into three datasets—train, validate, and test. See Fig. 11.8. The process of train-validate-test is repeated for the convergence of performance. This systematic approach evaluates how well a model is likely to perform on new, unseen data. Let's delve into the distinct phases of this process:

1. Training phase: Here, the model acquires knowledge from the training dataset. The training data serves as the primary set for refining the model's weights and biases through various algorithms, enabling it to grasp patterns and relationships within the data.
2. Validation phase: Utilizing a subset known as validation data, this phase fine-tunes the model's hyperparameters. Hyperparameters, essential for controlling the learning process, are configuration settings distinct from data-driven parameters. By employing a separate validation set, adjustments can be made to enhance the model's performance without succumbing to overfitting.
3. Test phase: Following training and fine-tuning, it is imperative to evaluate the model on entirely new data to assess its generalization capabilities. The test data, an unbiased measure of final performance, is used to apply the model and gauge its effectiveness.

The entire training, validation, and testing process is often iteratively repeated until the model reaches a satisfactory level of performance. This iterative nature allows for dynamic adjustments, such as modifying hyperparameters or incorporating additional data, to improve the overall effectiveness of the model (Fig. 11.13).

Post-evaluation, the model is applied to real-world data for practical use. This may involve tasks such as product recommendations, fraud detection, medical diagnosis, or autonomous decision-making in driving. Continuous monitoring of the model's performance in real-world scenarios is critical for its sustained effectiveness. As real-world data distributions evolve over time, models may need periodic updates or retraining. Continuous monitoring ensures the model remains effective and relevant in dynamic environments.

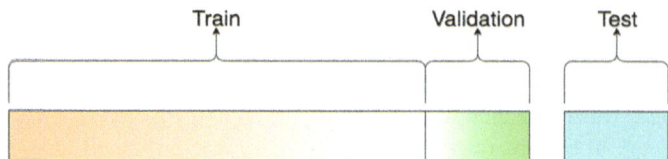

Fig. 11.13 Train, validate, test

In essence, the train-validate-test approach systematically develops machine learning models capable of generalizing new data. It ensures the model not only learns from training data but also undergoes fine-tuning and validation before practical deployment. The iterative nature of the process allows for ongoing adjustments, ensuring the model's adaptability and effectiveness in dynamic, real-world conditions.

11.6.6   Receiver Operating Characteristic: ROC/AUC Analysis

Receiver Operating Characteristic (ROC) analysis is commonly used with logistic regression models, especially in binary classification problems. In binary logistic regression, the model outputs probabilities that an observation belongs to the positive class. Observations with probabilities greater than or equal to 0.5 (threshold) are classified as positive, and those below 0.5 are classified as negative. ROC analysis involves changing the classification threshold to produce a range of true positive rates (sensitivity) and false positive rates (1 - specificity). The ROC curve is a graphical representation of this trade-off, where each point on the curve represents a different threshold. The Area Under the ROC Curve (AUC-ROC) is a scalar metric that summarizes the model's performance across all possible thresholds. It ranges from 0 to 1, with higher values indicating better discrimination between the positive and negative classes. Figure 11.14 illustrates ROC/AUC, comparing the performance of three classifiers on an arbitrary set of data.

Fig. 11.14 ROC/AUC analysis

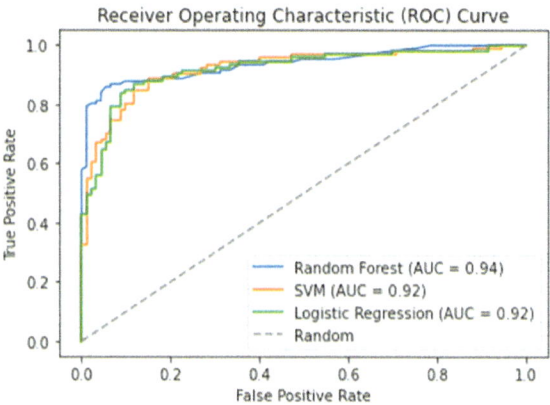

ROC/AUC analysis can also be extended to other classification problems. An illustration of an example where ROC/AUC analysis is extended to survival analysis is shown in Sect. 9.5.3.

Data Analytics in Action

A typical machine learning application: YouTube advertisements (Zappin et al. 2021)
Digital marketing refers to advertising campaigns on electronic gadgets, e.g., online videos, social media posts. It is more cost-effective and targeted than traditional marketing such as magazine ads, billboards, direct mail, and TV. YouTube has billions of users, and billions of hours of video are watched daily. Internet censorship refers to the control or suppression of publication or access of information enacted by Internet regulators. YouTube censors content using three primary methods—content removal, channel removal, and demonetization. In demonetization, the content creators are denied paid ads in their YouTube videos. Companies are more sensitive to the type of content to which they tie their ads. YouTube's censorship algorithm is not public.
The paper mentioned above proposes a methodology that employs four machine learning algorithms, i.e., C 4.5, Random Forest, Linear Regression, and Support Vector Machine, to predict if changes in the meta-data of the YouTube video will lead to censorship.

Summary

While humans learn from experience, machines learn from data and improve accuracy over time, without being programmed to do so. Machine learning involves identifying the correct model parameters that optimize the predictive capability and deriving information from available data. The machine learning process can be broadly categorized as supervised, unsupervised, semi-supervised, and reinforcement learning.

In supervised learning, the sample data consists of input features and the desired outputs (targets). The data sample is divided into two sets—the training dataset and the test dataset. The training dataset is used to train algorithms such as linear discriminant analysis, decision trees, and support vector machines. The model is then tested over the test dataset. The train-test process is repeated till an accurate model emerges. Algorithms for classification are trained to predict categorical class labels (nominal or ordinal). Regression algorithms are continuous-valued functions trained to predict numeric values (real numbers).

An exhaustive algebraic solution for optimization is computationally prohibitive with large data. An alternative is to use an iterative solution, giving due consideration to model fit. Gradient descent is an iterative optimization technique for finding the local minimum of a differentiable function. This technique can be used to find the model parameters (coefficients) that optimize a specified cost function. Variations include stochastic gradient decent, mini-batch gradient descent, batch gradient descent, Adagrad, Adam, etc.

Batch gradient descent uses each iteration's entire dataset (size 'n'). In Stochastic Gradient Descent (SGD), only one randomly selected data point is used per iteration from the 'n' available points. Mini-batch gradient descent, a compromise between SGD speed and batch gradient descent's goodness of fit, randomly selects a subset of 'm' datapoints ($1 < m < n$) in each iteration. It repeats the gradient descent process multiple times until a specified convergence criterion is met. SGD is memory efficient and converges rapidly, making it preferable for large datasets. While it provides a satisfactory solution quickly, it may not be optimal, yet it efficiently escapes local minima. The learning rate's influence on convergence speed is crucial, as too low or high rates result in slow convergence or oscillations, respectively. Scaling, especially in SGD, is sensitive, making standardized data input a common practice. SGD finds applications in large-scale and sparse machine learning, particularly in tasks like text classification and natural language processing.

Regularization techniques, including ridge and lasso regression, aim to reduce model complexity and prevent overfitting. Ridge regression minimizes the L2 norm with an added penalty, addressing multicollinearity. Lasso regression uses the L1 norm, adding the sum of coefficient magnitudes as a penalty, facilitating feature elimination. Elastic net regularization combines L1 and L2 regularization and is helpful when dealing with correlated features. These techniques contribute to improved model generalization and feature selection.

The components of prediction error include noise, bias, and variance. Noise, stemming from errors in data collection and outliers, contributes to unexplainable errors. Bias arises when a model makes simplistic assumptions about complex feature-to-target mappings, potentially resulting in an underfit model. High-bias models include linear regression, logistic regression, and linear discriminant analysis, known for speed but limited flexibility. Conversely, low-bias models offer more flexibility, such as decision trees, k-nearest neighbors, and support vector machines. Variance, representing sensitivity to data variations, is associated with model complexity. Complex models, like higher degree polynomials, may exhibit high variance, leading to overfitting. Nonlinear models such as decision trees, k-nearest neighbors, and support vector machines often have high flexibility and variance. Regularization techniques, like ridge regression and lasso regression, are employed to mitigate overfitting. The note emphasizes the need for a trade-off between model complexity and variance, advocating for an optimal balance.

Train-test convergence is an approach to ensure model generalization to new, unseen data. Randomly dividing the dataset into a large training set (80%) and a smaller test set, the model's performance is assessed by comparing training and test results. Overfitting is identified when a model excels on training but falters on testing, prompting techniques like cross-validation, regularization, early stopping, and hyperparameter tuning to achieve convergence. Train, validate, test split extends this concept by dividing sample data into three sets during training: train, validate, and test. The training phase focuses on learning model parameters, with validation data used to fine-tune hyperparameters. In the application phase, test data provides a final, unbiased estimate of the model's performance in real-world scenarios.

Hyperparameters are manually set parameters affecting model development. Crucial hyperparameters include train-test split ratio, cost or loss function choice, optimization algorithm, number of iterations in training, learning rate, regularization terms, and other model-specific parameters.

Ensemble methods, exemplified by techniques like bagging, boosting, and random forests, amalgamate multiple models to bolster accuracy and robustness. Bagging involves resampling, while boosting assigns increased weight to misclassified data, both contributing to improved accuracy despite potential overfitting challenges. The receiver operating characteristic curve and the AUC-ROC metric offer valuable insights into a model's ability to discriminate between classes at various thresholds.

Questions

Comprehension—Write Brief Notes on:

1. Cross-validation.
2. Bootstrapping.
3. Resampling.
4. Regularization.
5. Elastic net.
6. Mini-batch gradient descent.
7. Train-validate-test model.
8. Penalty versus learning rate.
9. Loss functions used in regression.
10. Loss functions used in classification.
11. Gradient descent algorithm.
12. Hyperparameter tuning.
13. Bias-variance trade-off.

Analysis

14. Explain the fundamental difference between supervised, unsupervised, semi-supervised, and reinforcement learning. Provide examples of real-world applications for each of these machine learning paradigms.

15. Regularization techniques like ridge regression and lasso regression are commonly used in linear models to prevent overfitting. Compare and contrast the key differences between ridge and lasso regression. In what situations would you choose one over the other, and why?

16. Discuss the concept of gradient descent optimization in machine learning. What are the various gradient descent techniques such as stochastic gradient descent and mini-batch gradient descent? How does the choice of gradient descent method impact the model's convergence and computational efficiency?

17. Explain the importance of hyperparameters in machine learning models. Please provide specific examples of hyperparameters used in different machine learning algorithms and describe their roles in model development and optimization.

18. Resampling techniques, like K-fold cross-validation and bootstrap, are essential for assessing the performance of machine learning models. Describe how K-fold cross-validation works and its benefits in evaluating model performance. In what situations would you prefer bootstrap resampling instead?

19. Define the bias-variance trade-off in the context of machine learning models. How does increasing model complexity affect bias and variance? Provide practical examples of how different modeling choices can impact the bias-variance trade-off in real-world applications.

20. Discuss the data standardization process in machine learning and why it is essential before training certain algorithms. Provide examples of machine learning models that benefit significantly from data standardization and explain why.

21. Ensemble methods are widely used in machine learning. Explain the concept of ensemble methods and provide examples of algorithms like Random Forest and Gradient Boosting. How do these methods improve predictive performance, and what types of problems are they well suited?

22. Explain the difference between bagging and boosting in the context of ensemble methods. How do these methods improve prediction accuracy, and what are the potential drawbacks of each approach?

23. Why must a model perform well on both the training and testing datasets when discussing test-train convergence? What does it indicate when a model performs well on the training data but poorly on the testing data?

24. Consider the concept of train-validate-test split. Why is it necessary to have a separate validation dataset in addition to training and testing datasets? What is the purpose of each of these subsets in the machine learning process?

25. In the context of random forests, explain how random sampling and feature randomness contribute to improving prediction accuracy. What are the advantages of using random forests over a single decision tree?

Exercises

I. Regression

IMDB Movie User Ratings for Regression

The goal here is to predict user ratings based on a set of features.

IMDB database is a collection of movie data supplied by studios and fans, probably the biggest movie database on the web, and run by Amazon. We are considering a movie user rating database with 58,788 rows and 24 variables.

The target that we would like to predict is movie 'rating'.

The features were selected to include most of the numeric variables. We will exclude non-numeric variables such as title and mpaa. We will also exclude the budget, which has numerous null entries.

Refer:

> https://vincentarelbundock.github.io/Rdatasets/articles/data.html—ggplot2movies
>> DOC https://vincentarelbundock.github.io/Rdatasets/doc/ggplot2movies/movies.html
>> CSV https://vincentarelbundock.github.io/Rdatasets/csv/ggplot2movies/movies.csv

Exercise 11.1 Cross-Validation
Perform regression analysis of diamond price versus features. Use the cross-validation method to predict the model performance.

Exercise 11.2 Regularization with L2 Penalty
Perform regression analysis of diamond price versus features, using L2 penalty for regularization.

Exercise 11.3 Regularization with L1
Perform regression analysis of diamond price versus features, using L1 penalty for regularization. Does this help in feature selection?

Exercise 11.4 Stochastic Gradient Descent
Perform regression analysis of diamond price versus features, using stochastic gradient descent. How do the results differ from linear regression without the gradient descent method?

Exercise 11.5 Grid Search to Identify the Best Model Parameters
Perform regression analysis of diamond price versus features. Use grid search to select the best model parameters.

II. Classification

The following exercises use the iris flowers dataset.

Exercise 11.7 Cross-Validation

Build a classification model for classifying iris species based on four features. Use the cross-validation method to predict the model accuracy.

Exercise 11.8 Regularization with L2 Penalty

Build a classification model for classifying iris species based on four features, using the L2 penalty for regularization.

Exercise 11.9 Regularization with the L1 Penalty for Feature Selection

Build a classification model for classifying iris species based on four features, using the L1 penalty for regularization. Does this help in feature selection?

Exercise 11.10 Stochastic Gradient Descent

Build a classification model for classifying iris species based on four features, using stochastic gradient descent. How do the results differ from the classification model without the gradient descent method?

Exercise 11.11 Grid Search to Identify the Best Model Parameters

Build a classification model for classifying iris species based on four features. Use grid search to select the best model parameters.

References

Chollet F (2019) Chollet—2018—Deep learning with Python. In: Manning, vol 53, Issue 9

Géron A (2019) Hands-on machine learning with Scikit-Learn, Keras, and TensorFlow (2019, O'reilly). In: Hands-on machine learning with R

Han J, Kamber M, Pei J (2014) Data mining. Concepts and techniques, 3rd edn. The Morgan Kaufmann series in data management systems. In: Proceedings—2013 international conference on machine intelligence research and advancement, ICMIRA 2013

Zappin A, Malik H, Shakshuki EM, Dampier DA (2021) YouTube monetization and censorship by proxy: a machine learning prospective. Procedia Comput Sci 198. https://doi.org/10.1016/j.procs.2021.12.207

Chapter 12
Artificial Intelligence and Deep Neural Networks

Learning Objectives

- Understand the basic concepts of artificial intelligence and artificial neural networks.
- Acquire familiarity with machine learning models—supervised, unsupervised, semi-supervised, and reinforcement learning.
- Get a detailed understanding of artificial neural network architecture, components, concepts, and associated software frameworks.
- Discuss feed-forward neural network (FFNN) and demonstrate its application.
- Discuss convolutional neural network (CNN) and demonstrate its application.
- Understand recurrent neural networks (RNN).
- Discuss Long Short-Term Memory (LSTM) and demonstrate its application.

Overview

This chapter starts with an introduction to artificial intelligence and its frontiers from a cognitive science perspective. Then we move on to an overview of artificial neural networks—concepts, components, architectures, and software packages. We perform a detailed exploration of a multilayered feed-forward network and its implementation using Python and Keras frameworks. A discussion on convolutional neural networks and their implementation follows. Finally, we explore recurrent neural networks, LSTM, and their implementation.

Supplementary Information The online version contains supplementary material available at https://doi.org/10.1007/978-981-99-0353-5_12.

391
S. Sundararajan, *Multivariate Analysis and Machine Learning Techniques*,
Transactions on Computer Systems and Networks,
https://doi.org/10.1007/978-981-99-0353-5_12

Definitions

Activation function: The activation function is a transformation function in a neu-ron, where the weighted sum of its inputs is transformed into the output using a function such as the sigmoid, the hyperbolic tangent, softmax, and rectified linear unit.

AdaGrad: AdaGrad, an optimization algorithm employing gradient descent, cus-tomizes the step size for each variable. This adjustment is determined by the par-tial derivative of the optimizing function with respect to that specific variable.

Artificial Intelligence (AI): AI is 'the science and engineering of making intelli-gent machines'. There are two approaches to AI—symbolic and connectionist. A medical expert system, which is a rule-based system, is an example of a symbolic system. Artificial neural networks are examples of connectionist systems.

Backpropagation: During the learning process of ANN, the deviation of the output from the target is computed using a loss function. The weights are adjusted based on their contribution to error, applying the chain rule of derivatives by an algo-rithm that traverses backward from the output layer. This is called backpropaga-tion in a multi-layer neural network.

Convolution: Convolution is applying a filter (F) iteratively over all the pixels of an image (I). A filter (F) carries a set of weights learned using the backpropagation algorithm. A filter can be considered as storing a single pattern. When we con-volve this filter across the corresponding input, we find the degree of similarity between the filter and different locations in the input. This principle is used for edge detection.

Convolutional Neural Network Architecture (CNN): CNN is a special case of feed-forward neural network suitable for image processing. CNN consists of the following segments—feature extraction and classification.

Cost function: Same as loss function.

Deep learning: Deep learning emphasizes learning by successive layers of increas-ingly meaningful representations using a connectionist design of neurons.

Feature detection: Feature detection is the first stage in image processing. There are various features—edges, corners, ridges, and blobs/regions. Small contiguous areas often describe these features. The recognition of features, such as edges, does not depend on their location within the edge.

Feature Extraction: The feature extraction segment consists of many convolutional layers, followed by pooled layers.

Feed-Forward Neural Network (FFNN): A feed-forward neural network is a sim-ple multilayered neural network architecture. The input data moves through hid-den layers to the output layer in one direction. The layers do not give feedback to any previous layers. The connections between nodes do not form a cycle at any stage.

Gradient descent: In a neural network, the learning process entails discovering a configuration of model parameters that minimize a specified loss function across a given set of training data samples. Random batches of data samples are selected, and the gradient of the model parameters concerning the loss is calculated. The parameters are then adjusted by a proportion determined by the learning rate, moving in the direction opposite to the gradient. The process is reiterated through multiple passes over the entire dataset, aiming for convergence based on predefined criteria.

Learning process: In a neural network, learning involves finding a combination of model parameters that minimize a loss function for a given set of training data samples.

A loss function, also known as a cost function or objective function, is a mathematical function that quantifies the difference between the actual target values and the values predicted by a machine learning model. The loss function measures model performance.

Machine learning: Machine learning involves identifying the correct model parameters that optimize the predictive capability based on the information in the data available.

Network layers: A deep learning neural network will have multiple layers, each with a set of neurons. A layer other than the input layer consists of (a) an activation function—which accepts a tensor input and generates a tensor output and (b) the current state—the (coefficient) weights held in TensorFlow variables.

Neural network: Comprising neurons arranged in layers, a neural network processes external inputs through its input layer and produces final outputs from its output layer. The intermediate, hidden layers are typically trained to identify intermediate concepts crucial for determining the ultimate output.

Neuron: A neuron takes one or more inputs, multiplies each input by a weight parameter, sums them up, and adds a bias term to the sum. The weighted sum is passed as input to an activation function, which transforms it. The output from a neuron is sent forward to the neurons of the next layer.

Optimization techniques: Optimization techniques help to minimize the loss function (or optimize the cost function). The optimizer specifies how the gradient of the loss will be used to update parameters. The following are some of the commonly used optimizers—Gradient Descent, AdaGrad, RMSProp, and Adam.

Recurrent Neural Network (RNN): Recurrent neural network (RNN) is a feedback neural network. It is also known as auto-associative network. RNN is designed to recognize sequences, such as the sentences in a natural language, which is composed of sequences of words, or the stock market position of a company, which is composed of a series of stock prices.

Regularization serves to counteract overfitting by incorporating various techniques. These techniques encompass penalizing substantial weights (coefficients), randomly excluding certain nodes during each iteration of gradient descent, and utilizing a validation set to facilitate stopping the training process when the loss on the validation set reaches convergence, etc.

RMSProp: RMSProp (root mean squared propagation) is an optimization algorithm that imbibes the properties of gradient descent and AdaGrad—it uses a decaying average of partial gradients to determine the step size for each variable.

The LSTM model is a refinement RNN for learning long-distance associations. LSTM can save selected information into the memory, forget information by purging it, and focus on memory that is relevant to the context.

12.1   Artificial Intelligence

Artificial Intelligence (AI) was coined in 1955 by John McCarthy, Emeritus Professor of Computer Science at Stanford University. He defined AI as 'the science and engineering of making intelligent machines' (Andresen 2002). Research in AI is associated with learning (memorization and generalization), reasoning (inductive and deductive), problem-solving, perception, and language.

AI research received a big boost in the 2010s with the emergence of powerful systems to process big data and support machine intelligence. Tracing the history of AI, two observations are worth mentioning, as they reveal profound thoughts in this area. Allan M Turing, a Mathematician, Computer Scientist, and Logician, considered one of the founders of AI, predicted that computers would one day play incredibly good chess. In 1997, IBM's chess computer Deep Blue beats the reigning world champion, Garry Kasparov, in a six-game match. Deep Blue was a 256-core machine that examined 200 million possible moves per second and looked ahead at as many as 14 turns of play. Noam Chomsky, Emeritus Professor of MIT, one of the world's greatest cognitive scientists and linguists, passed a thought-provoking comment—'a computer beating a grandmaster at chess is about as interesting as a bulldozer winning an Olympic weightlifting competition' (Copeland 2023).

However, many people on the other side of the spectrum, such as Yuval Noah Harari, are overawed at the possibilities of artificial intelligence. 'What will happen to the job market once artificial intelligence outperforms humans in most cognitive tasks? What will be the political impact of a massive new class of economically useless people? What will happen to relationships, families, and pension funds when nanotechnology and regenerative medicine turn eighty into the new fifty? What will happen to human society when biotechnology enables us to have designer babies and to open unprecedented gaps between rich and poor?' (Harari 2016). He continues to say that such developments may happen by about 2040.

While brushing aside profound questions from sociologists, we may need to acknowledge that AI raises some of the fundamental questions of philosophy—especially epistemology (theory of knowledge) and axiology (theory of principles and values). What is intelligence? How does it work? How does human intelligence differ from that of animals? What is the foundation of linguistic capability? What is 'reasoning'? What is ethics? These are all major questions of scientific research today. All living beings possess instincts that are genetically coded. Animals (including human beings) develop behavior patterns through interactions with the environment. However, human beings appear to have a distinguishable faculty, 'reasoning', that leads to self-consciousness and abstract quests such as 'meaning' and 'truth' of life. The constitution and rules of the human mind remain elusive and a matter of speculation starting from the time of Gautama Buddha and Zeno of Citium to Nobel Laureate Francis Crick (biophysicist and neuroscientist), who hypothesizes that all human feelings, thoughts, actions, and consciousness itself are products of neural activity in the brain. How far can AI emulate the human mind? How many more years of scientific investigations would lead us to such effective models? These are questions for posterity (Fig. 12.1).

Now let us examine the state of the art of machine intelligence today. As we know, AI finds application in a multitude of domains such as bioinformatics, health care, natural language processing, digital image processing, automation, business, government and public services, and so on.

Exploring genetic data facilitates detecting abnormalities, discerning causal genotype-phenotype connections, anticipating potential issues, and tailoring personalized drug recommendations to enhance medical interventions. The data underlying a single human genome sequence is about 200 gigabytes! Researchers construct probabilistic models such as Markov chains and hidden Markov models to study DNA sequences. Pharma companies use AI for cost-effective drug discovery. AI systems support medical diagnosis, preventive care, and prediction of failures based on historical data.

Natural language processing (NLP) techniques help to understand human language in written or spoken form and translate it to another language or respond to commands. Some applications include voice-operated GPS systems, translation tools, speech-to-text dictation, and customer service chatbots.

Convolutional Neural Network (CNN) architectures are used extensively to build models for digital image processing. One of the applications is biomedical

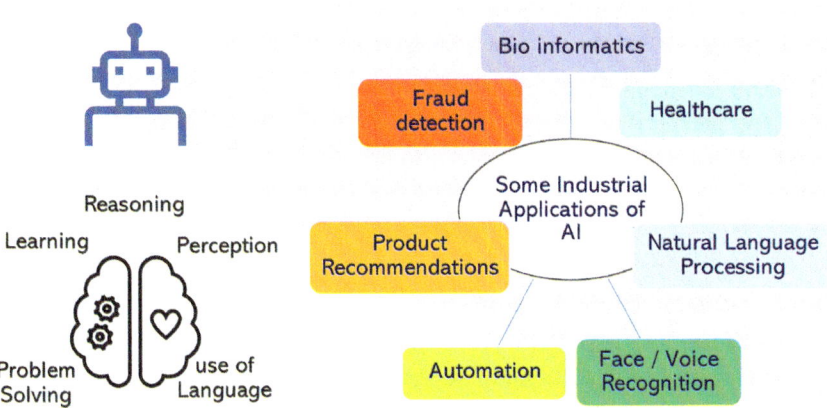

Fig. 12.1 Machine intelligence

image processing, which assists medical practitioners in visualizing, interrogating, identifying, and treating deformities in internal organs and systems. The investigation into remote sensing focuses on the automated and efficient analysis of extensive information that is challenging to acquire directly or evaluate manually, such as satellite images, geosensor data, storms, wildfires, and more. Computer vision enables computing devices and systems to gather meaningful information from digital images and act based on that.

One of the flourishing domains in AI is generative AI, encompassing algorithms that leverage reinforcement learning, generative adversarial networks (GANs), and transformer-based models for content creation across various mediums such as text, code, audio, images, and videos. An exemplary application in this realm is ChatGPT by OpenAI, which adeptly codes computer programs, composes music and crafts short stories and essays. This application is powered by a Large Language Model (LLM), representing one of the most successful applications of transformer models. LLMs acquire knowledge from extensive datasets and are being explored in a wide spectrum of topics, ranging from enhancing linguistic capabilities in AI to decoding proteins.

12.2   The Machine Learning Model

Machine learning became a buzzword in the late 1990s. While AI is a broader concept of creating intelligent machines, Machine Learning (ML) is a specific approach within AI that involves using algorithms to enable machines to learn from data. Machine learning involves identifying the correct model parameters that optimize the predictive capability and deriving information from available data. The data sample is expected to consist of a set of features and possibly the corresponding targets. From the data sample, subsets are selected randomly and further partitioned into training and test datasets. The training and testing are repeated iteratively to develop a model with accurate performance. We also saw that machine learning can be broadly categorized as follows:

- Supervised learning: Training data includes desired outcomes.
- Unsupervised learning: Training data does not include desired outcomes.
- Semi-supervised learning: Training data includes a few desired outcomes.
- Reinforcement learning: Rewards from a sequence of actions.

12.2.1   Supervised Learning

We had a detailed discussion of supervised learning and its application in the last chapter. As we know, in this method, the training data includes input features and desired targets. Based on the data samples, an algorithm learns to predict the output (target) given a set of features. The task is called classification if the target is a set of discrete categories. If the target is a continuous variable, the task is called regression. Some supervised learning applications include image classification, medical diagnosis, language translation, speech recognition, sentiment analysis, autonomous driving, customer churn prediction, stock price prediction, recommendation systems, weather forecasting, market segmentation, etc.

12.2.2   Unsupervised Learning

In this learning method, the training data includes input features but does not include the desired outcomes. Some unsupervised learning methods aim to discover groups with similar features within the data. Called cluster analysis, it offers several techniques to segregate objects into homogeneous groups called clusters. The objects in each cluster tend to be similar while being dissimilar to those in the other clusters (Han et al. 2014). Some of the applications of unsupervised learning include social network analysis, recommendation systems (collaborative filtering), customer segmentation, market basket analysis, image compression, genomic sequencing, etc.

ChatGPT and other models utilizing the GPT architecture undergo pre-training through unsupervised learning. The pre-training phase entails predicting the next word in a sentence based on the context of preceding words, leveraging extensive and diverse publicly available text from the Internet. However, the following must be noted. After the pre-training stage, these models undergo fine-tuning for specific tasks using supervised learning or alternative methods. In the case of ChatGPT, the fine-tuning process incorporates custom datasets and reinforcement learning from human feedback, enhancing the model's control and safety for interactive applications.

12.2.3   Semi-supervised Learning

Semi-supervised enables us to build systems where labeled training data may be sparse or costly. In this learning method, the training data comprises a small amount of labeled data and a large amount of unlabeled data. At first, a neural network model is trained on a limited set of data, where clustering algorithms group and label similar items. These labels are subsequently used to classify the other data points.

An example is the Google application—'Smart Reply for Inbox'. This application developed by Google's 'Expander machine learning system is used for massive graph building '(containing billions of nodes and trillions of edges)' and processing to recognize 'concepts in natural language, images, videos, and queries. Google products for applications such as reminders, question answering, language translation, visual object recognition, dialogue understanding', etc. (Sujith 2016). make use of this unsupervised machine learning platform.

12.2.4   Reinforcement Learning

This learning method constitutes rewards from a sequence of actions. The problem is to find suitable actions to be taken in each situation to maximize a reward. Unlike supervised learning, the training data does not consist of the desired output in each situation. Instead, they are discovered by a process of trial and error.

Richard S. Sutton, a Distinguished Research Scientist at DeepMind and a Professor of computing science at the University of Alberta, is considered one of the founders of modern computational reinforcement learning. Reinforcement learning relies on the Markov Decision Process (MDP). MDP is a sequential decision problem for a fully observable, stochastic environment with a Markovian state transition model and associated rewards {Sutton}. Ideas from dynamic programming methods are used to build approximate solutions to deal with large stochastic state-action space {Sutton}. Reinforcement learning finds applications in various fields such as gaming, robotics, transportation, energy, health care, manufacturing, and finance.

Reinforcement Learning (RL) and Deep Learning (DL) are two powerful paradigms in machine learning. They are combined to create reinforcement learning with deep learning techniques, often called Deep Reinforcement Learning (DRL). In classical reinforcement learning, the emphasis is on using algorithms that explicitly represent and update the environment's value functions, policies, or models.

Deep reinforcement learning involves using deep neural networks to approximate complex functions, such as value functions or policies to handle high dimensional state spaces and complex decision-making scenarios. DRL typically enables end-to-end learning, where the agent directly learns from raw sensory input to action, with minimal need for manual feature engineering. DRL has successfully played complex games, such as DeepMind's AlphaGo (see the story at the end of the chapter). DRL is employed in computer vision and robotics. Such models adapt to various environments and continually changing conditions by self-learning. In NLP, DRL is used in dialogue systems, language understanding, and generation. DRL is used in transfer learning where ANN models are architected, pre-trained on related tasks, and then fine-tuned for specific applications, enhancing generalization.

12.3   Deep Learning

As stated earlier, intelligence is associated with learning, reasoning, problem-solving, perception, and language. Scientists have developed two approaches to AI—symbolic and connectionist. See Fig. 12.2. During the last century, researchers believed intelligence could be emulated by connecting facts and rules. This approach is known as symbolic AI. Expert systems of the 1980s (with the

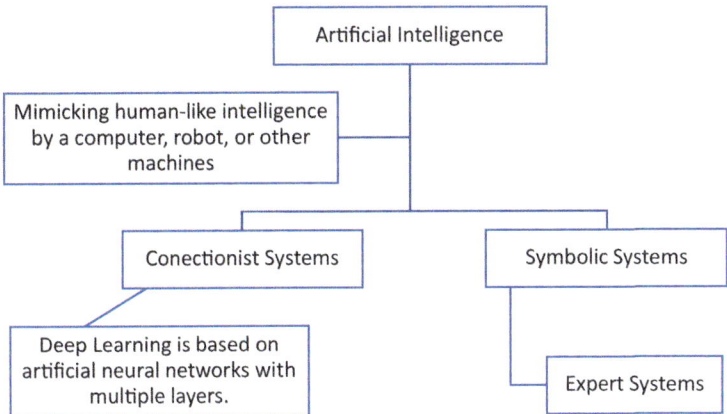

Fig. 12.2 Deep learning—a subset of AI

arrival of microprocessor chips) were the pinnacle of this movement. Though symbolic AI proved suitable for solving well-defined, logical problems, such as playing chess, it failed to tackle complex, fuzzy problems such as pattern recognition and language translation.

An artificial neural network (ANN) is founded on connectionist design that attempts to mimic the neurons and their interconnection in the human brain. Deep Learning (DL) constitutes a specialized variant of artificial neural networks, encompassing neural networks with multiple layers, commonly known as deep neural networks. Deep learning relies on learning by successive layers of increasingly meaningful representation of input data.

The origin of deep learning can be traced to cybernetics (1940s–1960s), and connectionism (1980s–1990s). The current resurgence of AI happened in 2006, with the availability of cheap and powerful systems that can manage big data. For example, graphics processing units (GPUs) have built-in circuits facilitating highly parallelized linear algebra operations. Some of the earliest learning algorithms in this area were intended to model neurobiological learning. As a result, deep learning got its popular pet name—artificial neural networks (ANNs). However, it may be noted that deep learning models are not models of the brain, and there is no evidence that the brain implements anything like the learning mechanisms used in modern deep learning models.

Deep learning is a specific type of machine learning, as it involves building models by learning patterns and relationships from data. Deep learning is different from other ML models by way of its architecture, viz., multilayered neural network architecture. Deep learning methods find use in all categories of advanced machine learning—supervised (e.g., FFNN, CNN, and RNN/LSTM that are covered in this chapter), semi-supervised (e.g., generative adversarial networks or GANs), unsupervised (e.g., cluster analysis in customer segmentation for targeted marketing), and reinforcement learning (games, robotics, autonomous, agents, etc.)

The rest of the chapter is devoted to the exposition of supervised learning using deep learning architectures such as feed-forward neural networks, convolutional neural networks, recurrent neural networks, and LSTM using the Keras framework.

12.4   The Artificial Neural Network

A neural network comprises neurons arranged in layers, with the outputs of one layer serving as inputs for the next (Goodfellow et al. 2016). The input layer receives external inputs, while the output layer generates the ultimate output. The intermediate layers between them, known as hidden layers, are typically trained to identify and understand intermediate concepts crucial for determining the final output.

12.4.1   Neuron

A neuron takes one or more inputs, multiplies each input by a weight parameter, sums them up, and adds a bias term to the sum. The weighted sum is passed as input to a nonlinear activation function, which transforms it. The output from a neuron is sent forward to the neurons of the next layer (see Fig. 12.3).

Propagating sequentially through layers, the input features undergo repeated transformations. The final outputs that emerge are expected to represent the target values. The deviation of the final output from the actual target is computed using a loss function. The weights are adjusted in all the layers to reduce the error. This is achieved by an algorithm that traverses backward through the network, starting from the output layer (François 2019). The weights are adjusted based on their contribution to error, applying the chain rule of derivatives.

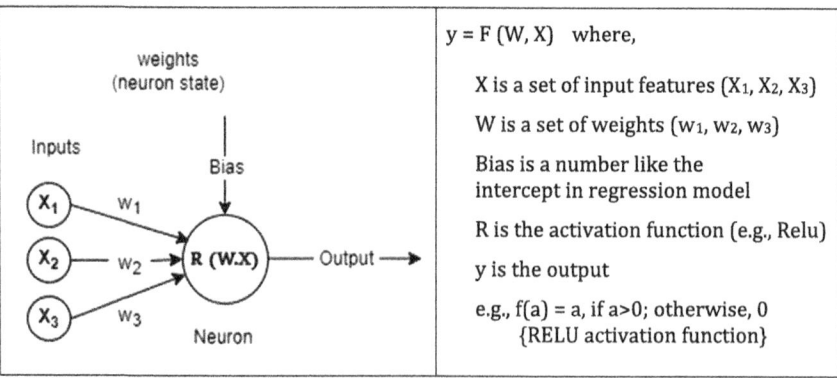

Fig. 12.3 Neuron, the building block of neural networks

The artificial neuron may be compared to the model of a biological neuron. In the biological neurons, the dendrites receive electrical signals (inputs) from the axons of other neurons. At the synapses between the dendrite and axons, electrical signals are modulated (weights are applied) in various amounts. The neuron fires when it reaches a threshold.

12.4.2   Activation Function

Determining a node's output in a neural network involves a two-step process. First, the system computes the weighted sum of the inputs assigned to that node. Note that these weights are dynamically adjusted through the learning process during training. This step develops the network's ability to assign varying importance to different inputs. In the second step, an activation function is applied to the weighted sum. The activation function is crucial in introducing nonlinearity to the model, allowing the neural network to learn and represent complex relationships within the data.

Several activation functions are commonly employed in neural networks, each serving specific purposes. The sigmoid function often produces an output distribution bounded by 0 and 1. The hyperbolic tangent function produces outputs between -1 and 1, aiding in mitigating the vanishing gradient problem. The SoftMax function is employed in the output layer for multi-class classification, providing a probability distribution over multiple classes. Finally, the rectified linear unit (ReLU) function, being computationally efficient, is commonly used for hidden layers, introducing nonlinearity by outputting the input for positive values and zero for negative values. ReLU and Sigmoid distributions are illustrated below (Fig. 12.4).

$$\text{Relu}(X) = \text{maximum}(0, X) \text{ see Fig. } 12.4(a)$$
$$\text{Sigmoid}(X) = 1/\left(1 + e^{-X}\right) - \text{ see Fig. } 12.4(b)$$

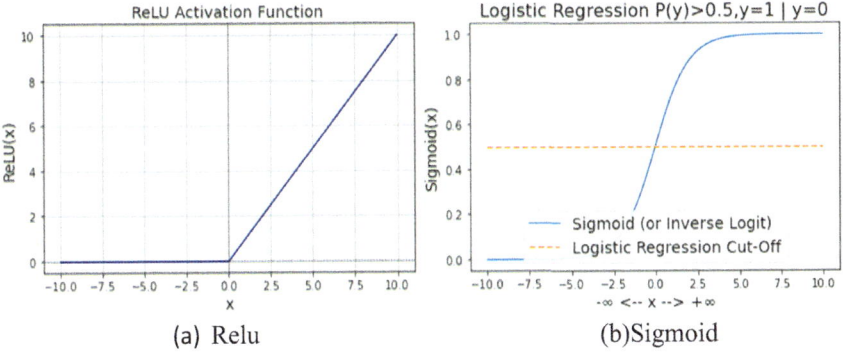

(a) Relu   (b)Sigmoid

Fig. 12.4   Activation functions—ReLU and Sigmoid

12.4.3 Network Layers

A deep learning neural network consists of multiple layers, each with a set of neurons. See Fig. 12.5. A layer other than the input layer consists of neurons that function as follows. Each neuron maintains its *current state* in the weight vector (W), accepts an *input* vector (X), and generates an *output (y)* using a nonlinear *activation function.*

12.4.4 The Learning Process

Three important features of deep learning are (a) the neural network architecture, (b) the learning method—gradient descent or its variants, and (c) the reduction of model complexity by using regularization techniques. In a neural network, learning involves finding a combination of **model parameters** that minimize a **loss function** for a given set of training data samples.

Let us assume that random batches of data samples are drawn and the **gradient** of the model parameters with respect to the loss (output—actual target)

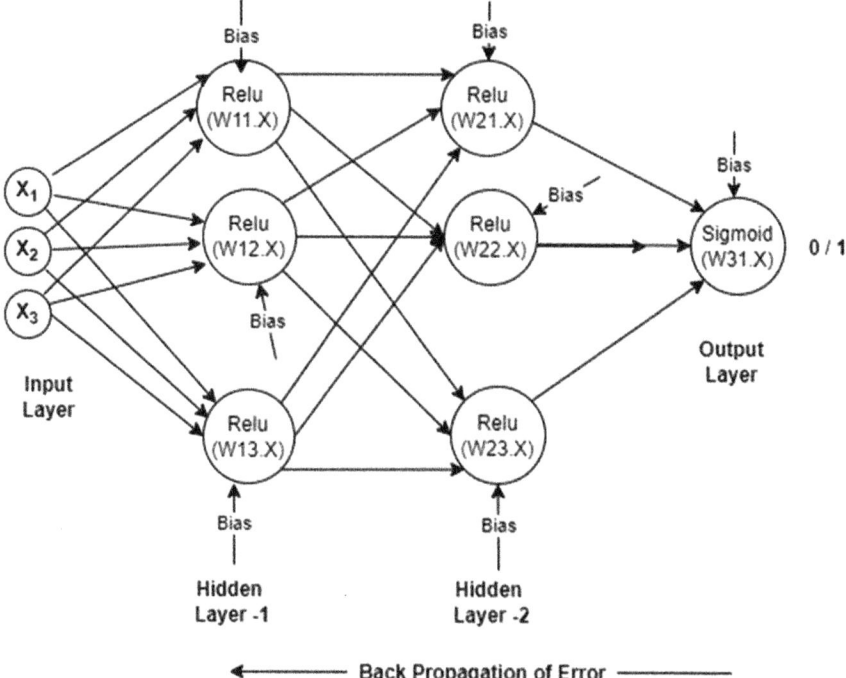

Fig. 12.5 Feed-forward network with backpropagation of error

is computed. The parameters are modified by a fraction defined by the **learning rate** in the direction opposite to the gradient. This process is repeated with several passes over the entire dataset, for convergence against set criteria. See Chap. 11 for a detailed description of the steps involved.

The learning process of ANN can be summarized as follows:

Assume, $y = W_0 + W_1X_1 + \cdots + W_kX_k$.

1. Draw a 'mini-batch' of training samples and the corresponding targets y.
2. Run the sample data through the network and obtain the predictions $\widehat{y}$.
3. Compute the loss $= (y - \widehat{y})^2$.
4. Compute the gradient of the loss with respect to W_i. This is done by a backward pass through each layer starting from the last layer and moving toward the first layer, in sequence. At any layer, the gradient is a set of partial derivatives of the loss $(y - \widehat{y})^2$ with respect to each weight W_i, i.e., $\delta(\text{loss function})/\delta(W_i)$.
5. Move the parameters (W_i) by a *step* in the opposite direction of the gradient to reduce the loss on the batch by a small step.

$$W_i = W_i - (step * \text{gradient}).$$

6. Repeat steps 1—5 until a termination criterion is reached. The criteria may be one or more of the following: the loss or mismatch is within an acceptable limit; there is no significant difference in the loss over a set of iterations; the number of iterations has reached a set limit; etc.

Neural networks are chains of differentiable tensor operations. This makes it possible to apply the **chain rule of derivation** to find the gradient function mapping the current parameters and current batch of data to a gradient value (Goodfellow et al. 2016). This forms the basis of learning in neural networks.

Assume the simple scenario of a purely sequential **feed-forward network** (Goodfellow et al. 2016). The input feature vector 'X' passes through the hidden layers sequentially to the output layer. In the learning process, we need to correct the prediction error, by minimizing the loss. Therefore, we propagate the error back from the output layer to the hidden layers backward in sequence to adjust the weights using partial derivatives in each node, in each layer. This is called **backpropagation** in a multi-layer neural network.

12.4.5 The Loss Function (Cost Function)

We discussed the loss function (or cost function) in Chap. 11. The loss function is a function of the error or the difference between the output that emerges from the network and the output predicted by the model. During training, our objective is to minimize prediction errors.

$$\text{Prediction error} = \frac{1}{n} \sum |y - \widehat{y}|$$

where

y is the observed value,
$\widehat{y}$ is the value predicted by the model,
n is the sample size.

$$\text{Loss Function} = \text{f(Prediction Error)}$$

Numerous loss functions, such as absolute error, squared error, and Huber loss, apply to regression models. The classification models use loss functions such as cross entropy, log loss, exponential loss, and hinge loss. Please check Chap. 11 for a detailed description.

12.4.6 Optimization Techniques (for Learning)

We had a discussion of optimization techniques in Chap. 11. Optimization techniques help to minimize the loss function or optimize the cost function (Géron 2019). Assume the following regression equation:

$$y = W_0 + W_1 X_1 + \cdots + W_k X_k$$

where k is the number of features.

The optimizer specifies how the gradient of the loss will be used to update parameters. The following are some of the commonly used optimizers (Goodfellow 2016):

- Gradient Descent.
- Adaptive learning rate methods—AdaGrad, RMSProp, Adam.

Gradient descent is an optimization algorithm that uses the negative gradient of an objective function to locate the minima (Chollet 2017). It uses the same learning rate (step size) for each feature (input variable). We had a detailed discussion on gradient descent in the last chapter. AdaGrad uses gradient descent. However, the step size (of change in the coefficients w_i) is different for each variable (X_i). The step is based on the partial derivative of the optimizing function with respect to that variable. The parameters with the larger partial derivatives of the loss are assigned a rapid decrease in their learning rate. RMSProp (root mean squared propagation) imbibes the properties of gradient descent and AdaGrad—it uses a decaying average of partial gradients to determine the step size for each variable. Adam is another adaptive learning rate optimization algorithm that is very popular. Adam stands for adaptive moments, an improvement over RMSProp. It uses stochastic gradient descent based on adaptive estimation of first-order and second-order moments. The gradient is different for each parameter. It takes advantage of momentum by using the moving average of the gradient (Chollet 2017).

12.4.7   Regularization Techniques (for Smoothening)

We had a detailed discussion of regularization techniques in Chap. 11. Regularization helps to prevent overfitting. Specialized techniques are used to avoid overfitting a deep neural network. These include penalizing large weights (coefficients or Wi's), randomly dropping some nodes each time we apply a step of gradient descent, and the use of a validation set to enable us to stop training when the loss on the validation set converges (Géron 2019).

12.4.8   NN Architectures, Training Challenges, and Transfer Learning

Neural network architectures form the foundation of deep learning models. Various architectures cater to different tasks (Goodfellow 2016). Convolutional Neural Networks (CNNs) excel in image-related tasks, Recurrent Neural Networks (RNNs) handle sequential data, and Transformers dominate natural language processing. Autoencoders and Generative Adversarial Networks (GANs) find applications in unsupervised learning and data generation, respectively. Choosing the right architecture depends on the nature of the data and the problem at hand. We will discuss some of the architectures subsequently.

Training neural networks poses several challenges. Vanishing and exploding gradients can impede convergence, especially in deep networks. Choosing appropriate weight initialization techniques, activation functions, and optimization algorithms addresses these issues. Overfitting is another challenge, prompting regularization methods such as dropout and weight decay. Selecting an optimal learning rate and managing computational resources effectively are additional challenges. Continuous experimentation and fine-tuning are often necessary to overcome these obstacles.

Optimization and weight initialization techniques, dropout, regularization, and batch normalization are critical components in training deep neural networks.

Xavier Initialization proposed in 2010 is designed to address the challenges of training deep neural networks with various activation functions, including sigmoid and hyperbolic tangent (tanh). Its objective is to keep the scale of the gradients roughly the same across all layers. Kaiming Initialization, proposed in 2015, is designed for ReLU activation functions. Both methods help to address the vanishing gradient problem associated with traditional weight initialization methods.

They are important tools for improving the convergence and performance of deep neural networks, and the choice between them often depends on the specific activation functions used in the network.

Dropout is a regularization technique where randomly selected neurons are ignored during training. It helps prevent overfitting by adding noise to the network. Regularization methods like L1 and L2 add penalty terms to the loss function, discouraging overly complex models. Batch normalization normalizes the inputs of each layer, reducing internal covariate shift and potentially accelerating training.

Transfer learning leverages pre-trained models to enhance performance on a specific task. Models trained on large datasets, such as ImageNet, can serve as feature extractors for tasks with limited labeled data. Fine-tuning allows adapting pre-trained models to task-specific nuances. Domain adaptation extends transfer learning to different but related domains, addressing the challenge of domain shift. This approach is particularly beneficial when training a deep model from scratch is impractical due to limited data or computational resources.

12.4.9 Keras and TensorFlow

Keras, a deep learning API implemented in Python, operates seamlessly on the TensorFlow machine learning platform. It was designed to facilitate rapid experimentation. TensorFlow, an end-to-end open-source machine learning platform, originated from the Google Brain Team's efforts, utilizing Python, C++, and CUDA. Its initial release occurred in 2015, with CUDA being a parallel computing platform and programming model by NVIDIA for general GPU-based computing.

TensorFlow simplifies the process of creating machine learning models for both novices and experts. Keras, operating as a high-level API, is built on top of TensorFlow and supports multiple backends, including TensorFlow and Theano. Originating from MIT, Keras emphasizes expeditious experimentation in deep learning. Moreover, Keras provides a versatile application library containing various pre-built deep learning models. These models are invaluable for feature extraction and prediction tasks, equipped with pre-trained weights that offer the flexibility for fine-tuning to suit specific requirements.

12.5 Simple Feed-Forward Neural Network (FFNN)

A feed-forward neural network is a simple multilayered neural network architecture. The input data moves through hidden layers to the output layer in one direction. The layers do not give feedback to any previous layers. The connections between nodes do not form a cycle at any stage. For example, see Fig. 12.5.

12.5.1   FFNN—Basic Operation

Let us elaborate on the neural network shown in Fig. 12.5, including weights associated with each node. The network architecture is shown in Fig. 12.6 and described below. There are three input variables—X_1, X_2, and X_3. Each node has an additional input—the Bias—a number, similar to the intercept in the regression model.

There are two hidden layers—hidden layer-1 and hidden layer-2. Each of these hidden layers has three nodes each. The hidden layers use ReLU activation functions, represented, by Relu (W, X). Take a look at the weight vectors w_{LNF}. Here L indicates the hidden layer, N indicates the node, and F indicates the feature. The weights differ from feature to feature, node to node, and layer to layer.

The ReLU activation function generates exactly one output. In the output layer, the activation function used is sigmoid. In the output layer, sigmoid ($W_{31} \cdot X$) represents the sigmoid function applied to the dot product of the feature vector X and the corresponding weight vector W, which generates a binary output {0, 1}.

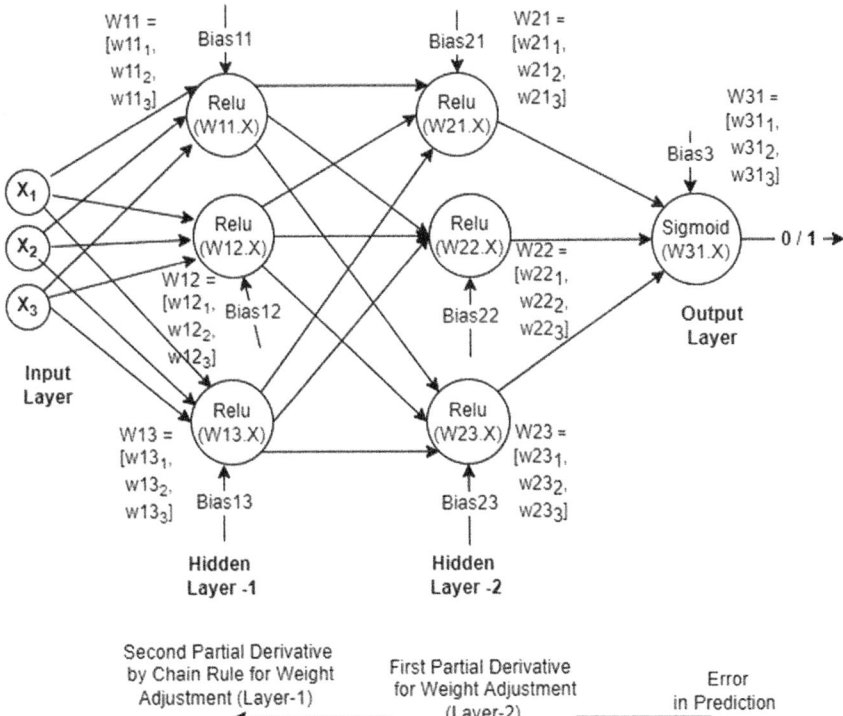

Fig. 12.6 Feed-forward neural network with weight vectors

The Operations at the Hidden Layer-1, Node-1

Look at hidden layer-1, node-1, which shows Relu(W11, X). Here, the operations involved are

- Take the dot product of feature vector X with the corresponding weight vector W11.

$$W11 \cdot X = W11_1 X_1 + W11_2 X_2 + W11_3 X_3$$

- Add the Bias term, Bias11.
- Apply ReLU function.

The operations can be summarized as follows:

$$\text{Relu} ([W11_1 X_1 + W11_2 X_2 + W11_3 X_3] + Bias11)$$

The Operations at the Output Layer

In the output layer, we have sigmoid activation function, which generates a binary outcome 0 or 1. The operations can be summarized as follows:

- Sigmoid (W31, X): Function Description
- Sigmoid (W31.X + Bias3): Dot Product W.X + Bias
- Sigmoid ($[W31_1 X_1 + W31_2 X_2 + W31_3 X_3] + Bias3$)

12.5.2 An FFNN for Diamond Price Prediction

Let us demonstrate a simple multilayered neural network, for predicting the price of diamonds. We are going to use the 'diamonds' dataset. The major steps to be followed in the tutorial are listed below:

1. Data preprocessing/setup.
2. Setup neural network architecture.
3. Train the neural network.
4. Measure the model performance.

1. Data Preprocessing/Setup

We consider nine features of diamonds for price prediction. A regression model requires numeric variables. Three of the features, cut, color, and clarity are categories. We will re-code them as integers to the variables cuti, colori, and clarity. The features that we choose as input to the neural network are (X1 ... X9),

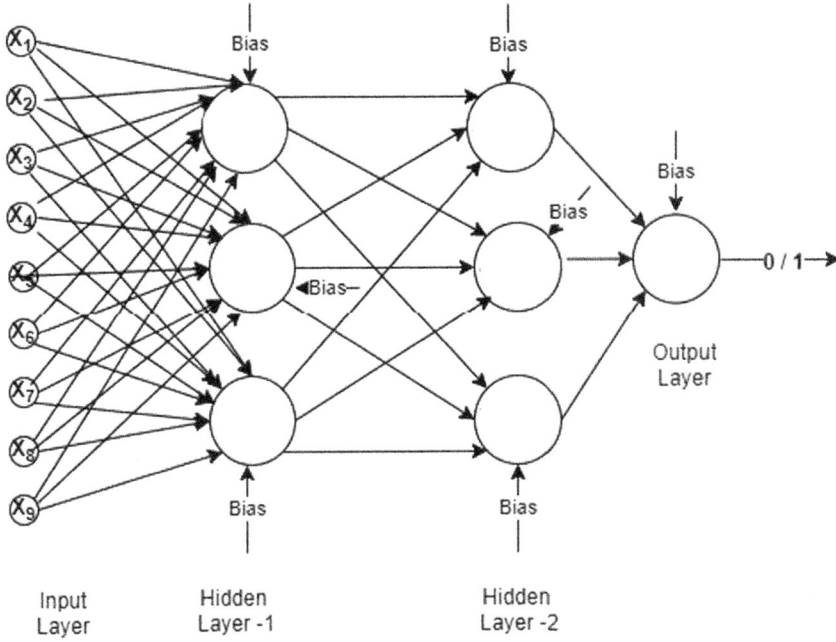

Fig. 12.7 A neural network for diamond price prediction

indicating 'carati', 'cuti', 'colori', 'clarityi', 'depth', 'table', 'x', 'y', 'z'. We will standardize these features, as gradient descent is sensitive to scale. Diamond price (y) is categorized into binary categories {0, 1} into variable pCat. Here, 0 stands for 'low-price' and '1' for 'high-price'. Figure 12.7 shows a neural network model for diamond price prediction (low or high), based on the above nine input features.

2. **Setup Neural Network Architecture**

Start with setting a random seed. This helps in the reproducibility of results on repeated trials. Set the count of input features—we have nine features. Build the neural network. Compile the neural network. The neural network architecture summary is shown in Table 12.1. The summary shows that 46 parameters are required. This implies that the network needs 46 variables for storing the weights and bias terms.

The layers in the neural network architecture are listed below:

- Add a fully connected layer with a ReLU activation function.
- Add a fully connected layer with a ReLU activation function.
- Add a fully connected layer with a sigmoid activation function in the last layer to facilitate binary classification.

Table 12.1 Neural network architecture summary

Layer	Nodes per layer	Inputs/node	Bias/node	Parameter description	Parameter count	Outputs/node
Hidden layer-1	3	9	1	3 x (9 + 1)	30	3
Hidden layer-2	3	3	1	3 x (3 + 1)	12	3
Output layer	1	3	1	1 x (3 + 1)	4	1
Total					46	

3. **Train the Neural Network**

   - Train the neural network.
   - Summarize the training history for accuracy (plot).
   - Summarize the training history for loss (plot).

4. **Measure the Model Performance**

   - Predict the classes over the entire dataset.
   - Print Confusion Matrix.
   - Print Classification Report.

Tutorial 12.5   A Neural Network for Diamond Price Category Prediction

Develop a multilayered feed-forward neural network for predicting the price of diamonds. The dataset diamonds is described in Chapter 1.

Let us consider nine features of diamonds, for price prediction. A regression model requires numeric input variables. Three of the features, cut, color, and clarity are categories. We will re-code them as integers into the variables cuti, colori, and clarityi. The features that we choose as input to the neural network are (X1 .. X9), indicating 'carati', 'cuti', 'colori', 'clarityi', 'depth', 'table', 'x', 'y', 'z'. We will standardize these features, as gradient descent is sensitive to scale.

Diamond price (y) is categorized into binary categories {0, 1} into variable pCat. Here, 0 stands for 'low-price' and '1' stands for 'high-price'.

Tutorial 12.5.1 Diamond Price - Data Preprocessing / Setup

```
import seaborn as sb
import numpy as np
from sklearn.preprocessing import StandardScaler
from sklearn.model_selection import train_test_split
```

Read Diamonds Data Set from seaborn Library
```
d = sb.load_dataset('diamonds')
d = d.dropna() # drop rows with null values
```

Recode Categorical Input Features to Integer

```
d['cuti'] = d.cut.astype("category").cat.codes
d['colori'] = d.color.astype("category").cat.codes
d['clarityi'] = d.clarity.astype("category").cat.codes
```

Choose the Input Features

```
X = d[['carat', 'cuti', 'colori', 'clarityi',
       'depth','table', 'x', 'y', 'z']]
```

Standardize Features

```
Xz = StandardScaler().fit_transform(X)
X.columns

d['pCat'] = np.zeros(d.shape[0])
for i in d.index:
    if    d.loc[i, 'price'] < 2000: d.loc[i, 'pCat'] = 0
    else: d.loc[i, 'pCat'] = 1
```

Setup target data - y, as binary categorical variable (pCat = 0,1)

```
y = d.pCat       # Target Vector
y[y==0].count() # Label-0: 24203
y[y==1].count() # Label-1: 29737
```

Test - Train Split; 80% Train, 20% test

```
X_train, X_test, y_train, y_test = train_test_split(
        Xz, y, train_size= 0.8, random_state=1)
```

Tutorial 12.5.2 Diamond Price - Setup Neural Network Architecture

Take a look at Figure 12-7: A Neural Network for Diamond Price Prediction

Load libraries

```
from keras import models
from keras import layers
```

Random seed - helps in the reproducibility of results on repeated trials

```
np.random.seed(0)
```

Set the count of input features

```
input_features_count = 9
```

Sequential class: groups a linear stack of layers into a tf.keras.Model.

```
ann = models.Sequential()
```

Add a fully connected layer with a ReLU activation function. Fully connected or 'Dense' layer, implies that, each input to the layer is connected to every node in that layer
In the first hidden layer, we must specify the input feature dimensions

```
ann.add(layers.Dense(units=3, activation="relu", input_shape=(
        input_features_count,)))
```

Add another fully connected layer with a ReLU activation function
```
ann.add(layers.Dense(units=3, activation="relu"))
```

Add fully connected layer with a sigmoid activation function in the last
layer, to facilitate binary classification
```
ann.add(layers.Dense(units=1, activation="sigmoid"))
```

Compile the Neural Network
With Root Mean Square Err Propagation, and Accuracy as a performance metric;
and Binary_crossentropy (a binary classifier).
```
ann.compile(loss="binary_crossentropy",
            optimizer="rmsprop",
            metrics=["accuracy"])
```

```
ann.summary()
Layer (type)    OutputShape Parameters Remarks - Weights
dense_1 (Dense) (None, 3)      30        3 Nodes x (9 inputs + 1 Bias Term)
dense_2 (Dense) (None, 3)      12        3 Nodes x (3 inputs + 1 Bias Term)
dense_3 (Dense) (None, 1)       4        1 Nodes x (3 inputs + 1 Bias Term)
Total params: 46
See Table 12-1 for details
```

Tutorial 12.5.3 Diamond Price - Train the Neural Network

Train Neural Network
```
history = ann.fit(X_train, # Features
    y_train,         # Target vector
    epochs=5,        # Number of epochs
    verbose=2,       # Describe epoch
    batch_size=100, # Number of observations per batch
    validation_data=(X_test, y_test)) # Test data
```

```
Epoch 1/5 - 2s - loss:0.431 - acc:0.852 - val_loss:0.307 - val_acc: 0.947
Epoch 2/5 - 1s - loss:0.266 - acc:0.954 - val_loss:0.220 - val_acc: 0.962
Epoch 3/5 - 1s - loss:0.198 - acc:0.964 - val_loss:0.169 - val_acc: 0.967
Epoch 4/5 - 1s - loss:0.156 - acc:0.968 - val_loss:0.138 - val_acc: 0.968
Epoch 5/5 - 1s - loss:0.129 - acc:0.968 - val_loss:0.117 - val_acc: 0.970
```

Let us get familiar with history variable
```
print(history.history.keys())
# dict_keys(['loss', 'accuracy', 'val_loss', 'val_accuracy'])
```

Plot the summary of training history, for accuracy, over the epochs
```
import matplotlib.pyplot as plt
plt.title('model accuracy', fontsize=16)
plt.plot(history.history['accuracy'])
plt.plot(history.history['val_accuracy'])
plt.ylabel('accuracy', fontsize=16)
```

```
    plt.yticks(np.arange(0.85, 1, 0.05), fontsize=16)
    plt.xlabel('epoch', fontsize=16)
    plt.xticks(np.arange(0, 6), fontsize=16)
    plt.legend(['train', 'test'], loc='best', fontsize=16)
    plt.show()
    # See Figure 12-8 (a): ANN Iterations Performance- Accuracy and Loss
```

Plot the summary of training history, for loss, over the epochs
```
    import matplotlib.pyplot as plt
    plt.title('train/test performance convergence', fontsize=16)
    plt.plot(history.history['loss'])
    plt.plot(history.history['val_loss'])
    plt.ylabel('loss', fontsize=16)
    plt.yticks(np.arange(0.05, 0.61, 0.1), fontsize=16)
    plt.xlabel('epoch', fontsize=16)
    plt.xticks(np.arange(0, 6), fontsize=16)
    plt.legend(['train', 'test'], loc='best', fontsize=16)
    plt.show()
```
See Figure 12-8 (b): ANN Iterations Performance- Accuracy and Loss

Tutorial 12.5.4 Diamond Price - Measure the Model Performance

```
    from sklearn.metrics import confusion_matrix, classification_report
```

Display the Confusion Matrix
```
    y_pred = ann.predict(Xz)
    print('Confusion Matrix')
    print(confusion_matrix(y, np.round(y_pred,0)))
Confusion Matrix
[[23510   693]
 [  945 28792]]
```

Display the Classification Report
```
    print('Classification Report')
    category_labels = ['low-price', 'high-price']
    print(classification_report(y, np.round(y_pred,0),
                        target_names = category_labels))
Classification Report
             precision    recall  f1-score   support
  low-price       0.96      0.97      0.97     24203
 high-price       0.98      0.97      0.97     29737
avg / total       0.97      0.97      0.97     53940
```

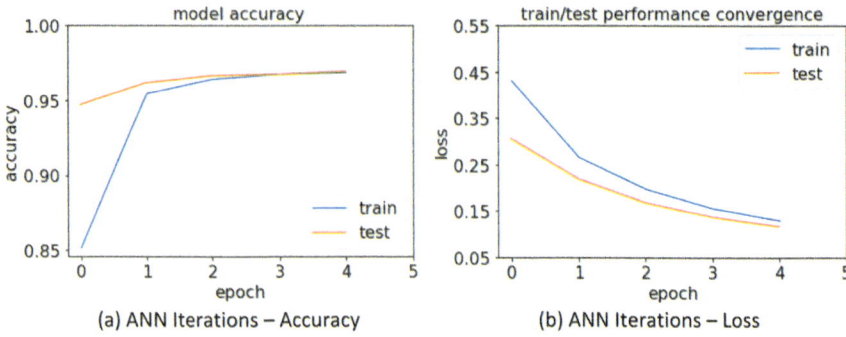

Fig. 12.8 ANN iterations performance—accuracy and loss

12.6   Convolutional Neural Network (CNN)

CNN is a special case of a feed-forward neural network suitable for image processing. CNN consists of the following segments: feature extraction and classification. Figure 12.9 shows a simple CNN architecture for image detection, demonstrated in the Tutorial 12.2. Prominent CNN architectures include LeNet-5, AlexNet, VGGNet, GoogLeNet (Inception), ResNet, and DenseNet.

LeNet-5 (year 1998) consists of seven layers, including three convolutional layers and two fully connected layers. The layers are arranged sequentially: C1 (convolutional), S2 (subsampling or pooling), C3 (convolutional), S4 (subsampling or pooling), C5 (convolutional), F6 (fully connected), and the output layer. Sigmoid activation functions are used in the convolutional layers. The output layer used SoftMax function.

AlexNet (year 2012) consists of eight layers, including five convolutional layers and three fully connected layers. It was one of the first deep neural networks to

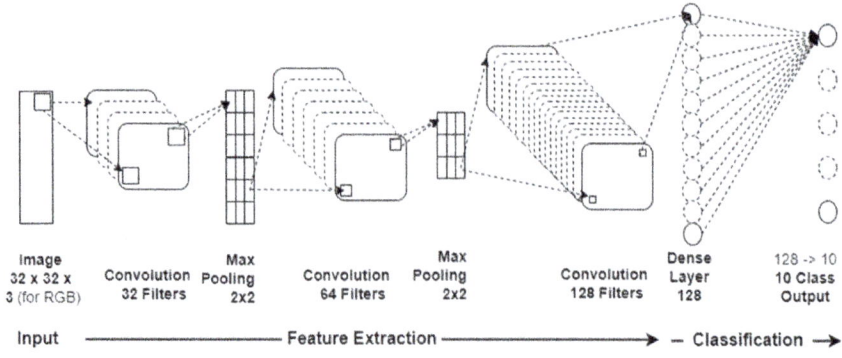

Fig. 12.9 CNN architecture

utilize many parameters, contributing to its success. Rectified Linear Unit (ReLU) activation functions are used after each convolutional and fully connected layer. ReLU helped address the vanishing gradient problem and accelerated the training of deep neural networks.

VGGNet uses small 3×3 convolutional filters, with multiple stacked layers throughout its architecture. GoogLeNet is characterized by its deep and wide architecture, with multiple inception modules stacked on each other. ResNet's key feature is the residual block, which allows gradients to flow more easily during training, enabling the training of extremely deep networks.

Convolution—Basic Concept

Feature detection is a challenging task in image processing. There are various types of features—edges, corners, ridges, and blobs/regions. Small contiguous areas often describe these features. The recognition of features such as edges does not depend on their location within the edge. Convolution makes use of these concepts.

Convolution is applying a filter (F) iteratively over the entire pixels of an image (I). Mathematically, a convolution is an integral that expresses the amount of overlap of one function 'F' (filter) as it is shifted over another function 'I' (image) (Leskovec et al. 2020). A filter scans the image similar to raster scan. The filter may move in steps of one or more pixels, left to right (across columns) and top to bottom (across rows). These steps are called strides.

A filter (F) carries a set of weights learned using the backpropagation algorithm. A filter can be considered as storing a single pattern. When we convolve this filter across the corresponding input, we find the degree of similarity between the filter and different locations in the input. This principle can be used for edge detection.

In CNN, several filters carrying different patterns are applied over the entire image. Initial convolutional layers help to detect simple features like the sections of edges. The subsequent layers help progressively assemble complex structures, such as legs or eyes.

12.6.1 Convolution

Let us consider an example shown in Table 12.2. The input image 6×6 is shown in Table 12.2a; the filter (or kernel) vector F [1, 0, 0, −1] in Table 12.2b, and the output from convolution in Table 12.2c. Table 12.2d shows three examples of convolution operation.

Let us consider the convolution operation on image segment B [0, 1, 0, 1]. It is computed as the dot product of these vectors B (image segment) and F (the convolution filter). This is shown in Table 12.2d as follows:

$$Q = B.F,$$

Table 12.2   Applying 2×2 filter to an image 6×6, Stride 1

| (a) Image 6x6 | (b) Filter | (c) Output 5x5 | (d) Convolution operation |

where Q is the result of convolution of B and F

$B = [0,1,0,1]$

$F = [1, 0, 0, -1]$

$Q = B.F = [0,1,0,1] . [1, 0, 0, -1] = 0 \times 1 + 1 \times 0 + 0 \times 0 + 1 \times (-1) = -1.$

See Table 12.2a. Consider an image 'A' measuring 6×6. Assume a window 'F' with dimensions (2×2) sliding horizontally across the image. It moves from left to right over the first row and continues this process for each subsequent row. This will result in the output P, which has $5 \times 5 = 25$ windows. See Table 12.2c.

$$A (6 \times 6) - > P (5 \times 5).$$

So, the convolution operation on an input of 6×6 results in an output of 5×5.

Note that, in practice, the entire set of 25 convolutions can be done in parallel.

In the above discussion, we assumed that we slide the window in strides of one step. However, in practice, the stride (s) can be a different number. If the stride $= 1$, an image of 6×6 will convolve into 5×5.

If the stride is 2, an image of 6×6 will convolve into 3×3. See Table 12.3. In this case, the filter moves 2 pixels at a time, from left to right and top to bottom. The 2×2 filter makes nine movements to cover the 6×6 image. The filter positions are shown by color—violet, green-blue, blue, green, yellow, light orange, dark orange, red, and finally shaded white, in order. This results in mapping 6×6 input to 3×3 output as shown below:

$$A (6 \times 6)^{stride=2} - > P (3 \times 3).$$

Table 12.3   Applying 2×2 filter to an image 6×6, Stride 2

| (a) Filter on Row-1 | (b) Filter on Row-2 | (c) Filter on Row-3 | (d) Output Image (A->P) |
| Row 1: 3 windows | Row 2: 3 windows | Row 3: 3 windows | Convolved Image 3x3 |

As we saw, for an input image A of size (6×6), the output is P(5×5). To maintain the image size (m) constant over the convolution, we may add extra columns and rows of zeros. This is called **zero-padding**. For a padding size of p, we add 'p' rows at the top and bottom each, and 'p' columns at the left and right.

Let us summarize our discussion.

Assume that the input image is of size $m \times m$ pixels, the output is $n \times n$, the filter size is (f x f), the stride is s, and padding is p. The output size 'n' can be expressed by the following formula:

$$n = (m - f + 2p)/s + 1 \tag{12.1}$$

Assume that we use k filters (and that we constrain all filters to have the same size, stride, and padding). Then the output contains k activation maps. The output layer will be.

$$n \times n \times k \text{ where the formula for n is given in Eq. (12.1)}$$

In the above discussion, we have assigned {0, 1} as the possible values of a pixel of the input image, assuming black/white. A common format for an image pixel is a set of three values representing the color channels R, G, B, the values of R, G, or B varying from 0 … 255. Considering (RGB), a 2×2 filter will have a dimension $2 \times 2 \times 3$. The input image and the output will both have one more dimension included—input $(6 \times 6 \times 3)$ and output $(5 \times 5 \times 3)$. Considering RGB, the output layer will be

$$n \times n \times k \times 3$$

12.6.2 Feature Extraction

The feature extraction segment consists of a large number of convolutional layers, followed by pooled layers. A simple example is given below:

- The input layer feeds the feature matrix, e.g., a handwritten image represented by [32 × 32] pixel matrix.
- In the convolutional layer, a set of filters (e.g., 3×3 window of pixels; $f = 3$) slide through the input feature matrix in strides of size 1 ($s = 1$); convolve the window and transform it using a nonlinear function (ReLU). We apply multiple filters and take several convolutions at a time (possibly in parallel), from which the vital ones are identified and given appropriate weightage during the training process.
- In the maxpooling layer, the data after convolution are divided into small segments (e.g., 2×2 window of pixels), and the maximum value in the segment is chosen as the representation of the segment.

12.6.3   Classification

This layer receives the extracted features and flattens them into fewer dimensions. Then it uses a fully connected neural network for classification (e.g., using SoftMax multi-class activation function). The output will be the class labels.

Tutorial 12.2   CIFAR-10 image feature extraction and classification

The CIFAR-10 dataset consists of 60,000 32×32 color images in 10 classes, with 6000 images per class. There are 50,000 training images and 10,000 test images. The image classes are airplane, automobile, bird, cat, deer, dog, frog, horse, ship, and truck (Refer: https://www.cs.toronto.edu/~kriz/cifar.html). Develop a convolutional neural network to extract the features and classify the images. Note: Fig. 12.9 shows the CNN architecture for CIFAR-10 image feature extraction and classification.

Tutorial 12.6   CIFAR-10 image feature extraction and classification

Tutorial 12.6.1 Data Setup

The following code was run on google colab.

```
from tensorflow.keras import datasets, layers, models, losses
import matplotlib.pyplot as plt
```

Download and prepare the CIFAR10 dataset. The dataset contains 60,000 color images in 10 classes. Data loading may take time
```
(X_train, y_train), (X_test, y_test) = datasets.cifar10.load_data()
```
Image shape X(32, 32, 3): (image_height, image_width, color_channels). Color_channels have 3 components (R,G,B), each with a value 0..255

Get familiar with the data
```
X_train[0].shape   # (32, 32, 3)    one image
X_train[0][0][0]   # [59, 62, 63]  (R,G,B) values of one pixel of the
first image
y_train[0].shape   # (1,)
y_train[0][0]      # (6) the target label of one object
# target labels are 0..9; corresponding names are given below
target_labels = ['airplane', 'automobile', 'bird', 'cat', 'deer',
                 'dog', 'frog', 'horse', 'ship', 'truck']
```

Normalize pixel values to be within 0 and 1
```
X_train, X_test = X_train / 255.0, X_test / 255.0
```

```
Display 12 sample images # See Figure 12-10
    plt.figure(figsize=(12,12))
    for i in range(12):
        plt.subplot(6,6,i+1)
        plt.xticks([])
        plt.yticks([])
        plt.grid(False)
        plt.imshow(X_train[i])
        plt.xlabel(target_labels[y_train[i][0]])
    plt.show()
    See Figure 12-10 Sample Images
```

Tutorial 12.6.2 Build the CNN Model - Feature Extraction

Typically, CNN is a sequential model built using a linear stack of layers
```
    model = models.Sequential()
```

Convolution: Add 32 filters, with kernel size = (f,f) = (3,3)
Choose the activation function relu. The input image is 32x32 pixels. Each pixel has 3 data points to represent color in RGB. So the input size is 32x32x3
```
    model.add(layers.Conv2D(32, (3, 3), activation='relu',
    input_shape=(32, 32, 3)))
```
Calculating the Image Size After Convolution
```
    m = 32 (input image is 32x32)
    f = 3  (3x3 filter)
    p = 0  (no padding)
    s = 1  (stride by 1 pixel)
    output shape is n.n
    n = (m - f + 2p)/s + 1 = (32 - 3 + 2x0)/1 + 1 = 30 (equation 12.1)
    Output image from one filter after convolution (n, n) = (30, 30)
```
Calculating the number of parameters in the Convolution layer

Fig. 12.10 Sample images

```
     Filter size = (3x3)
     Filter size including colour channel (R,G,B) =  = 3 x (3x3)
     Bias = 1
     Parameters per filter = 3 x (3x3) + 1 = 27 + 1 =28
     Number of filters  (current layer) = 32
     Total number of parameters = 28 x 32 = 896
```

Max pooling, with pool size = (w,w) = (2,2); padding = 0; stride = 1
   model.add(layers.MaxPooling2D((2, 2)))
Output image after max-pooling (n/w, n/w) = (30/2, 30/2) = (15, 15).
Max pooling does have any parameters, as it is a plain arithmetic

Convolution: Add 64 filters, with kernel size = (f,f) = (3,3)
   model.add(layers.Conv2D(64, (3, 3), activation='relu'))
Calculating the Image Size After Convolution
     m = 15 (input image is 15x15, from the previous maxpooling layer)
     f = 3 (3x3 filter)
     p = 0 (no padding)
     s = 1 (stride by 1 pixel)
     Output shape is nxn
     n = (m - f + 2p)/s + 1 = (15 - 3 + 2x0)/1 + 1 = 13 (equation 12.1)
     Output image from one filter after convolution (n, n) = (13, 13)
     Number of filters = 64
     Total number of parameters = (13, 13) x 64 = 18496
Calculating the number of parameters in the Convolution layer
     Number of filters from previous layer = 32
     Filter size = (3x3)
     Bias = 1
     Parameters per filter = 32 x (3x3) + 1 = 288 + 1 =289
     Number of filters (current layer) = 64
     Total number of parameters = 289 x 64 = 18496

Max pooling, with pool size = (w,w) = (2,2); padding = 0; stride = 1
   model.add(layers.MaxPooling2D((2, 2)))
Output image after max-pooling (n/w, n/w) = (13/2, 13/2) = (6, 6)
Max pooling does have any parameters, as it is a plain arithmetic

Convolution: Add 64 filters, with kernel size = (f,f) = (3,3)
   model.add(layers.Conv2D(128, (3, 3), activation='relu'))
Calculating the Image Size After Convolution
     m = 6 (input image is 15x15, from the previous maxpooling layer)
     f = 3 (3x3 filter)
     p = 0 (no padding)
     s = 1 (stride by 1 pixel)
     Output shape is nxn
     n = (m - f + 2p)/s + 1 = (6 - 3 + 2x0)/1 + 1 = 4 (equation 12.1)
     Output image from one filter after convolution (n, n) = (4, 4)
     Number of filters = 128

Calculating the number of parameters in the Convolution layer
    Number of filters from previous layer = 64
    Filter size = (3x3)
    Bias = 1
    Parameters per filter = 64 x (3x3) + 1 = 576 + 1 = 577
    Number of filters (current layer) = 128
    Total number of parameters = 577 x 128 = 73856

Flatten the 3D array to 1D
    model.add(layers.Flatten())
Output from the previous convolution layer = (4, 4, 128)
Output of flatten = 4 x 4 x 128 = 2048
Flatten just reshapes the array. So, there are no parameters

You may use dropout, a technique to prevent overfitting, after every convolution layer. Randomly selected neurons are ignored during training only. The
% of the nodes to ignore is specified as a parameter.
Dropout 20% of the nodes at random.
    model.add(layers.Dropout(0.2))

Tutorial 12.6.3 Build the CNN Model - Classification

Add Dense layers
    model.add(layers.Dense(128, activation='relu'))
    Output of dense layer = 128
Output from the previous flatten layer = 2048
Bias = 1
Dense layer: each node in dense layer will receive 2048 inputs + 1 bias
Therefore, parameters per node, in the dense layer = 2048 +1
Number of nodes in the dense layer = 128
Total number of parameters = (2048+1) * 128 = 262272

The images has 10 labels. So we will use a final Dense layer with 10 outputs
    model.add(layers.Dense(10))

Output of current dense layer = 10
Output from the previous dense layer = 128
Bias = 1
Dense connection: each of the 10 nodes will receive 128 inputs + 1 bias
Therefore, parameters per node, in the dense layer = 128 +1
Total number of parameters = (128+1) * 10 = 1290

```
    model.summary()
Layer (type)                        Output Shape       Param #
-----------------------------------------------------------
conv2d (Conv2D)                     (None, 30, 30, 32)      896
max_pooling2d (MaxPooling2D)        (None, 15, 15, 32)        0
conv2d_1 (Conv2D)                   (None, 13, 13, 64)    18496
max_pooling2d_1 (MaxPooling2D)      (None, 6, 6, 64)          0
conv2d_2 (Conv2D)                   (None, 4, 4, 128)     73856
flatten (Flatten)                   (None, 2048)              0
dense (Dense)                       (None, 128)          262272
dense_1 (Dense)                     (None, 10)             1290
-----------------------------------------------------------
Total params:                                          356,810
Trainable params:                                      356,810
Non-trainable params:                                        0
```

Tutorial 12.6.4 Compile and Run the Model

Compile the CNN model. Use, Optimizer: adam; Loss function for classification: cross entropy

```
    model.compile(optimizer='adam',
        loss=losses.SparseCategoricalCrossentropy(from_logits=True),
         metrics=['acc'])
```
Run the CNN model for 5 epochs
```
    run_history = model.fit(X_train, y_train, epochs=5,
                        validation_data=(X_test, y_test))
```
Save the CNN model for future use
```
    model.save('Cifar10_CNN_Model_version01')
```
Load the saved CNN model
```
    from keras.models import load_model
    saved_model=load_model('Cifar10_CNN_Model_version01')
```

Tutorial 12.6.5 Plot Train-Test Convergence History

```
    plt.plot(run_history.history['acc'],
            label='model accuracy on training dataset',linestyle='--')
    plt.plot(run_history.history['val_acc'],
            label = 'model acuracy on test dataset',linestyle='-.')
    plt.xlabel('Epoch',fontsize=14)
    plt.ylabel('Accuracy',fontsize=14)
    plt.ylim([0.5, 1])
    plt.legend(fontsize=14)
    See Figure 12-11: Train-Test Convergence
```

Fig. 12.11 The cross-validation history

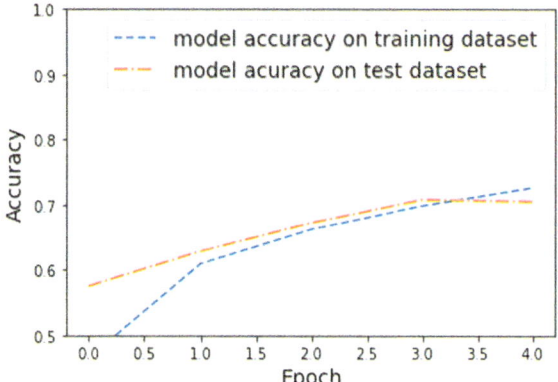

12.7   Recurrent Neural Network (RNN) and LSTM

A recurrent neural network (RNN) is a feedback neural network, unlike CNN. RNN is also known as auto-associative network. RNN is designed to recognize sequences, such as the sentences in a natural language which is composed of sequences of words, or the stock market position of a company, which is composed of a series of stock prices.

The LSTM model is a refinement RNN for learning long-distance associations. LSTM can save selected information into the memory, forget information by purging it, and focus on memory that is relevant to the context.

12.7.1   RNN

RNN is commonly used for natural language processing (Leskovec et al. 2020). Consider a sentence in the English language. After processing the initial set of words in a sentence ('prefix'), we may predict the next word in the sentence. The next word can be drawn from a set of words with probabilities associated with each word.

RNN is a special type of neural network. Figure 12.12 depicts RNN architecture. It has three types of layers: the input layer x (e.g., a set of words in a sentence), the hidden layer h, and the output layer o. The instance of time is indicated by the superscript (t). U, V, and W are weight vectors, and the hidden layer consists of an activation function (e.g., sigmoid).

RNN takes a set of words as input. Assume a fixed-length sequence of n words, x_1, x_2,..., x_n as input to the RNN. Each input word is usually one hot encoded (OHE), as a vector of size equal to the number of words in our dictionary. The output is also a sequence of words o_1, o_2,..., o_n. Each output o_i is a vector of words with probabilities assigned to them, from which we predict the next word in the

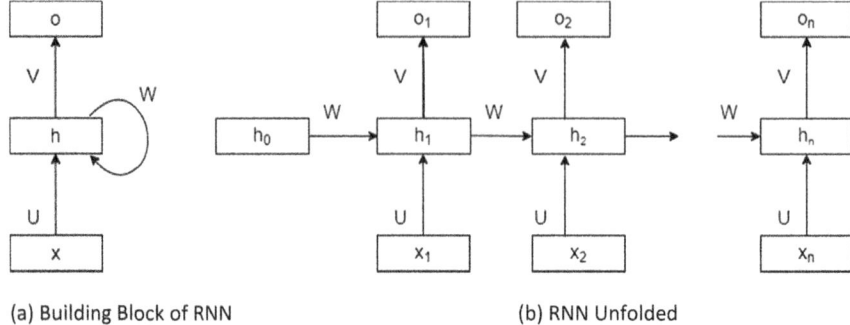

(a) Building Block of RNN   (b) RNN Unfolded

Fig. 12.12 RNN architecture

sentence. At each step t, we have the memory vector, h_t, generated in the hidden layer, based on the sequence of words it has already observed, and the input at time t, x_t. The memory vector can be expressed as

$$h_t = f(Wh_{t-1} + UX_t + b_h)$$

where

f may be a sigmoid function with output (0 ... 1) or tanh function with output $\{-1 ... +1\}$,

$W \text{ and } U$ are weight matrices,

h_{t-1} is the current hidden state,

$X_{t,}$ is the input,

b_h is the bias vector.

Similarly, the output at time t is

$$o_t = g(Vh_t + b_0)$$

where

g could be a SoftMax function to generate a vector of probabilities that sums up to 1,

V is the weight matrix,

h_t is the new hidden state,

b_o is the bias vector.

The main differences between a general neural network and RNN can be summarized as follows:

- The output at each point depends on the entire prefix of the sentence until that point. Therefore, the network needs to retain the prefixing words in memory.
- A language follows a given grammar, and it does not change across the words in the sentence. Therefore model parameters (weights—U, V, W) are the same across the nodes.

12.7.2   Long Short-Term Memory (LSTM)

LSTM is an evolution of RNN, capable of learning and remembering dependencies in a sequence and using that for prediction (Leskovec et al. 2020). This is useful for applications such as language translation and speech recognition. For example, consider Google's predictive search feature to predict a user's search query. If someone types 'what is pop ...', it may be suffixed by 'corn' (what is popcorn) or 'music' (what is pop music). LSTM applications are very much beneficial in such contexts.

The LSTM model is organized as a chain structure with three gates—the forget gate, input gate, and output gate. A typical LSTM network comprises memory blocks called cells. In each instance of time t, two states will be computed and transferred to the next cell—the hidden state (h_t) corresponding to the working memory and the cell state (c_t) corresponding to the long-term memory content. Sigmoid gates add (new) data or remove (old) data. These concepts are described below with the help of Fig. 12.13 LSTM architecture.

Current State and Inputs

Current cell state	c_{t-1} {Current Long Memory}
Current hidden state	h_{t-1}{Current Working Memory}
Input	x_t
Weight matrix at forget gate, input gate, and output gate	W {W_f, W_i, W_o}
Bias, at forget gate, input gate, and output gate	b {b_f, b_i, b_o}

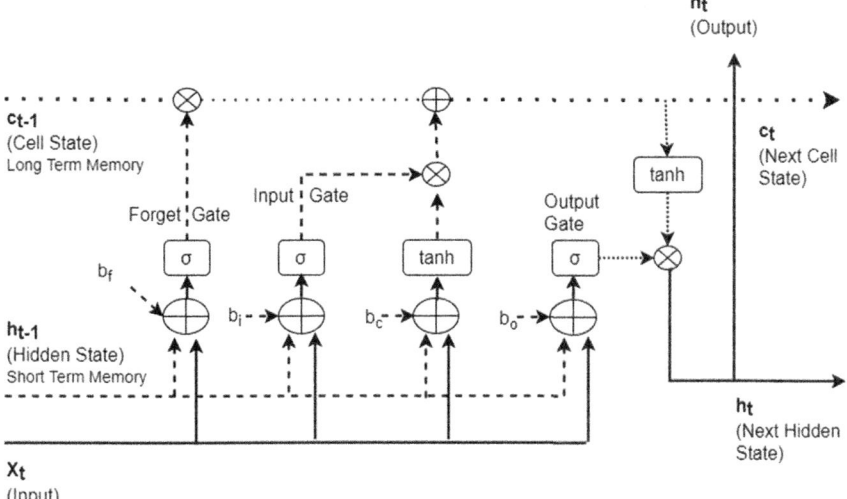

Fig. 12.13   LSTM architecture

The Forget Gate

The forget gate determines the data to be discarded from long-term memory. Note that the sigmoid function σ generates an output (0 … 1).

$$f_t = \sigma(W_f h_{t-1}, W_f X_t + b_f)$$

The Input Gate

The input gate determines the updates to the long-term memory. The sigmoid function σ generates an output (0 … 1)

$$i_t = \sigma(W_i h_{t-1}, U_i X_t + b_i)$$

Cell State (C_t)

Cell state (c_t) corresponds to the long-term memory content of the network. The long-term memory is updated, based on the input gate i_t and forget gate f_t, as shown below:

$$c_t = c_{t-1}.f_t + h_t.i_t)$$

The Output Gate

The output gate extracts meaningful data from the current cell state (h_{t-1}).

$$o_t = \sigma(W_0 h_{t-1} + U_0 X_t + b_0)$$

Then updates the working memory:

$$h_t = \tanh(c_t.o_t)$$

Note that the tanh function generates an output in the range of $\{-1 … +1\}$.

12.7.3   LSTM for Power Demand Prediction

The dataset 'PowerConsumption_2013_19.csv' consists of data from the state electricity board, from 2013 to 2019, of a state in India. The features include daily power demand, shortage, consumption, load, OD/UD, and date. Download the file from GitHub—(Sundararajan 2023). Develop an LSTM to predict power demand.

Tutorial 12.7   LSTM for Power Demand Prediction

Tutorial 12.7.1 Import Libraries

```
import numpy as np
import matplotlib.pyplot as plt
import pandas as pd
import math

from keras.models import Sequential
from keras.layers import Dense
from keras.layers import LSTM
from sklearn.preprocessing import MinMaxScaler
from sklearn.metrics import mean_squared_error
```

Tutorial 12.7.2 Data Setup

The dataset consists of daily power demand, shortage, consumption, load
Download the data (Refer: {(Sundararajan, 2023)} and provide the path in the
following instruction

```
d = pd.read_csv(r'PowerConsumption_2013_19.csv')
d.columns
['Max_Demand_Day_MW', 'Shortage_During_Max_Demand_MW','Energy_Met_MU',
'Draw_Schedule_MU', 'OD_UD_MU', 'Max_OD_MU', 'Year', 'Month', 'Date']
```

Convert 'Draw_Schedule_MU' to floating point format
```
   d = d.Draw_Schedule_MU.astype('float32')
```

Drop rows with null values, if any
```
   d.dropna(inplace=True)
```

Convert the array of values into 2D array in Numpy - 'ElectricPowerLoad'
```
   def create_ElectricPowerLoad(ElectricPowerLoad, look_back=1):
       dataX, dataY = [], []
       for i in range(len(ElectricPowerLoad)-look_back-1):
             a = ElectricPowerLoad[i:(i+look_back), 0]
             dataX.append(a)
             dataY.append(ElectricPowerLoad[i + look_back, 0])
       return np.array(dataX), np.array(dataY)
```

Normalize the ElectricPowerLoad
```
   scaler = MinMaxScaler(feature_range=(0, 1))
```

Reshape the array to 1 dimension
```
   d=np.array(d).reshape(-1,1)
```
If we give -1 as a parameter, eg., reshape (-1,1), numpy will compute the
number of rows implicitly

```
ElectricPowerLoad = scaler.fit_transform(d)
len(d) # 2161
```

Random seed helps reproducible results on repeated trials
```
np.random.seed(0)
```

Split into Train and Test sets
```
TrainSize = int(len(ElectricPowerLoad) * 0.67)
TestSize  = len(ElectricPowerLoad) - TrainSize
TrainSplit, TestSplit = ElectricPowerLoad[0:TrainSize,:],
ElectricPowerLoad[TrainSize:len(ElectricPowerLoad),:]
```

Reshape into X=t and Y=t+1
```
look_back = 1 # See Table 12-4
X_Train, Y_Train = create_ElectricPowerLoad(TrainSplit, look_back)
X_Test, Y_Test   = create_ElectricPowerLoad(TestSplit, look_back)
X_Train.shape # (1445, 1)
```

Reshape input to the format required for LSTM:
From [samples, features] to [samples, time steps, features]
From (1445, 1) to (1445, 1, 1)
```
X_Train = np.reshape(X_Train, (X_Train.shape[0], 1, X_Train.shape[1]))
X_Test.shape # Reshape from (712, 1) to (712, 1, 1)
X_Test  = np.reshape(X_Test, (X_Test.shape[0], 1, X_Test.shape[1]))
```

Tutorial 12.7.3 Configure and Compile the LSTM network

```
model = Sequential()
```
The IDE may throw an error regarding GPU usage. However, we do not need GPU
to execute this tutorial as the data volume is small)

Add LSTM layer with 1 input and 4 LSTM blocks
```
model.add(LSTM(4, input_shape=(1, look_back)))
```

We need 1 output
```
model.add(Dense(1))
```
The default activation function for LSTM is sigmoid; Loss function is MSE;
optimizer is adam.

The network is trained for 100 epochs; batch size of 1 is used
```
model.compile(loss='mean_squared_error', optimizer='adam')
model.fit(X_Train, Y_Train, epochs=100, batch_size=1, verbose=2)
```

Tutorial 12.7.4 Predict Power Demand

Make Predictions - using the Train data
```
TrgPredictions = model.predict(X_Train)
# [[0.2129], [0.2405], [0.2401], ..., [0.5037], [0.5392], [0.4887]]
```

De-normalise or get the predictions in the original scale
```
   TrgPredictions = scaler.inverse_transform(TrgPredictions)
   # [[ 61.91], [70.05], [69.95], ..., [147.79], [158.29], [143.38]]
```

Train-Y values: de-normalize or get the values in the original scale
```
   Y_Train = scaler.inverse_transform([Y_Train])
```

Make Predictions - using Test Data
```
   TestPredictions  = model.predict(X_Test)
```

Test X values- de-normalise or get the predictions in the original scale
```
   TestPredictions = scaler.inverse_transform(TestPredictions)
```

Test Y values - de-normalize or get the values in the original scale
```
   Y_Test = scaler.inverse_transform([Y_Test])
```

Calculate RMSE Error for Train and Test Data Separately

Calculate RMSE Error for Train Data
```
   TrainScore = math.sqrt(mean_squared_error(Y_Train[0],
   TrgPredictions[:,0]))
   print('Train Score: %.2f RMSE' % (TrainScore)) # Train Score: 10.61 RMSE
```

Calculate RMSE Error for Test Data
```
   TestScore = math.sqrt(mean_squared_error(Y_Test[0],
   TestPredictions[:,0]))
   print('Test Score: %.2f RMSE' % (TestScore))  # Test Score: 11.54 RMSE
```

Tutorial 12.7.5 Power Demand Plot

Shift Train predictions for plotting
```
   TrainingPredForPlot = np.zeros(ElectricPowerLoad.shape)
   TrainingPredForPlot[look_back:len(TrgPredictions)+look_back, :] =
   TrgPredictions
   TestPredForPlot = np.zeros(ElectricPowerLoad.shape)
   TestPredForPlot[len(TrgPredictions)+(look_back*2)+1:len(ElectricPowerLoad
   )-1, :] = TestPredictions
```

Plot baseline and predictions
```
leg = ['entire data',' training predictions','test predictions']
plt.title('Electrical Load Schedule 2013-2018')
plt.plot(scaler.inverse_transform(ElectricPowerLoad),
         linestyle='dotted',color='black')
plt.plot(TrainingPredForPlot,linestyle='dashdot',color='blue')
plt.plot(TestPredForPlot,linestyle='dashed',color='green')
plt.xlabel('Time Step in Days')
plt.ylabel('Million Units')
plt.legend(leg)
plt.show()
```
See Figure 12-14 LSTM for Power Demand Prediction

Table 12.4 Predicting the next value (Y_{t+1}) from a sequence ($X_0 \ldots X_t$)

Row#	X_Train (value at t)	Y_Train (value at t+1)
0	0.2207	0.2464
1	0.2464	0.2461
2	0.2461	...
	...	0.5297
1442	0.5297	0.5795
1443	0.5795	0.5101
1444	0.5101	0.5304

Fig. 12.14 LSTM for power demand prediction

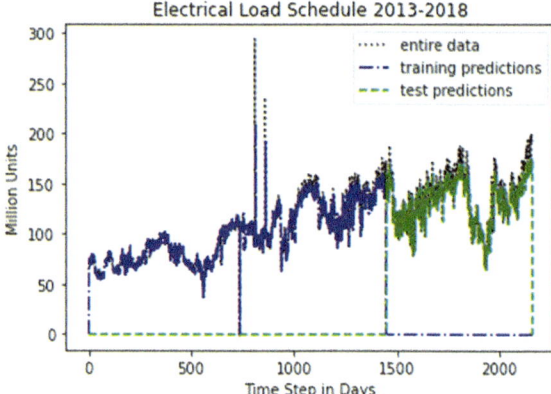

Data Analytics in Action

'Go' is a 3000-year-old Chinese strategy game. Two players with white or black stones take turns placing their stones on a board. The goal is to surround and capture their opponent's stones or strategically create spaces of territory. Once all possible moves have been played, the stones on the board and the empty points are tallied. The highest number wins (brainpool.ai 2021). Go has about 10 simple rules. However, this gives rise to 10^{170} possible board configurations!

AlphaGo is a computer program developed by DeepMind Technologies, a Google subsidiary. AlphaGo's research work started in 2014. It defeated the reigning European champion in 2015. After defeating many world champions in several games, it was retired in 2017. AlphaGo used the tree search to evaluate positions and deep neural networks to select the moves. These neural networks were trained by supervised learning from human experts. Then AlphaGo trained itself through self-play, using reinforcement learning.

Summary

The term Artificial Intelligence (AI) is defined AI as 'the science and engineering of making intelligent machines'. Research in AI is associated with learning (memorization and generalization), reasoning (inductive and deductive), problem-solving, perception, and language. AI research greatly boosted in the 2010s with the emergence of powerful systems to process big data and support machine intelligence.

While humans learn from experience, machines learn from data and improve their accuracy over time, without being programmed to do so. Machine learning involves identifying the right model parameters that optimize the predictive capability and deriving information from available data. Machine learning can be broadly categorized into four groups—supervised learning, unsupervised learning, semi-supervised learning, and reinforcement learning.

Scientists have developed two approaches to AI—symbolic and connectionist—expert systems are symbolic; deep learning is connectionist. Deep learning emphasizes learning by successive layers of increasingly meaningful representations, using a connectionist design of layers of neurons.

A neural net is a collection of neurons organized in layers. The outputs from one layer provide inputs to the next layer. The input layer receives external inputs. The final output emerges from the output layer. The other layers in the middle are called hidden layers and generally are trained to recognize intermediate concepts needed to determine the output. A neuron takes one or more inputs, multiplies each input by a weight parameter, sums them up, and adds a bias term to the sum. The weighted sum is passed as input to an activation function, which transforms it. Common activation functions include the sigmoid function, the hyperbolic tangent, SoftMax, and rectified linear unit functions.

A deep learning neural network will have multiple layers, each with a set of neurons. A layer other than the input layer consists of (a) an activation function—which accepts a tensor input and generates a tensor output and (b) the current state—the (coefficient) weights held in TensorFlow variables.

In a neural network, learning involves finding a combination of model parameters that minimize a loss function for a given set of training data samples. Random batches of data samples are drawn and the gradient of the model parameters with respect to the loss (output—actual target) is computed. The parameters are modified by a fraction defined by the learning rate in the direction opposite to the gradient. This is achieved by an algorithm traversing the entire network backward, starting from the output layer. The weights are adjusted based on their contribution to error, applying the chain rule of derivatives. This is called backpropagation in a multi-layer neural network. This process is repeated with several passes over the entire dataset, for convergence against set criteria.

Optimization techniques help to minimize the loss function (or optimize the cost function). The optimizer specifies how the gradient of the loss will be used to update parameters. Some of the commonly used optimizers are Gradient Descent,

AdaGrad, RMSProp, and Adam. Gradient descent is an optimization algorithm that uses the negative gradient of an objective function to locate the minima. Other methods are improvements over gradient descent.

Regularization helps to prevent overfitting. Specialized techniques are used to avoid overfitting a deep neural network. These include penalizing large weights (coefficients), randomly dropping some nodes each time we apply a step of gradient descent, and using a validation set to stop training when the loss on the validation set converges.

Feed-Forward Neural Network (FFNN) is a simple multilayered neural network where the input data moves through hidden layers to the output layer in only one direction. The layers do not give feedback to any previous layers. Convolutional Neural Network (CNN) is a feed-forward neural network, commonly used for image processing. CNN consists of the following segments—feature extraction and classification. The feature extraction segment consists of many convolutional layers, followed by pooled layers. Convolutional layers help detect features like edges and thereby progressively put together complex structures, such as eyes. These features are passed on to the classification layers to determine the class labels. A recurrent neural network (RNN) is a feedback network, commonly used for sequence/natural language processing. Long Short-Term Memory (LSTM) improves RNN architecture for learning long-distance associations. LSTM can save selected information into the memory, forget information by purging it, and focus on memory that is relevant to the context.

Questions

Comprehension

Write brief notes on:

1. How is learning achieved in an ANN?
2. Unsupervised learning.
3. Reinforcement learning.
4. The architecture of an artificial neuron.
5. Activation function.
6. Loss functions used in regression models.
7. Loss functions used in classification models.
8. Various optimization techniques.
9. Feed-forward neural network.
10. Convolution in convolutional neural network.
11. Feature extraction method.
12. Describe deep neural network with a suitable example.
13. Describe regularization techniques.

14. Describe convolutional neural network (CNN) architecture.
15. Describe a recurrent neural network (RNN) architecture.
16. Describe long short-term memory (LSTM) architecture.
17. Write a note on the capabilities of AI today.
18. Write a note on current applications of AI.

Analysis

19. How did historical predictions about AI, such as Turing's chess-playing computers and Chomsky's skepticism, shape the perception and development of AI over the years?
20. In what ways does AI raise fundamental questions in philosophy, especially in the realms of epistemology and axiology? How might the developments in AI influence our understanding of intelligence, reasoning, and ethics?
21. Explore the diverse applications of ML in bioinformatics, health care, natural language processing, and image processing. How has ML contributed to advancements in these domains?
22. Compare and Contrast Artificial Intelligence with Human Intelligence capabilities.
23. Explore the challenges in training deep neural networks, including issues like vanishing/exploding gradients and overfitting. How do techniques like dropout and regularization address these challenges?
24. What are the main segments of a Convolutional Neural Network (CNN) for image processing? Explain the purpose of each segment in a CNN, such as feature extraction and classification.
25. Compare and contrast LeNet-5, AlexNet, VGGNet, GoogLeNet, ResNet, and DenseNet in terms of their key features and contributions.
26. Describe the basic concept of convolution in image processing. Explain how feature extraction works in CNNs and its significance in image analysis.
27. Explain the process of applying a filter (kernel) to an image using convolution.
28. What distinguishes RNNs from other types of neural networks, especially in handling sequential data?
29. Describe the Long Short-Term Memory (LSTM) model and its purpose compared to traditional RNNs.
30. Explain the roles of the forget gate, input gate, and output gate in an LSTM. How do these gates contribute to the memory management of the LSTM?

Application

31. How has the emergence of powerful systems for processing big data in the 2010s impacted real-world applications of AI in various sectors such as business, health care, and government services?
32. Explore the societal implications raised by Yuval Noah Harari, such as the impact on the job market and the emergence of economically useless individuals. How might these scenarios play out by the year 2040?
33. Explore how Convolutional Neural Networks (CNNs) are applied in digital image processing, especially in fields like biomedical image processing and remote sensing.
34. Explore the practical applications of generative AI, focusing on models like ChatGPT. How is it utilized in coding, music composition, and content creation across various mediums?
35. Explore success stories in the application of Reinforcement Learning (RL) in various fields such as gaming, robotics, and health care. What are the specific achievements and challenges faced?

Exercises

Exercise 12.1 Build a Neural Network for Term Deposit Prediction

Refer to the Bank Marketing dataset described in Chap. 1. Build a Neural Network Model to predict whether a client will subscribe to a term deposit.

For the dataset, refer: https://archive.ics.uci.edu/ml/datasets/bank+marketing/

Exercise 12.2 Build a Neural Network for Hand-Written Character Recognition

The MNIST database of handwritten digits houses 60,000 training samples and 10,000 test samples of handwritten digit images in black and white $\{0, 1\}$. The digits have been size-normalized and centered in a fixed-size image of 28×28. Develop a convolutional neural network to classify the images. Describe the network structure and parameters.

For the dataset, refer: http://yann.lecun.com/exdb/mnist/

Exercise 12.3 Build a Neural Network for Visual Image Detection

We have used CIFAR-10 in a previous tutorial. The CIFAR-100 dataset is similar to the CIFAR-10. It has 100 classes containing 600 images each. There are 500 training images and 100 testing images per class. The 100 classes in the CIFAR-100 are grouped into 20 superclasses. Each image comes with a 'fine' label (the class to which it belongs) and a 'coarse' label (the superclass to which it belongs).

Refer: https://www.cs.toronto.edu/~kriz/cifar.html

Exercise 12.4 Develop a Neural Network for Energy Demand Prediction

The dataset 'PowerConsumption_2013_19.csv' consists of data from the electricity board for a state in India from 2013 to 2019. There are nine features and 2161 rows. Download the file from GitHub. (# Data File—Sundararajan 2023). Develop an LSTM to predict power shortages.

References

Andresen SL (2002) John McCarthy: father of AI. IEEE Intell Syst 17(5):84–85

brainpool.ai. (2021) Reinforcement learning: optimising data-driven decision-making by rewarding positive Outcomes. https://blog.brainpool.ai/reinforcement-learning-optimising-data-driven-decision-making-by-rewarding-positive-outcomes/

Chollet F (2017) Deep learning with Python. Simon and Schuster

François C (2019) Chollet—2018—Deep learning with Python. In: Manning, vol 53, issue 9

Copeland BJ (2023) Artificial intelligence. https://kids.britannica.com/scholars/article/artificial-intelligence/9711

Géron A (2019) Hands-on machine learning with Scikit-Learn, Keras, and TensorFlow (2019, O'reilly). In: Hands-On Machine Learning with R

Goodfellow I, Bengio Y, Courville A (2016) Deep learning. MIT press

Harari YN (2016) Homo Deus: a brief history of tomorrow. random house

Han J, Micheline Kamber JP (2014) Data mining. concepts and techniques, 3rd edn (The Morgan Kaufmann Series in Data management systems). In: Proceedings—2013 international conference on machine intelligence research and advancement, ICMIRA 2013

Leskovec J, Rajaraman A, Ullman JD (2020) Mining of massive datasets. Cambridge. In: Biometrics, issue 4. Cambridge University Press. https://doi.org/10.1111/biom.12982

Sujith R (2016) Graph-powered Machine Learning at Google. Google Research. https://blog.research.google/2016/10/graph-powered-machine-learning-at-google.html

Sundararajan S (2023) MVA-ML. https://github.com/sun-sri/MVA-ML

The manufacturer's authorised representative in the EU is Springer
Nature Customer Service Centre GmbH, Europaplatz 3, 69115 Heidelberg,
Germany. If you have any concerns regarding our products, please
contact ProductSafety@springernature.com

Printed and bound by CPI Group (UK) Ltd, Croydon, CR0 4YY
29/04/2026
02099466-0007